Energy Modelling

Second Edition

Energy Modelling
Advances in the Management of Uncertainty
Second Edition

Edited by Vincent Kaminski

Published by Risk Books, a Division of Incisive Financial Publishing Ltd

Haymarket House
28–29 Haymarket
London SW1Y 4RX
Tel: +44 (0)20 7484 9700
Fax: +44 (0)20 7484 9800
E-mail: books@riskwaters.com
Sites: www.riskbooks.com
       www.incisivemedia.com

Every effort has been made to secure the permission of individual copyright holders for inclusion.

© Incisive Media Investments Limited 2005

ISBN 1 904339 42 5

British Library Cataloguing in Publication Data
A catalogue record for this book is available from the British Library

Managing Editor: Laurie Donaldson
Development Editor: Tamsine Green
Copy Editor: Andrew John
Senior Designer: Matthew Hadfield

Typeset by Mizpah Publishing Services, Chennai, India

Printed and bound in Spain by Espacegrafic, Pamplona, Navarra

Conditions of sale
All rights reserved. No part of this publication may be reproduced in any material form whether by photocopying or storing in any medium by electronic means whether or not transiently or incidentally to some other use for this publication without the prior written consent of the copyright owner except in accordance with the provisions of the Copyright, Designs and Patents Act 1988 or under the terms of a licence issued by the Copyright Licensing Agency Limited of 90, Tottenham Court Road, London W1P 0LP.

Warning: the doing of any unauthorised act in relation to this work may result in both civil and criminal liability.

Every effort has been made to ensure the accuracy of the text at the time of publication, this includes efforts to contact each author to ensure the accuracy of their details at publication is correct. However, no responsibility for loss occasioned to any person acting or refraining from acting as a result of the material contained in this publication will be accepted by Incisive Financial Publishing Ltd.

Many of the product names contained in this publication are registered trade marks, and Risk Books has made every effort to print them with the capitalisation and punctuation used by the trademark owner. For reasons of textual clarity, it is not our house style to use symbols such as TM, ®, etc. However, the absence of such symbols should not be taken to indicate absence of trademark protection; anyone wishing to use product names in the public domain should first clear such use with the product owner.

Contents

	List of Contributors	vii
	Introduction *Vincent Kaminski*	xiii

SECTION 1: MODELLING ENERGY MARKETS AND PRICING DERIVATIVES

	Introduction *Vincent Kaminski*	3
1	Selecting Stochastic Processes for Modelling Electricity Prices *Blake Johnson, Graydon Barz* Stanford University	9
2	Fundamentals of Electricity Derivatives *Alexander Eydeland; Hélyette Geman* Morgan Stanley; University Paris IX Dauphine	59
3	Pricing, Modelling and Managing Physical Power Derivatives *Corwin Joy* Baylor College of Medicine	75
4	Valuing Power and Weather Derivatives on a Mesh Using Finite Difference Methods *Craig Pirrong; Martin Jermakyan* Olin School of Business and Washington University; Vernadun, LLC	99
5	Modelling Energy Prices and Derivatives using Monte Carlo Methods *John Putney* National Power plc	121
6	Fundamental Analysis of Power Price Modelling *Roman Kosecki* MAK Energy Consultants	153
7	Management of Transmission in the Electricity Markets *Martin Lin*	169

SECTION 2: MODELLING AND MARKET REALITIES

    Introduction   211
    Vincent Kaminski

8   The Importance of Market Structure and Incentives in Determining Energy Price Risk   217
    Giulio Federico; Adam Whitmore
    CRA International (UK) Ltd; Deloitte

9   Impacts of the Weather on Energy Demand and Supplies   247
    Daniel Guertin
    Sempra Energy Trading

10   Full-Requirement Contracts   279
    Yan Gao; Harald Ullrich; Krzysztof Wolyniec
    Progress Energy; Constellation Energy Commodities Group; Sempra Commodities

11   Heat Rate Options   303
    Boris Chibisov, Alexander Eydeland; Krzysztof Wolyniec
    Morgan Stanley; Sempra Commodities

12   Credit Risk Management for the Energy Industry – Some Perspectives   331
    Vincent Kaminski; Vasant Shanbhogue
    Citigroup; AIG Financial Products Corp

13   Capital Adequacy for Companies Transacting in US Electric Power Markets   359
    Laura L. Brooks
    PSEG

14   Generator Bid Strategies in Deregulated Markets: an Empirical Approach   391
    Paul Flemming
    Energy Security Analysis, Inc

    Index   417

List of Contributors

Graydon Barz, at the time of the first edition, was a recent graduate of the Engineering-Economic Systems & Operations Research Department at Stanford University. His doctoral research was on stochastic financial models for electricity derivatives. Graydon previously worked as an associate in the research group at Enron Corporation and as an associate at Edison Enterprises, a subsidiary of Edison International. In addition to a PhD, Graydon holds an MS degree in mathematics as well as a BSE in mechanical engineering from Princeton University.

Laura L. Brooks was appointed vice president – risk management, and chief risk officer (CRO) for PSEG in November of 2002, where she refines and implements PSEG's global risk management strategy and recommends methodologies for assessing and evaluating risk across all PSEG businesses. She is an integral member of the risk management committee. Laura came to PSEG from PG&E Corporation where she had been vice president – risk management, since 2000. She also is currently treasurer for the Committee of Chief Risk Officers. Laura previously was employed at Deloitte & Touche as a senior manager; at Equitable Resources as director of corporate risk management and vice president of reservoir engineering at Equitrans an interstate pipeline, and as a senior engineer at Southern California Gas Company. Laura has MS degrees from Carnegie Mellon University and Stanford University, and an MA and BA from the University of Colorado.

Boris Chibisov has a PhD in particle physics from the University of Minnesota and later joined the research department at Mirant Corporation. His work is focused on modelling joint distribution of commodity prices, valuation and risk management of commodity derivatives and structured products. Boris is currently working in the commodities analytical modelling department at Morgan Stanley.

Alexander Eydeland is executive director at Morgan Stanley, where he is in charge of global commodities analytic modelling. His previous positions include head of research at Mirant Corp, vice president with Lehman Brothers and Fuji Capital Markets, and associate professor of mathematics at the University of Massachusetts. Alexander holds a PhD degree in mathematics from Courant Institute of Mathematical Sciences. His papers on risk management, scientific computing, optimisation and mathematical economics have appeared in a number of major publications and he has lectured extensively on these subjects throughout the US, Europe, and Japan. Alexander is a co-author (with K. Wolyniec) of the book *Energy and Power Risk Management*, published in 2002.

Giulio Federico is a principal in the European competition practice of CRA International (UK) Limited, an economics, finance and business consulting firm. He is an expert of economic models of industrial organisation, with particular reference to electricity markets. His previous consultancy experience includes work at Lexecon Ltd and in the energy team of London Economics. He holds a BA Hons (PPE), MPhil and a DPhil in economics from the University of Oxford. Giulio has published a number of papers on the economics of the electricity industry.

Paul Flemming manages ESAI's power and gas team and has been with ESAI since 1999, after over six years as trading manager with both Koch Industries and Caltex Petroleum. Paul has expertise in the dynamics of Northeast Power Market, he has been directly involved in numerous plant evaluation feasibility studies and performs regular analysis to determine the factors influencing the pricing dynamics of the deregulated Northeast power markets. Paul also has direct responsibility for California and WECC market analysis and for the firm's natural gas capability. He has over 22 years of global experience in the international energy arena. At ESAI, Paul has also been responsible for global oil refining analysis, Asia product market analysis, risk management support services and project management for consulting projects, and he plays a lead role in expert witness work. Paul is a registered CTA with the National Association of Securities Dealers.

Yan Gao is currently principal structure analyst in the research and structure group at Progress Energy Ventures, the unregulated marketing and trading arm of Progress Energy. Before joining Progress, he was a lead quantitative analyst in the research group at Mirant Corporation. Prior to Mirant, he held faculty position at Pennsylvania State University and visiting faculty position at Yale University. Yan holds a PhD in mathematics from Yale University.

Hélyette Geman, at the time of the first edition, was professor of finance at the University Paris IX Dauphine and at ESSEC Graduate Business School. In 1993, Hélyette received the first prize of the Merrill Lynch awards for her work on exotic options and in 1995 was awarded the first Actuarial Approach for Financial Risk (AFIR) international prize. She was the co-founder and editor of *European Finance Review*, associate editor of *Mathematical Finance, Geneva Papers on Insurance* and other international journals, co-chair of the French branch of the International Association of

Financial Engineers and president elected of the Bachelier Finance Society. Hélyette is a graduate of the Ecole Normale Supérieure, holds a masters degree in theoretical physics, a PhD in mathematics from the University Paris VI Pierre et Marie Curie and a PhD in finance from the University Paris I Panthéon Sorbonne.

Daniel (Dan) Guertin is currently employed by Sempra Energy Trading Corp. Dan has been employed as the chief meteorologist at Sempra since October 1997. His responsibilities include short-term and long-term weather forecasting and analysing the impacts of weather patterns on energy demand and supply. Prior to joining Sempra, Dan completed his masters degree in meteorology from the Pennsylvania State University in 1997, and graduated with honors from Purdue University in 1995.

Martin Jermakyan is currently the president of Vernadun, LLC. Previously, he worked as a principal at Bank of America's energy trading department and as a vice president at Altra Energy Technologies. He was also one of the principals at ElectraPartners.com. Martin has been on the faculty at UCLA and University of Michigan Mathematics Departments, and the University of Michigan Business School. He has authored proprietary technologies and models in the areas of estimation of volatility surfaces, covariance matrices and developed models and techniques for quantifying such attributes of power and weather derivatives products as the market price of risk, and its impact on derivatives pricing. Martin has consulted with major market makers and utilities, and received his PhD in mathematics from Moscow State University.

Blake Johnson is consulting professor at Stanford University, where in recent years his work has focused on modelling power prices, valuing power assets and derivatives, and on the development of integrated risk management tools for the power industry. He has authored numerous articles and book chapters on these and related topics, and is also the founder of Options, Markets & Analytics, a provider of price modelling software and real options based valuation tools to the power industry. Previously, Blake was an associate in the corporate finance department at the investment-banking firm CSFB in New York. He holds BS, MS, and PhD degrees in engineering from Stanford University.

Corwin Joy first became interested in quantitative energy finance while working at Enron under the direction of Vincent Kaminski, where he worked on sophisticated VAR tools, credit risk engines, and derivative pricing models. In 1996, Corwin founded a successful energy software and consulting practice, Positron Energy Consulting, which was bought two years later by Caminus and then Sungard, whereupon Corwin led the energy modeling efforts at Sungard. In 1998, Corwin was recognised by Hart Energy Markets as one of the "100 Most Influential People in Gas and Electricity" in the last century. He received an MS degree in applied mathematics from Carnegie-Mellon University in 1993, and a BA degree in mathematics and statistics from Rice University in 1991. Corwin currently works for the Human Genome Sequencing Center at Baylor College of Medicine where he is helping the centre financially optimise their sequencing pipeline and data mine genomic sequences.

Vincent (Vince) Kaminski is currently working as a managing director at Citigroup, Houston. Before assuming this role, he served as a managing director and a consultant at Sempra Energy Trading and a senior vice president of commercial analytics at Reliant Resources, Inc in Houston. Before joining Reliant, Vince was a managing director of Citadel Investment Group LLC. Prior to this, Vince was the head of the quantitative modelling group at Enron Corp (from 1992 to 2002) and a vice-president in the research department, bond portfolio analysis group, of Salomon Brothers in New York (from 1986 to 1992). Vince is an adjunct associate professor at Rice University in Houston (Jones Graduate School of Management) and he serves on the executive committee of the Global Energy Management Institute at Bauer College of Business, University of Houston. Vince holds an MS degree in international economics, a PhD degree in theoretical economics from the Main School of Planning and Statistics in Warsaw and an MBA from Fordham University in New York. He is a recipient of the 1999 James H. McGraw award for Energy Risk Management (Energy Risk Manager of the Year). His recent publications include: *Managing Energy Price Risk* (all three editions) and "The Challenge of Pricing and Risk Managing Electricity Derivatives" in *The US Power Market*, both from Risk Books; and *Energy Derivatives: Pricing and Risk Management*, Lacima Publications.

Roman Kosecki is founder and president of MAK Energy Consulting, specialising in weather, emissions and energy analytics and trading. Roman has extensive experience in the pricing and modelling of cross-commodity derivatives, asset valuation and VAR methodologies and implements trading strategies in new and existing markets including weather, emissions, power, natural gas, crude oil and products. Prior to independent consulting, Roman was a senior trader at PSEG, expanding their weather and emissions trading presence as well as advancing the analytical capabilities throughout their trading organisation. Roman also worked at MIECO and Southern Company Energy Marketing, as a leader in energy research, risk management and structured product analysis. Before entering the energy markets, Roman was involved with portfolio optimisation procedures for a money management fund, and was a professor in the area of partial differential equations. He holds a PhD in mathematics from the Courant Institute of Mathematical Sciences.

Martin Lin is currently with the marketing, analysis and strategy group at Shell Trading Gas & Power, where he develops fundamentals-based forecasts and analyses of gas, power, and transmission markets across North America. He has also participated in the development of congestion management policy and market design in the ERCOT market. Prior to Shell, Martin held quantitative and fundamentals analyst positions at RWE Trading Americas, Citadel Investment Group, UBS, and Enron. Martin holds a BS from the California Institute of Technology and an MSE and PhD in electrical engineering from The University of Texas.

Craig Pirrong is professor of finance and energy markets director for the Global Energy Management Institute at the Bauer College of Business at the University of Houston. He was previously Watson Family Professor of commodity and financial risk management at Oklahoma State University, and a faculty member at the University of Michigan, the University of Chicago, and Washington University. Craig's research focuses on the economics of commodity markets, the relation between market fundamentals and commodity price dynamics, and the implications of this relation for the pricing of commodity derivatives. He has published over 30 articles in professional publications and is the author of three books. He holds a PhD in business economics from the University of Chicago.

John Putney, at the time of the first edition, was a risk analyst in the energy risk management group at National Power. He was responsible for developing pricing and risk analysis methods for power and gas derivatives and the implementation of these methods in the Company's energy trading and risk management systems. He spent his early career modelling multiphase flow and heat transfer for reactor safety studies. He has also been involved in developing models of the UK Power Pool and bidding decision support systems. John holds a PhD in nuclear reactor physics from Imperial College.

Vasant Shanbhogue has been a quantitative analyst in the energy industry for ten years. After getting an MBA from the University of Chicago, Vasant worked in Enron's Research Group, focusing on risk analysis and option pricing. He was then, as vice president of quantitative research, part of the core team that set up the infrastructure for Citadel Investment Group's energy funds. He is a Chartered Financial Analyst (CFA), and is now a vice president at AIG Financial Products Corp, and is active in structuring and pricing energy transactions. Vasant also has a PhD in computer science from Cornell University.

Harald Ullrich is currently vice president in the strategies department at Constellation Energy Commodities Group. Before joining Constellation, he worked as a senior analyst in the research group at Mirant Corporation and as a financial manager for Procter & Gamble. Harald holds a masters degree in management science from the University of Rochester, NY and graduated cum laude with a degree in economics and computer science from the University of Darmstadt in Germany. Harald has published in the journals *IEEE Transactions on Knowledge and Data Engineering*, *Lecture Notes in Computer Science*, and *Data and Knowledge Engineering*. He also co-authored papers winning "Best Conference Paper" awards at Informs 1996 and ER 1997.

Adam Whitmore has 20 years experience of the energy sector. He is currently a director in the economic consulting team at Deloitte. He previously worked at both Shell and London Economics. He has extensive experience of applying economics to modelling a range of energy markets. He has a first class honours degree and an MPhil in chemistry from the University of Southampton.

Krzysztof Wolyniec is currently vice president of quantitative analysis at Sempra Commodities. Before joining Sempra, he was the director of research at Mirant Corporation. Krzysztof holds graduate degrees in management science from the University of Rochester and theoretical physics from the University of Gdansk in Poland. He has published extensively on modelling, pricing and hedging in energy markets. Among other publications, he is the co-author of *Energy and Power Risk Management* with Alexander Eydeland.

Introduction

Vincent Kaminski

Risk management in any trading organisation is a challenging task that requires strong analytical skills and the ability to act as a critical interface between the traders and originators and senior management. This function tends to be as much an exercise in financial engineering as in social engineering. The energy business creates additional complications that result from a combination of several factors, unique to this industry.

Imperfect price discovery and the unconventional behaviour of energy prices is the first obvious problem any energy markets practitioner will encounter. The historical price data are often available from the vendors who collect the transaction data provided by the counterparties on the voluntary basis and with varying frequency. The rapid evolution of the industry and the changing regulatory landscape make historical data irrelevant to the current problems. Another set of problems results from the limited applicability of the stochastic processes used widely in the financial markets to the modelling of the dynamics of energy prices. A combination of seasonality, frequent jumps and dependence of price behaviour on the environmental variables, such as weather and the condition of the physical industry infrastructure, creates serious challenges to any trader, quant or risk manager.

Additional complications arise from the growing integration of all the energy markets. Fifteen years ago, a trader could spend a career trading one energy commodity, such as coal or natural gas, without worrying about the developments in other markets. Today, the shocks from one commodity or geographical market are transmitted to other parts of the energy complex through a web of complicated and constantly evolving links. In order to develop an understanding of the interactions between different energy markets one has to accumulate knowledge of the physical processes underlying production and transportation of energy and of the exogenous influences, ranging from the weather to the legal and regulatory framework.

Many skills required to function in this business are typically acquired through participation in the industry, through a process of trial and error and learning from more seasoned players, in the same way folk wisdom was transmitted through the generations in the past. The academic literature tends to be several years behind industry practice or is too technical for the average practitioner. This book is designed to close to some extent the existing gap and offer insights into most recent developments in financial engineering as applied to the energy markets. The book contains also several chapters that explain how the physical infrastructure of the industry affects market operations. We hope that the book will not only become a useful reference but also will introduce the reader to the leading practitioners of the merchant energy business.

The book is focused on electricity markets but the insights it offers and analytical techniques it explains are applicable across the entire energy complex. The common features of the commodity markets are found in their most extreme form in the power industry.

The chapters contributed by some of the most respected practitioners and academics are organised into two sections. The first section of the book introduces some basic concepts and models that can be applied to a wide range of practical problems. This part of the book reviews also the salient features of the power markets that have to be recognised and addressed in the daily practice of the merchant power industry. The second section offers examples demonstrating how different theoretical tools are used to solve the problems of valuation of specific transactions. The chapters included in the second section show also how important it is to supplement highly theoretical models with the studies that analyse how behaviour of market participants can affect the outcome of the competitive process.

Section 1

Modelling Energy Markets and Pricing Derivatives

Introduction

Vincent Kaminski

Modelling the dynamics of energy prices is a critical component of any system of valuation and risk management tools designed and used for a portfolio containing both linear energy-related instruments, such as forwards, and nonlinear instruments, such as options or transactions with volumetric risk exposures. The modelling difficulties are common to most commodity markets but no industry offers greater challenges than the electricity business. The volatility of electricity prices is very high compared with the volatility of other energy commodities and dwarfs the volatility observed in the financial markets. As several chapters in this section point out, this is due primarily to non-storability of electricity and high variability of demand that is driven mostly by weather fluctuations. The shocks that take place in the fuel markets (such as natural gas or coal) are quickly transmitted to the power market and amplified by its special characteristics.

The first chapter, written by **Blake Johnson** and **Graydon Barz**, combines empirical analysis of power prices based on the data from a cross-section of different power pools in the UK, California, Australia and Scandinavia and reviews analytical tools that can be used for price modelling. The properties of the electricity prices they identify include extremely high volatility combined with mean reversion tendencies, dependence of volatility on the price level, strong seasonal effects, the tendency of prices to gap, or spike, both up and down, and also significant differences in the behaviour of prices across different geographical markets.

These differences can be explained by variability of demand patterns, differences in the level and structure of the installed generation capacity, differences in the developments of the transmission network and the power pool design. Their discussion of the empirically observed characteristics of electricity prices is followed by a review of widely used stochastic processes used for modelling energy prices and the discussion of their required properties and estimation methods. The authors find the geometric mean reversion process with jumps superior to other commonly used alternatives. The authors mention, although do not discuss in detail, an alternative model that is likely to attract a lot of attention from industry professionals in the future. The model is identified with the initialism LMRDJ, which stands for "locally mean-reverting-diverting", possibly augmented with jumps. The detailed description of the model can be found in the patent application.

Alexander Eydeland and **Hélyette Geman** review analytical tools for pricing options offered in the power markets. They start with the review of different types of options popular in the power markets and link them to some unique characteristics of the electricity markets. As do Barz and Johnson in the previous chapter, they emphasise non-storability of electricity and regional differences between different markets. Another complicating factor is market incompleteness that makes it difficult, if not impossible, to create replicating portfolios, the foundation stone of the Black–Scholes theory of option pricing. Non-storability of electricity invalidates the relationship between the spot and forward prices based on the familiar cost-of-carry argument. The usefulness of models based on the dynamics of the spot process is also undermined, as the well-defined link between the spot and forward prices does not exist in the case of electricity. This observation brings to the forefront the importance of modelling the dynamics of the forward prices. The approach the authors are proposing is based on the production-related framework. The dynamics of the forward prices are determined through the interaction of forward prices for fuels, the composition of the supply stack and demand fluctuations. The authors discuss certain restrictive assumptions under which the dynamics of power prices can be modelled using the traditional geometric Brownian motion process.

Corwin Joy emphasises, like the previous contributors, the limitations of the traditional Black–Scholes model and the importance of capturing physical reality in the modelling of electricity markets. Unlike the previous authors, Joy uses the system lambda, historical data representing the marginal cost of electricity and available from the FERC filings. The starting point for his model is the technique developed originally by Heath, Jarrow and Morton (HJM) for evolving the term structure of interest rates. This approach, adopted for the commodity markets by Cortazar and Schwartz, allows modelling the behaviour of the forward price curves treated as unified objects. A realistic implementation of this model has to capture the autocorrelation of hourly prices of electricity, the tendency of prices to mean reversion, the propensity of prices to jump under the impact of the demand and/or supply shocks, as well as intraday and long-term seasonality patterns. The generalisation of the price model proposed by Corwin Joy combines the HJM framework with jumps and the additional assumption that conditions volatility on the price levels. The practical application of the model requires estimation of the parameters of the model from the historical information, including the system lambdas. Additional complication arises from the fact that electricity contracts often contain volumetric optionality in addition to price optionality. Corwin Joy proposes to address this problem using the framework developed for the so-called quanto options.

The chapter contributed by **Craig Pirrong** and **Martin Jermakyan** describes a numerical technique designed to handle a number of difficulties encountered in pricing of energy options and identified in the previous chapters. The authors recognise highly non-standard behaviour of electricity prices (such as jumps, seasonality and mean reversion) and market incompleteness that make hedging difficult, if not impossible, under many conditions. The solution proposed by the authors is based on the equilibrium model that uses the demand (load), and possibly the fuel price, as the state variables. The price of power at the option contract expiration is related to the level of the state variable that determines the option valuation. The option value is a solution of a partial differential equation that the authors solve using the finite difference method. Additional complications arise from the fact that the markets that could allow transferring the load-related risk to another party don't currently exist and, therefore, option valuation requires determination of the market price of risk associated with the load variable. The market price of risk is not observable and has to be estimated from the market data (forward prices). The authors find that the market price of risk is rather significant but also, as one might expect, has a strong seasonal pattern. This approach offers an elegant framework that allows us to combine the methods of financial economics with the fundamentals of the power markets. The numerical techniques proposed by the authors lend themselves to the valuation of the swing options (options containing volume optionality) and path-dependent options.

One of the very powerful numerical techniques used in the valuation and risk management of electricity derivatives is the Monte Carlo approach. **John Putney** reviews the basics of this technique and focuses next on the particular problems related to the practical applications in the energy business and valuation of the specific contracts. The list of contract types he covers includes Asian options, swaptions, spark spread options and swing options. One advantage of the Monte Carlo approach is that this technique allows us to use an arbitrary price process to describe the price behaviour of the commodity underlying an option contract. The author reviews the applications of the Monte Carlo approach for risk management. One shortcoming of the Monte Carlo method is its relatively slow speed. Putney reviews different numerical tricks that can be used to increase the speed of convergence and accuracy of the results of the Monte Carlo-based models.

The next two chapters, contributed respectively by **Roman Kosecki** and **Martin Lin**, attack the fundamental problem facing any power trader and risk manager, and that is the dependence of prices on the physical infrastructure of the industry: generation units and transmission grid. The condition of the infrastructure and the environmental and economic factors, such as weather and the levels of economic activity, determine jointly the level and behaviour of electricity prices. Kosecki concentrates on the unit characteristics and the load dynamics, driven in the short run primarily by weather conditions. Interaction of the units' characteristics, captured by the generation stack and the load volatility, explains the nature of electricity prices (the properties described and documented in the earlier chapter by Johnson and Barz). The discussion of the properties of the power prices is followed by the review of the typical option structures offered in the electricity markets, such as monthly options (with a forward price as the underlying) and daily options (based on the spot price such as, for example, the day-ahead or real-time price).

The electricity traders use a number of technical tools in their decision-making process, including so-called generation stack and a load forecast. The generation stack is a supply curve of electricity, constructed based on the information about the available generation units, their thermal efficiency, installed capacity and unit output cost. Of course, the hydro and nuclear units require special treatment in construction of a stack. The stack, in conjunction with forecast load, allows predicting the level of market prices, given by the intersection of the stack and expected load. The main assumption behind the generation stack is that the units with lower unit costs are dispatched first. The generation stack is a relatively straightforward tool that ignores the subtleties of units' characteristics and optimal dispatch logic as well as the possibility of market prices diverging from the marginal costs due to exercise of market power or strategic bidding into power pools. One of the main limitations of the generation stack tool is that it ignores the importance of transmission in determination of the power prices. This is equivalent to building the theory of international trade that ignores the transportation costs. This omission is understandable, given the highly technical and complex nature of the theory explaining power flows in the integrated generation and transportation grids. The chapter on transmission contributed by Lin provides an extensive review of the physics and technology of electricity transmission and of the financial instruments available in the marketplace to manage the exposure to congestion (a technical term for transmission cost) in the electricity markets.

1

Selecting Stochastic Processes for Modelling Electricity Prices

Blake Johnson, Graydon Barz
Stanford University

The ability to model power prices in an effective way provides a necessary foundation for the principal activities of electricity market participants. These activities include the structuring, pricing, trading and risk management of physical and financial contracts, and the valuation, risk management, and choice of operating policies for generation and transmission assets. The observed behaviour of electricity prices over the available history of deregulated electricity markets, however, suggests that there are significant challenges associated with modelling power prices. For example, the volatility seen in power prices is unprecedented in other commodity markets, with annualised volatilities of 1,000% common. The mean price of power also varies substantially by time of day, week and year. The frequency, magnitude and complexity of these fluctuations are also unprecedented in other commodity markets, and their importance is compounded by the fact that price volatility is strongly correlated with price level. As a result, price volatility follows similar daily, weekly and annual patterns. Both positive and negative price "spikes" (defined as sudden upward or downward movements in price that can be extremely large and are quickly reversed), are commonly observed. Finally, basis relationships between power prices at different geographic locations show similar volatility, and also frequently exhibit cycles by time of day, week and year.

This chapter addresses the challenge of building models capable of capturing this complex behaviour in the following way. First, the actual behaviour of electricity prices in a cross-section of the world's electricity markets is carefully reviewed. This review serves to identify the most important characteristics of the behaviour of electricity prices, and to consider how this behaviour varies by time of day, week and year and across markets. Next, these observed characteristics of electricity price behaviour are linked to the economic fundamentals of electricity markets. These fundamentals, which also vary from market to market, include:

- the pattern of demand over the course of the day, week and year;
- the characteristics of the regional generation supply stack, including its fuel type, operating costs and constraints, and ownership structure;
- the available transmission network and;
- the rules governing the regional power market and the behaviours of its participants.

A strong link between these market fundamentals and the characteristics of electricity price behaviour will be demonstrated, both within individual markets and in the cross-sectional behaviour of prices across markets, and this is traced principally to the non-storable nature of electricity. Finally, the relative ability of a range of possible models of electricity price behaviour to adequately represent the key characteristics of electricity prices is evaluated. The models considered include those commonly used to model the price processes of financial instruments and other commodities, as well as models designed specifically to reflect the unique properties of electricity prices. The relative performance of each of the models is evaluated by fitting it to price data from a cross-section of the world's electricity markets using maximum likelihood methods.

Two principal conclusions are generated by the analysis. The first is that electricity's non-storable nature is the principal cause of its complex and volatile behaviour. Non-storability is important because it rules out the substitution of electricity at one point in time for electricity at other points in time. As a result, inventories cannot be used to smooth price fluctuations over time, and the cost-of-carry-based methods that have traditionally been used to bound commodity price fluctuations are unavailable. In their absence, prices are free to adjust, and in fact must adjust, on a real-time basis to the level necessary to balance real-time electricity production and consumption. Consequently, the stochastic characteristics of real-time electricity production and consumption are reflected directly in electricity prices over time, rather than in the "smoothed" form characteristic of storable commodities. The cyclical patterns of electricity demand over the course of the day, week and year therefore account for the cyclical pattern of electricity prices over the same periods, whereas temporary generation and transmission asset outages or capacity constraints trigger sudden changes in electricity price levels and locational-basis relationships.

The second principal conclusion of the analysis is that price-process models tailored to electricity prices are necessary to adequately represent these characteristics of electricity prices. The need for such specialised models is made intuitively clear by the analysis of the unique characteristics of electricity price behaviour presented in the first two parts of the chapter, and the links between this behaviour and electricity market fundamentals. This intuition is verified empirically in the third section of the chapter, where alternative models of electricity price behaviour are constructed and calibrated, and their performance evaluated. The fourth section addresses the unique challenges of modelling intra-day price behaviour, including performance requirements and necessary model capabilities. The fifth section describes how available data about future supply, demand, and market conditions can be used to improve model calibration and performance.

THE BEHAVIOUR OF ELECTRICITY PRICES

To gain insight into the behaviour of power prices and the challenges associated with modelling them effectively, their behaviour in a cross-section of markets around the world is examined. Given the unique and complex behaviour of power prices, the maxim "a picture is worth a thousand words" applies. Accordingly, graphics make up an important part of the analysis. Although data from a range of markets are examined, much of the analysis focuses on four markets selected for the diversity of their supply-and-demand characteristics, market structure and regulation, and associated price behaviour. These markets are the power pools in the UK, Scandinavia, the Australian state of Victoria, and California.

Figure 1 shows a sample time series of hourly prices over the first 10 days in April and August 1998, respectively, in each of the

Figure 1 Sample time series of prices (1998 data)

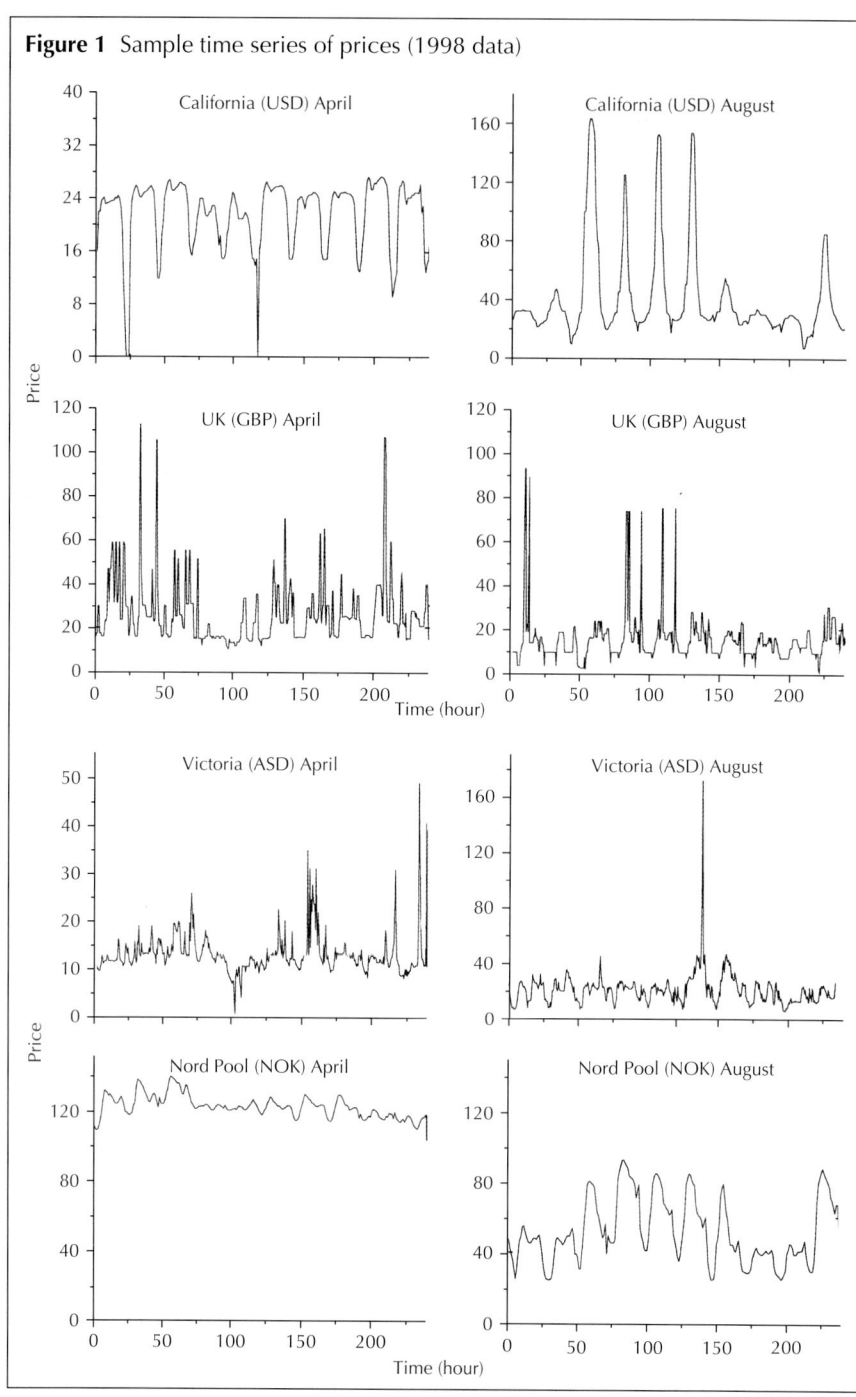

California, UK, Victoria and Scandinavia markets. The series illustrate the cyclical and volatile nature of power prices in each market, as well as important differences in price behaviour across markets and seasons.

The figure illustrates many of the many distinctive properties of electricity prices, including:

❑ the highly volatile but also mean-reverting nature of hourly and daily prices in all markets;
❑ the strongly seasonal nature of prices;
❑ the presence of both positive and negative price "spikes" as visible, for instance, in the California market in August and April, respectively; and
❑ the substantial differences in price behaviour across markets.

Figure 2 shows the path of prices over the course of individual days in the California market for a sample of days in the months of April–May and August–September 1998. The sample paths show strong serial correlation over the course of the individual days, but only limited serial correlation between days.

Figure 3 shows the mean price by time of day in each of the California, Scandinavia, UK and Victoria markets. Due to the highly seasonal nature of electricity prices, the mean price over the course of the day at each location is shown for three separate two-month time blocks: April–May, August–September, and November–December. Figure 3 illustrates:

❑ the pronounced fluctuation in mean price over the course of the day in each market, particularly during the high demand season;
❑ the seasonal nature of the levels and patterns of mean prices; and
❑ that substantial differences exist in the shape and magnitude of these fluctuations across the markets.

Note, in particular, the smaller size of the fluctuations in mean price over the course of the day in the Scandinavia market but the larger seasonal fluctuation.

Figure 4 shows the standard deviation of prices in each market over the course of the day for the same three two-month time blocks. On an annualised basis, the volatility of the associated hourly or

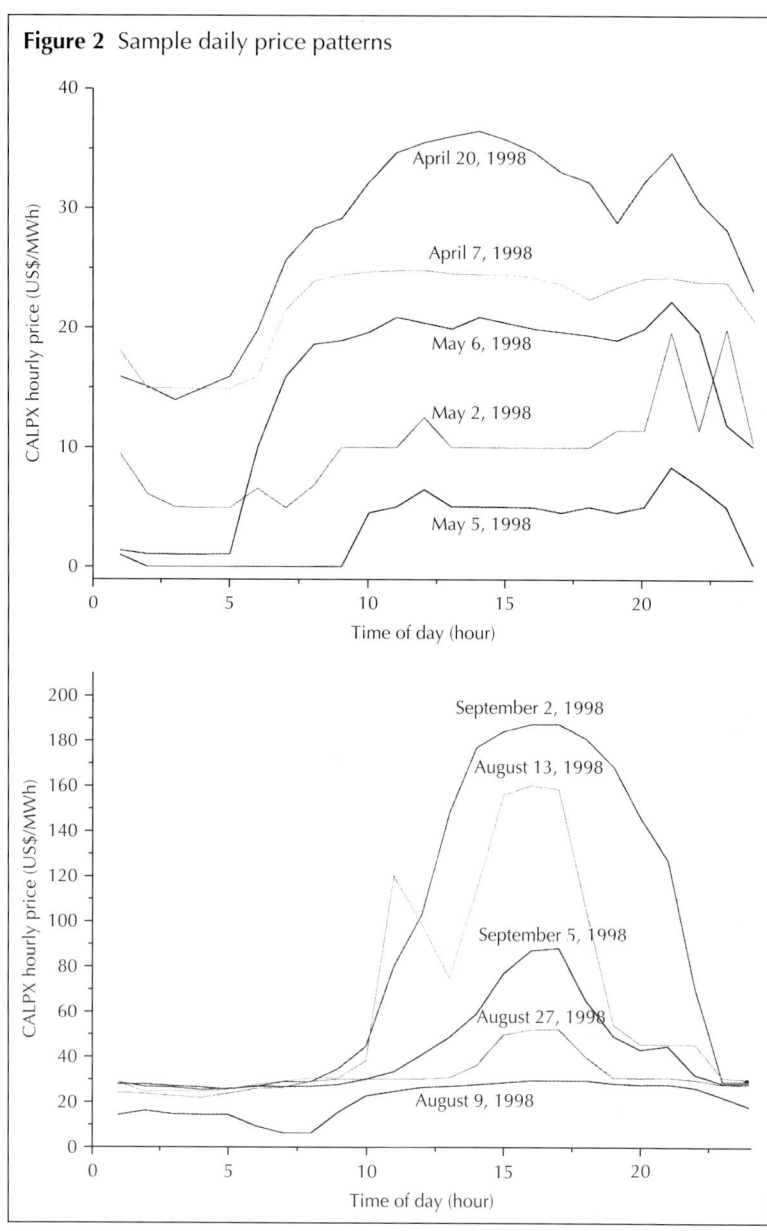

Figure 2 Sample daily price patterns

half-hourly returns in the figures range from 120% to 2,600%. Like Figures 1–3, Figure 4 illustrates the pronounced fluctuations in price volatility that occur over the course of the day, the seasonal variation in these patterns, and the substantial differences that exist

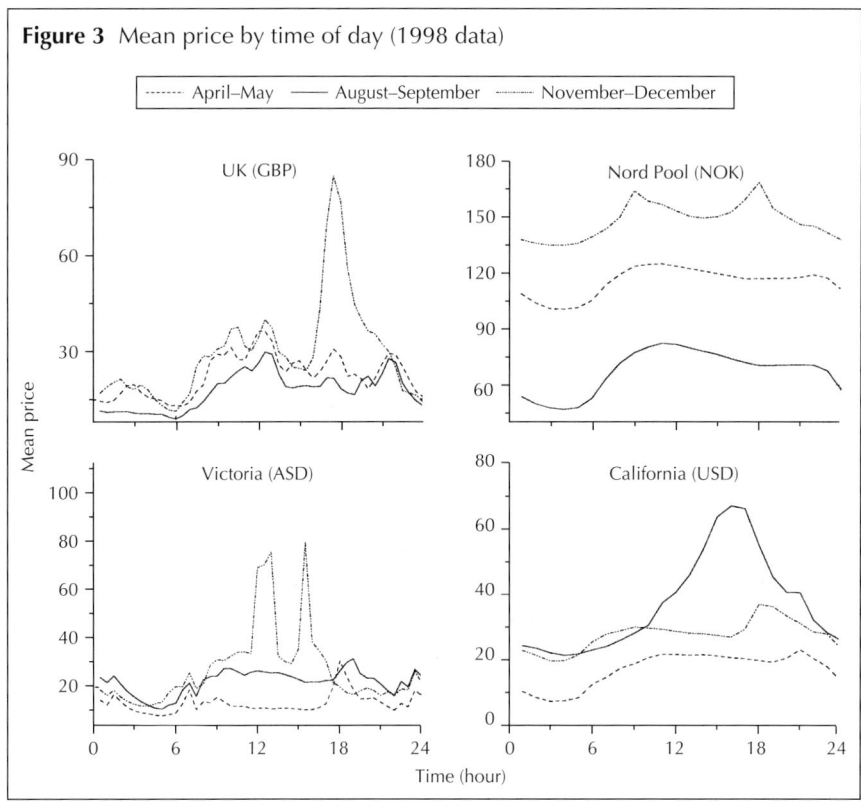

Figure 3 Mean price by time of day (1998 data)

in both these patterns and in the overall magnitude of volatility across markets.

Figure 5 shows, on a single graph, the mean price and volatility curves at each location during August–September 1998. It shows that a strong positive correlation exists between the mean price level and the level of price volatility. This pattern is strikingly consistent across markets, despite the substantial differences in overall price behaviour across the markets. Possible explanations for this relationship will be considered below.

Figure 6 shows the term structure of the volatility of returns in each of the four markets, presented as volatility per unit time for periods ranging from one to 100 hours in length. The top charts plot the actual volatility in each market, while the lower charts plot normalised versions of the same curves. The charts illustrate the

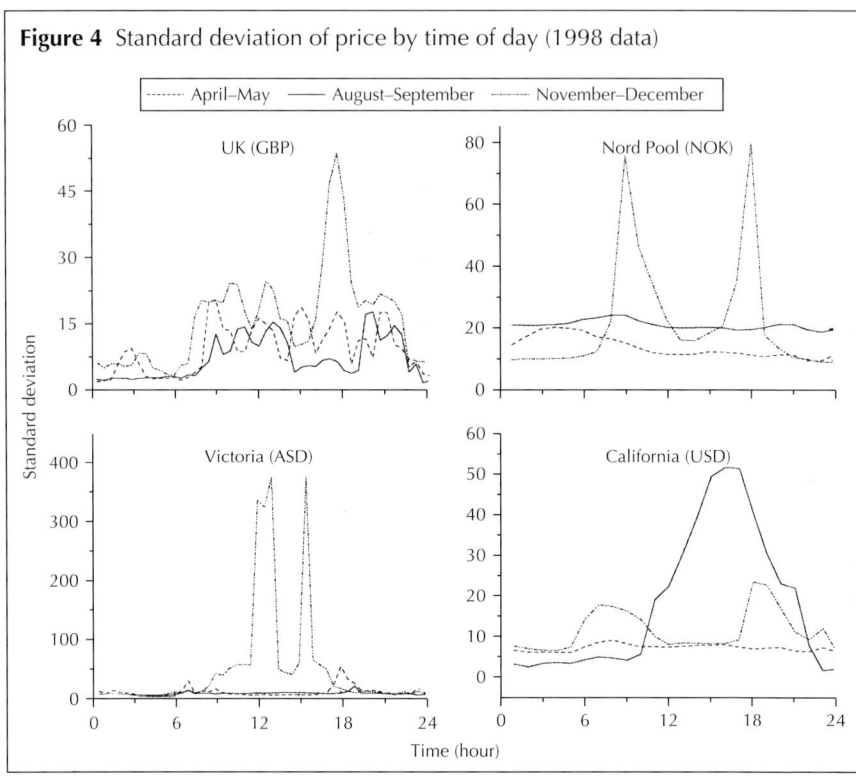

Figure 4 Standard deviation of price by time of day (1998 data)

strongly mean-reverting nature of power prices over even very short periods of time.

Figures 7–12 document a variety of basis relationships in the four markets. First, Figures 7–9 show a range of intertemporal basis relationships. Figure 7 provides histograms of the basis between the price of power at 4pm and 4am, and Figure 8 shows the basis between power at 4 pm and the daily average price of power. Figure 9 illustrates the basis between the daily and monthly average price, in all cases for each of the four markets during the months of August–September 1998. Together the figures illustrate the highly variable nature of the intertemporal basis relationships of electricity prices, particularly over short intervals of time. Also note the differences in the shape of the distributions, particularly between the California distribution and those of the other markets. An important implication of this variability is that hedges constructed with

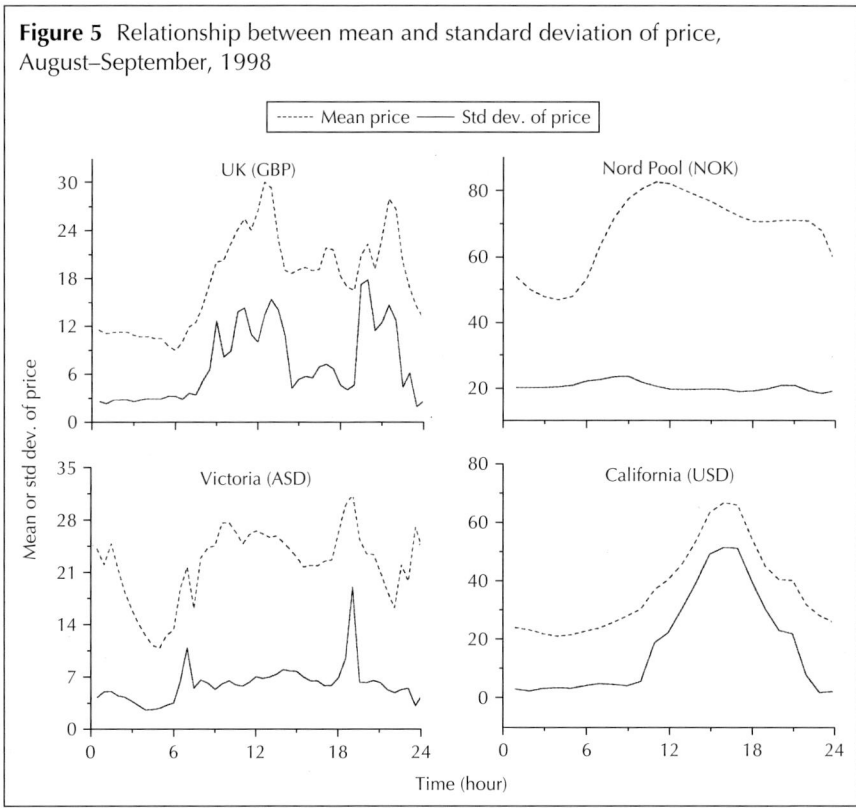

Figure 5 Relationship between mean and standard deviation of price, August–September, 1998

electricity to be delivered at a time other than precisely the same time as the underlying exposure may be highly ineffective. This may in part account for the low trading volumes of the existing electricity futures contracts, all of which are based on monthly average prices, given that the majority of electricity price volatility occurs over the course of the day and week.

Figures 10–12 show three locational basis relationships. Figure 10 shows both a histogram and a time series for the price difference between the Victoria and New South Wales markets in Australia. The power markets in these neighbouring states in Australia are connected by a 1,100 MW transmission line. When the transmission line is available and operating below capacity, prices in the two markets are within the transmission losses on the line. As Figure 10 shows, however, when the line is either out of service or at capacity, prices in the markets can diverge dramatically.

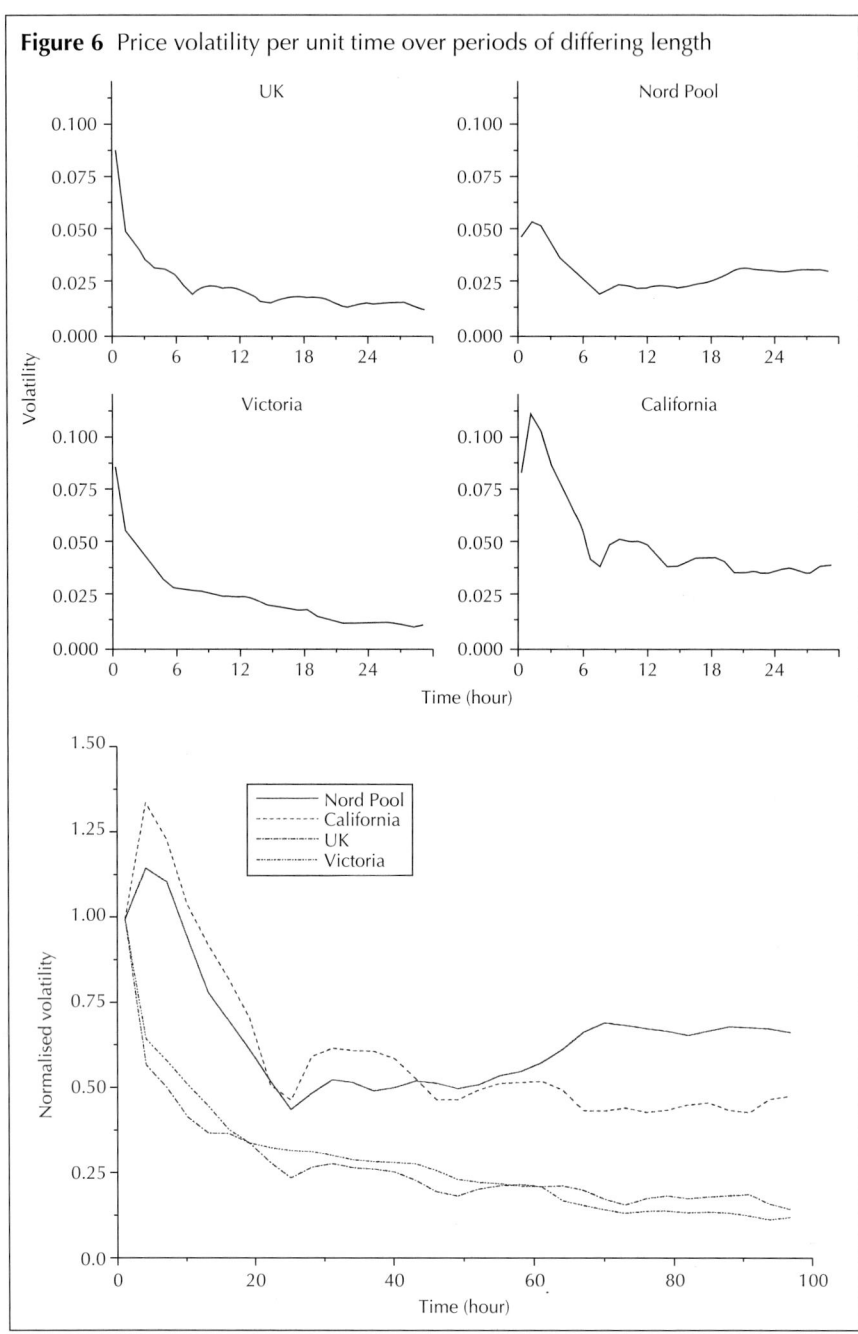

Figure 6 Price volatility per unit time over periods of differing length

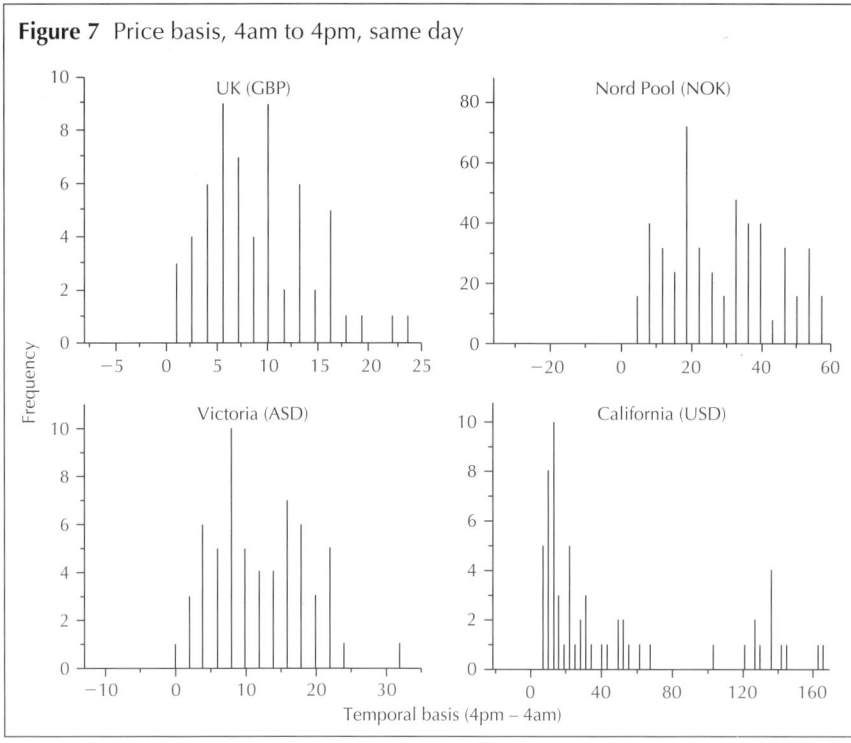

Figure 7 Price basis, 4am to 4pm, same day

Figure 11 shows the locational basis between the daily prices of power during 1998 at the California–Oregon border (COB) and at the Palo Verde (PV) high voltage switchyard in Arizona. COB and PV serve as the main trading locations for power flowing into and out of California, as well as for the western region of the US as a whole, and were chosen as the delivery points for the first two US electricity futures contracts. In contrast with the Victoria – New South Wales basis, the COB-PV basis is more consistent through time, but is also highly seasonal due to capacity constraints on the North–South transmission lines during the summer high demand period in Southern California.

Figure 12 shows the locational basis between the daily prices of power during June and July 1998 in the east-central region (ECAR) of the US, which includes the states of Indiana, Kentucky, Michigan, Ohio, West Virginia and parts of Pennsylvania, and the Pennsylvania, New Jersey and Maryland (PJM) power pool.

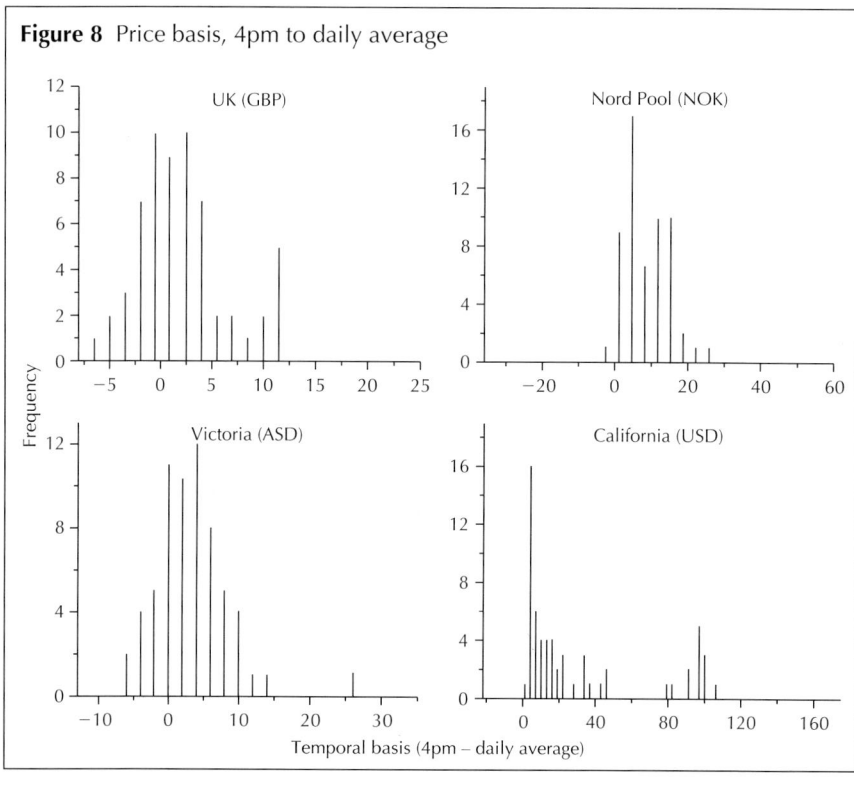

Figure 8 Price basis, 4pm to daily average

Although transmission capacity was generally available between these markets during this period, due to regulatory restrictions prohibiting the flow of power out of the PJM pool when prices in the pool exceed US$50/MWh, prices in these neighbouring regions diverged dramatically on several occasions when prices spiked in ECAR.

THE EVOLUTION OF ELECTRICITY PRICE DISTRIBUTIONS OVER TIME

As the deregulation of electricity markets only began in 1990 (in the UK market) and spread to other markets quite slowly, the available trading history in most electricity markets is limited. As a result, few data are available on how the distribution of

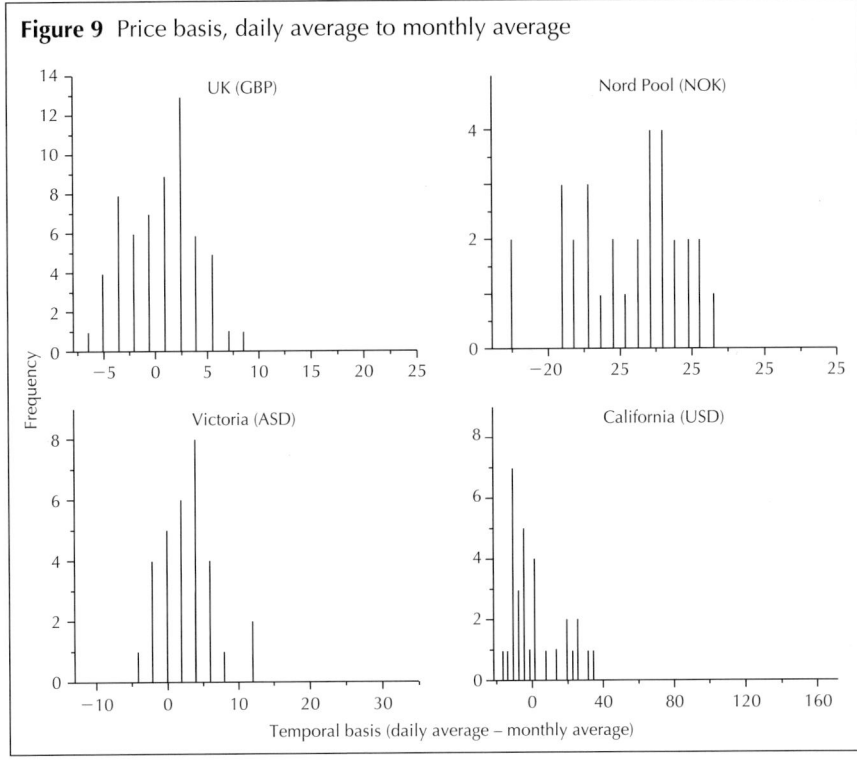

Figure 9 Price basis, daily average to monthly average

electricity prices varies from year to year, or on whether and how this distribution might evolve over time. The UK market, however, provides an exception in this regard, and a brief review of the history of price behaviour within it reveals several interesting trends. For example, Figures 13 and 14 show what appears to be an evolution in the distribution of electricity prices over time in the UK market. Figure 13 shows histograms for the price of power between 2pm and 4pm during August for the years 1991–6 in the UK, and Figure 14 shows histograms for the price of power between 2am and 4am for the same time periods. Note that afternoon power prices have become substantially more dispersed over the six-year period, but prices during the night have remained concentrated and have shifted downwards. Possible reasons for these changes include the construction of new assets,

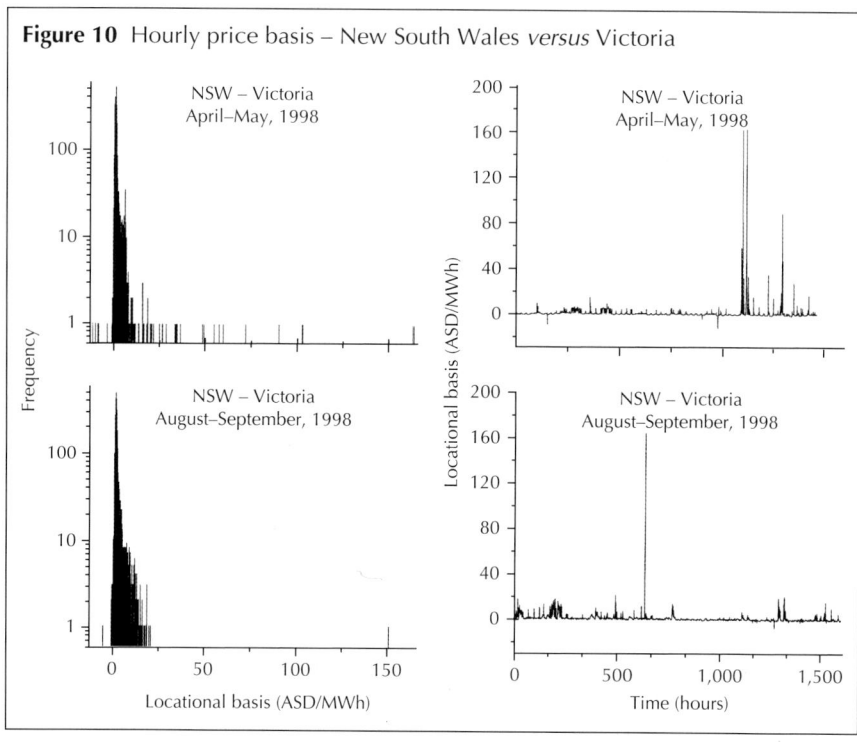

Figure 10 Hourly price basis – New South Wales *versus* Victoria

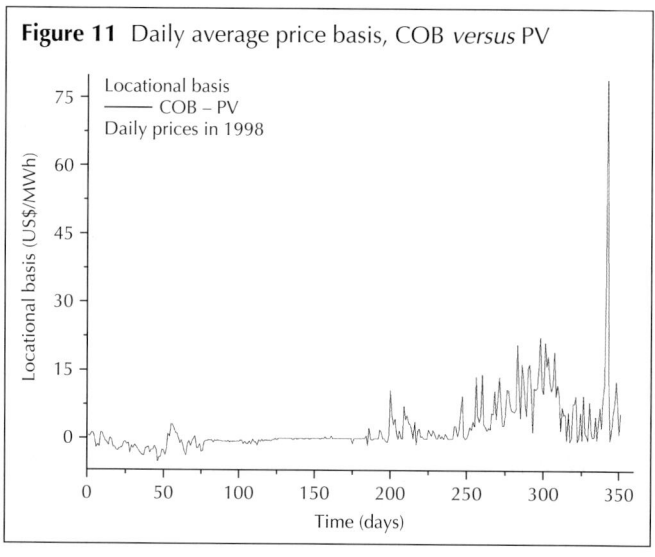

Figure 11 Daily average price basis, COB *versus* PV

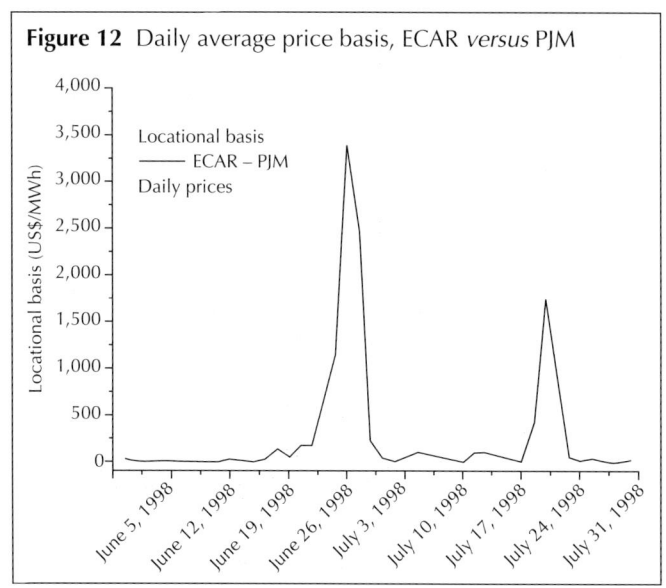

Figure 12 Daily average price basis, ECAR *versus* PJM

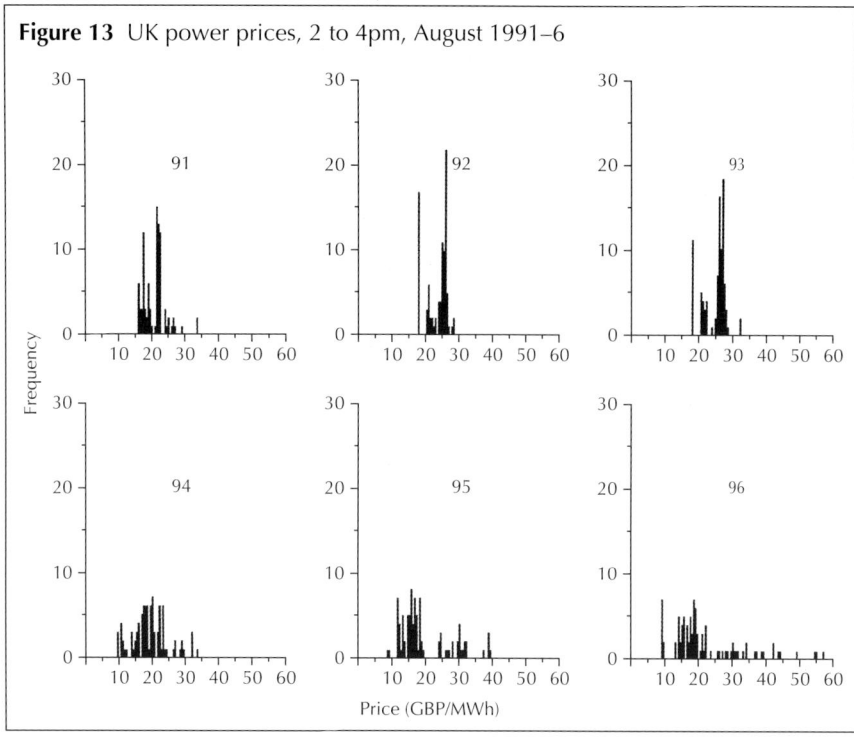

Figure 13 UK power prices, 2 to 4pm, August 1991–6

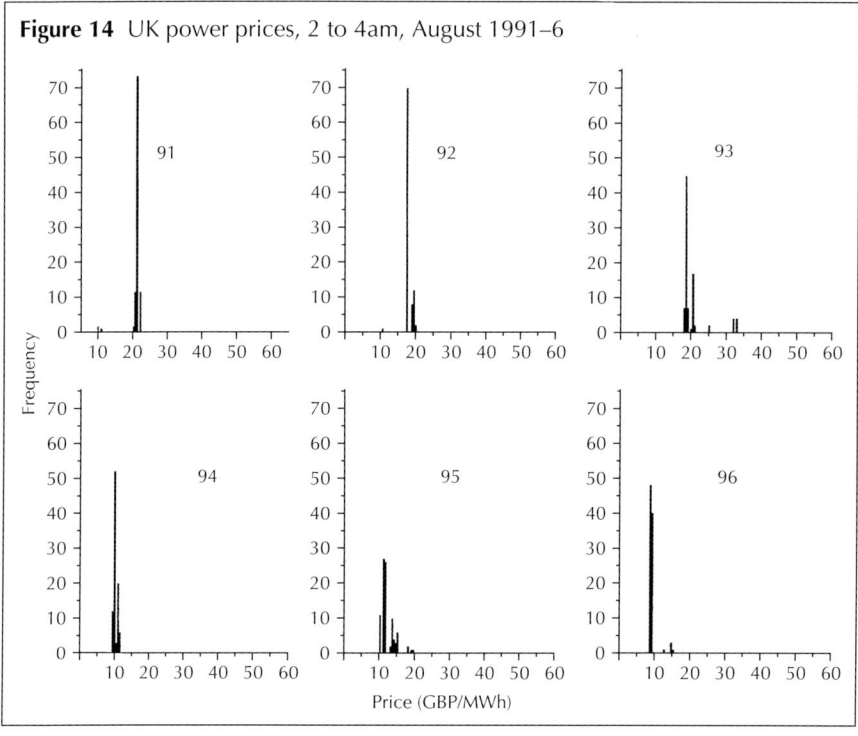

Figure 14 UK power prices, 2 to 4am, August 1991–6

changes in the ownership and dispatch policies of existing assets, including attempts to exercise market power, the retirement of existing generation assets, and changes in load patterns in response to the proliferation of time-of-use pricing.

ELECTRICITY MARKET FUNDAMENTALS AND ELECTRICITY PRICE BEHAVIOUR

Many of the complex and unique characteristics of electricity price behaviour documented above can be traced to the physical properties of electricity, most importantly its non-storability, and to the properties of its supply and demand. The objective of this section is to explore these relationships, in particular the real-time market clearing made necessary by non-storability, and the characteristics and key drivers of electricity production and consumption.

PANEL 1 STORAGE COSTS AND COMMODITY PRICE BEHAVIOUR

The cost-of-carry based arbitrage bounds that can be placed on a commodity's price over time provide a direct link between a commodity's cost of storage and the available arbitrage bounds on its price behaviour. The nature of this cost-of-carry based relationship is reviewed briefly below.

Storage costs and cost-of-carry based bounds on price movements

Upper and lower bounds on the future price of a commodity can be imposed with standard cost-of-carry arguments. The upper bound is constructed by storing the commodity and the lower bound by borrowing it, and so the tightness of the resulting bounds depends crucially on the commodity's cost of storage and its closely related physical borrowing cost. For example, under the simplifying assumption that the interest rate r_t and the costs of storing S_t and of borrowing a commodity b_t (expressed as rates) are deterministic (but potentially time varying), the upper and lower bounds are

$$P_t \exp\left(\int_t^T (r_\tau + s_\tau) d\tau\right)$$

and

$$P_t \exp\left(\int_t^T (b_\tau - r_\tau) d\tau\right)$$

respectively (see Figure A).

Precisely where within these bounds the actual forward price will lie is determined by the balance between the supply of and demand for forward positions in the commodity, not by arbitrage arguments, with the implied holding cost that yields the current price defining the commodity's convenience yield. As the difference between these bounds grows with the length of the holding period and the magnitude of the commodity's storage and borrowing costs, relatively tight bounds can be placed on the forward price at relatively distant points in time for commodities with low storage costs (such as gold) whereas meaningful bounds can only be imposed on the forward price of commodities that are more costly to store, such as oil and most agricultural commodities, over relatively short periods. For an essentially non-storable commodity like electricity, the bounds lose their economic relevance in a matter of minutes.[i]

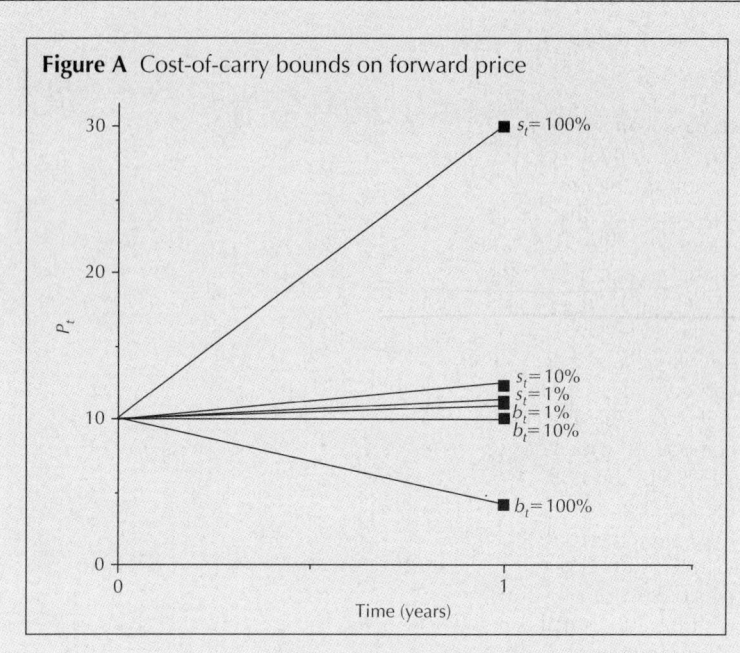

Figure A Cost-of-carry bounds on forward price

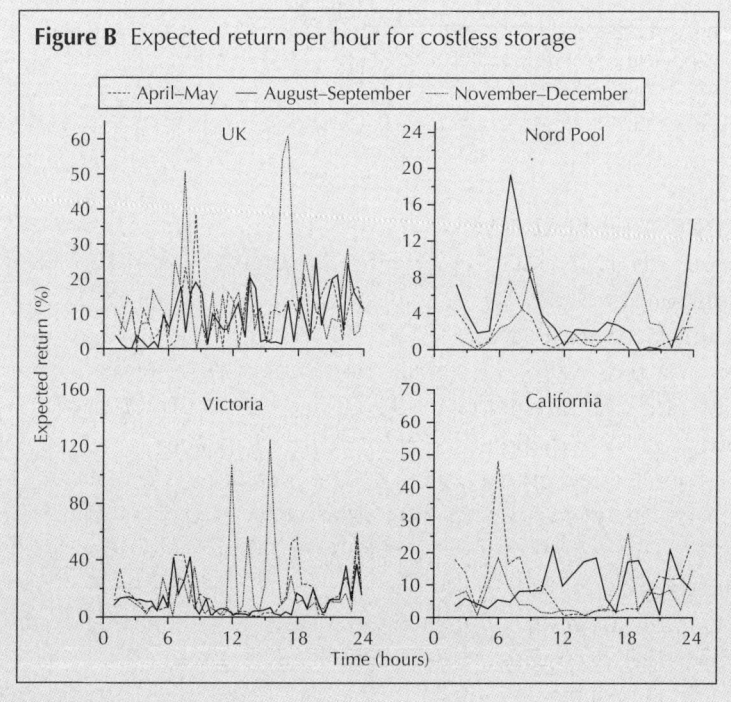

Figure B Expected return per hour for costless storage

SELECTING STOCHASTIC PROCESSES FOR MODELLING ELECTRICITY PRICES

> A second way of arriving at the same conclusion is to note that, as the cost of maintaining an inventory of a commodity grows, the magnitude of the fluctuations in the commodity's price that cannot be economically smoothed by buying into and selling out of an inventory position in the commodity also grows. As a result, the size of the fluctuations in the commodity's price that can be supported in equilibrium increases with the commodity's cost of storage. In the extreme case of a non-storable commodity, even predictable and dramatic price changes cannot be ruled out. As the electricity price data presented earlier illustrate, price changes of this kind are common in electricity markets. Figure B summarises this characteristic of electricity prices by showing the expected return *per hour* (by season) which could be earned in each of the four markets considered with a technology that makes possible the costless storage of electricity. The cumulative (non-compound) expected return *per day* ranges by season from 147% to 229% in California, 44% to 107% in Scandinavia, 417% to 759% in the UK and 494% to 966% in Victoria. It seems reasonable to view these numbers as the "shadow price" of storage in each market. When interpreted in this way, they suggest a much lower value of incremental storage in the Scandinavian market, for example. This is consistent with the substantial role that hydroelectric generation and its storage-like properties play in the Scandinavian market.
>
> [1] The absence of meaningful cost-of-carry based bounds rules out the traditional cost-of-carry approach as a method of valuing all types of electricity derivatives, not just forward contracts, as discussed in Panel 3.

THE ROLE OF REAL-TIME MARKETING CLEARING

Due to electricity's non-storable nature, inventory cannot be used to smooth fluctuations in production or demand over time. As a result, prices must adjust to the level necessary to balance production and consumption on a real-time basis (see panel 1, "Storage costs and commodity price behaviour"). Both predictable and unpredictable variations in production and demand over time will in general affect prices. For example, sudden, short-lived price movements, both positive and negative, may result from temporary imbalances in supply and demand, with their durations determined by asset availability, the time and cost required to ramp asset utilisation up or down, and the ability and lead time required to manage demand. As the real-time balancing of production and consumption must occur at each geographic location, the

ENERGY MODELLING

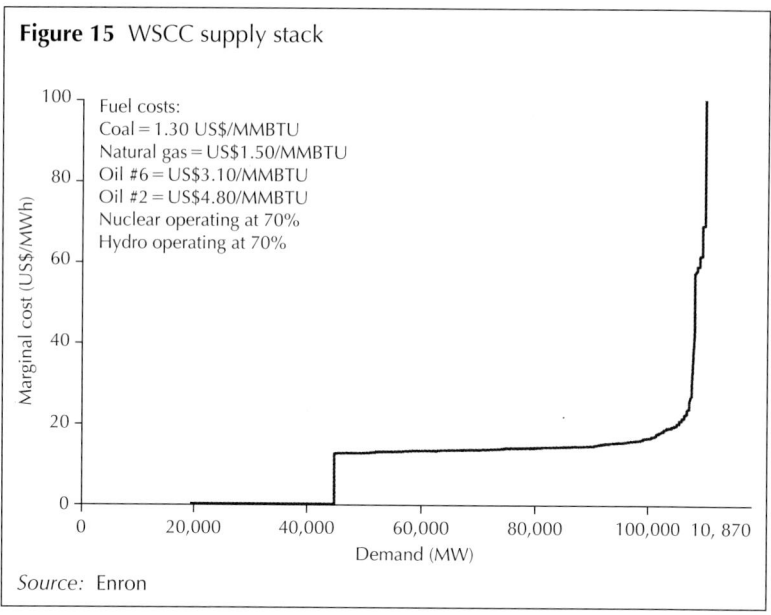

Figure 15 WSCC supply stack

Fuel costs:
Coal = 1.30 US$/MMBTU
Natural gas = US$1.50/MMBTU
Oil #6 = US$3.10/MMBTU
Oil #2 = US$4.80/MMBTU
Nuclear operating at 70%
Hydro operating at 70%

Source: Enron

relationship between prices over time at different locations will also depend on the cost and availability of real-time transmission between the locations.

THE CHARACTERISTICS OF ELECTRICITY PRODUCTION AND CONSUMPTION

Regional, real-time production and consumption play an important role in determining the behaviour of electricity prices. The most important characteristics of electricity production are that:

❑ it is accomplished by the discrete set of generally relatively large assets that collectively make up a region's electricity "supply stack" or "merit order"; and
❑ these assets are (generally) dispatched in order of their marginal cost of operation.

Figure 15 shows the supply stack for the Western States Coordinating Council (WSCC), which includes the states of Arizona, California, Colorado, Idaho, Montana, Nevada, New Mexico, Oregon, Utah and

Washington. Operationally, the dispatch of a generation asset is constrained by:

❑ the cost and time required to turn it on and off, which are generally quite small for hydroelectric and gas-fired facilities, but quite substantial for coal and nuclear facilities; and
❑ the production levels at which the asset can be operated, which commonly range between 70% and 100% of its rated capacity.

Generation assets are also subject to occasional planned and unplanned outages. Ideally, the electricity supply stack in a region will be optimised to minimise the expected cost of meeting the region's demand by selecting an appropriate mix of high-fixed-cost, low-operating-cost "base-load" units through low-fixed-cost, high-operating-cost "peaking units". In many cases, however, previous regulatory requirements and incentives, combined with the absence of active regional electricity markets, encouraged utilities to tailor their generation and transmission investments to local fuel sources and to the needs of their own customer base, rather than to the fuel stock and consumption patterns of the region as a whole.

The principal characteristics of electricity demand are that it is:

❑ highly inelastic, due to a combination of its non-storability, its status as a necessity in many uses, and the currently limited use of time-of-use based pricing;
❑ highly cyclical over short periods, due to the patterns of economic activity over the course of a day and week; and
❑ seasonal, due primarily to weather patterns.

Figure 16 shows the mean demand by hour of the day in each of the four markets under consideration during the months of April–May, August–September, and November–December, 1998. Figure 17 illustrates the relationship between demand, price and price volatility over the course of the day during August–September in each of the markets. In this figure, each series is normalised so that its initial value is one. Finally, Figures 18 and 19 plot hourly demand and price pairs in each market for the months of April–May and August–September. These figures show that a positive correlation exists between demand and price, but that the strength of this relationship varies substantially across markets.

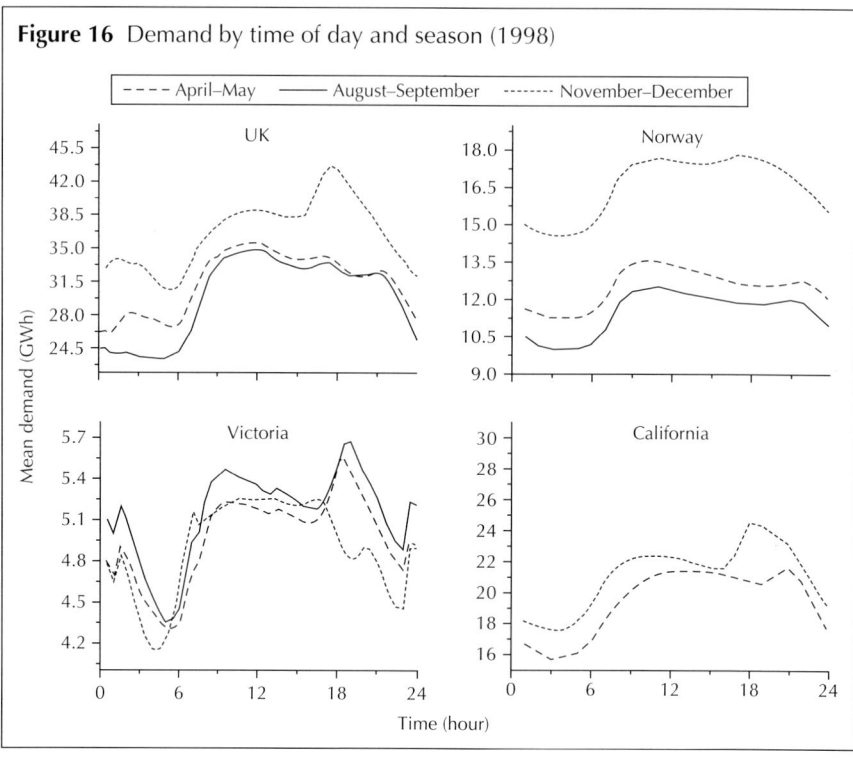

Figure 16 Demand by time of day and season (1998)

To better understand the relationships in the figures note that, in an economically efficient electricity market, the relevant supply stack will be dispatched in merit order, and the prevailing price of power will be the marginal cost of the last unit dispatched. When this is the case, the relationship between demand and price will trace out the curve corresponding to the marginal cost curve of the relevant supply stack. A relationship of this kind is in fact quite easy to identify in the California data. The relevant supply stack for the California market is the WSCC, shown in Figure 15. The non-linear shape of the demand – price relationship for California plotted in Figures 18 and 19 mirrors the shape of this supply stack quite closely. The steep downward slope of the plot at low levels of demand is due to the substantial start-up and shut-down costs of some base-load generators, such as nuclear and coal plants. To avoid these costs, operators of these plants frequently price power below cost during periods of very low demand, with prices of zero common.

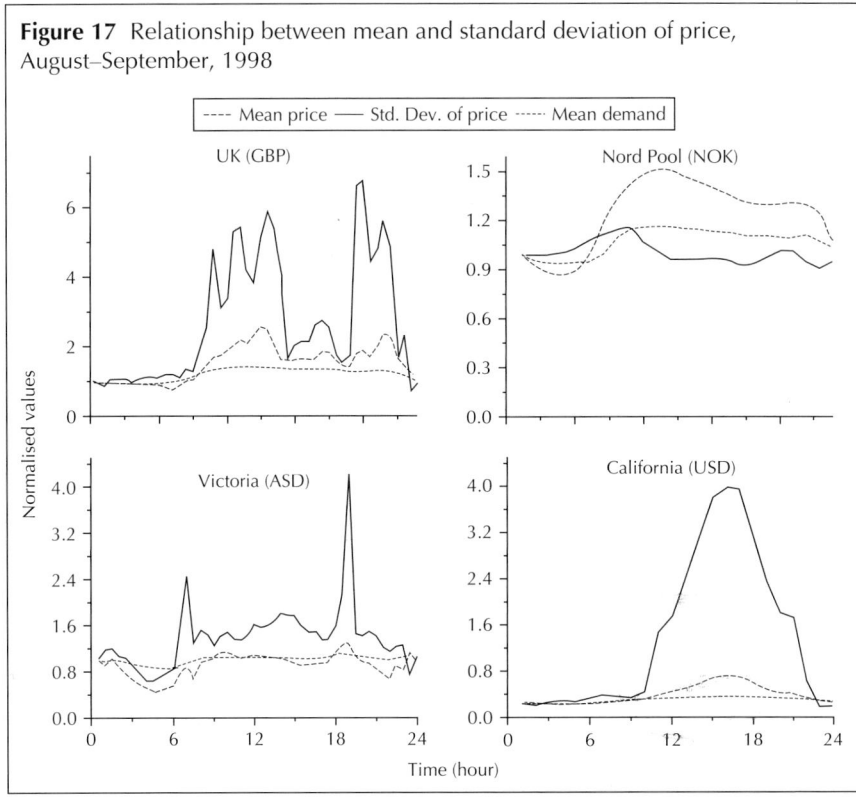

Figure 17 Relationship between mean and standard deviation of price, August–September, 1998

The relationship between the demand–price plots and the regional supply stacks in the other three markets are less easy to identify. The supply stack for Norway is dominated by quite similar hydro assets, but prices in the Scandinavia market are also influenced by transfers of power into and out of the market from neighbouring regions where nuclear and thermal assets are dominant. In Victoria, three large brown-coal generators, with nearly identical cost structures and operating constraints, together have generating capacity substantially in excess of the usual range of demand in the market. As a result, prevailing prices are determined as much by the strategic price and quantity decisions of the generators as by the prevailing demand level, with occasional exceptions during periods of high demand or the temporary outage of one or more of the assets. In the UK, two large generators are generally considered to have exercised market power quite

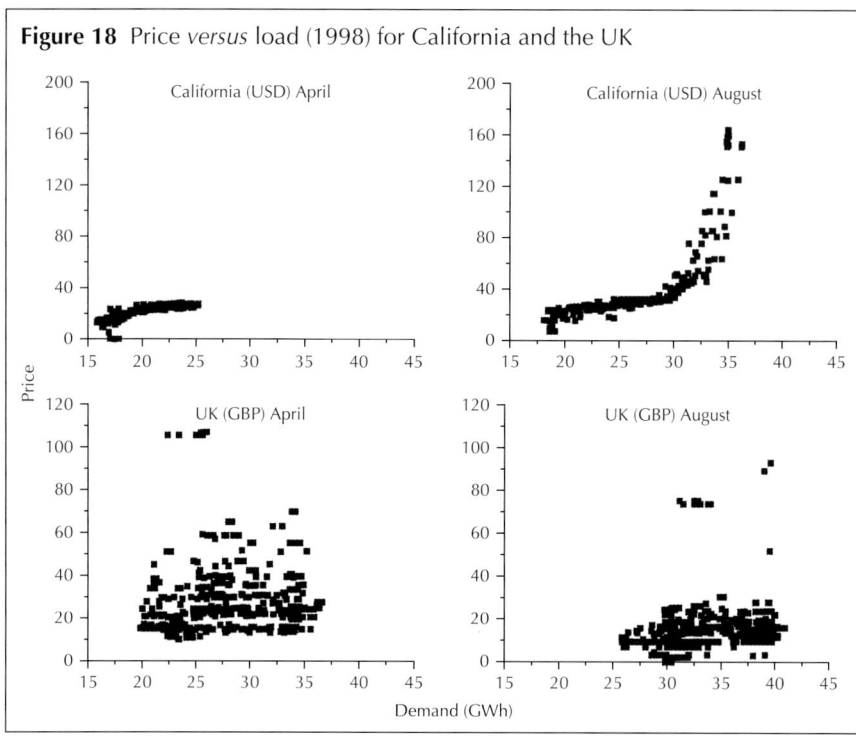

Figure 18 Price *versus* load (1998) for California and the UK

successfully on a consistent basis, causing divergences between price and marginal cost.

Taken together, the markets in California, Scandinavia, Victoria and the UK highlight the importance of the specific characteristics of the production base in a market, including the mix of generating assets (concentrated hydro in Scandinavia, brown coal in Victoria) and the ownership structure of the assets (concentrated ownership in the UK). In the WSCC, both the mix and the ownership structure of the supply stack are apparently sufficiently diverse to result in a market outcome that initially approximated that of an efficient market quite well, and later digressed dramatically as suppliers began to exhibit market power.

EVALUATION OF ALTERNATIVE PRICE MODELS

The unique behaviour of electricity prices identified above and the link between this behaviour and the non-storability of electricity

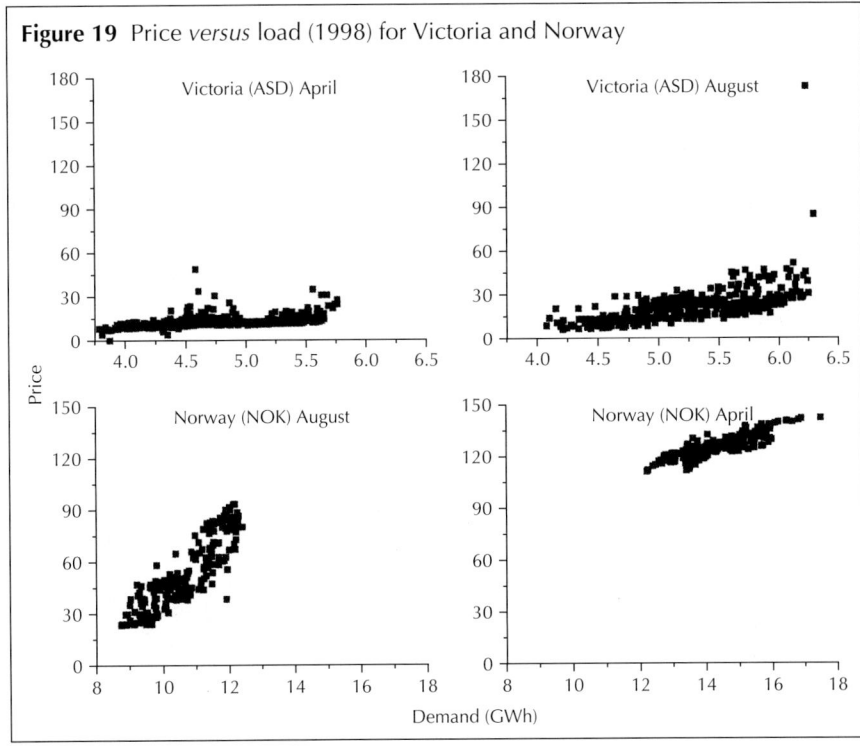

Figure 19 Price *versus* load (1998) for Victoria and Norway

and the underlying production and demand fundamentals of a particular market suggest that effective models of electricity prices must include variations in mean price and price volatility by time of day, week and year, and must be market-specific. They should also reflect the strongly mean-reverting nature of electricity prices, and may need to allow for brief positive and negative "spikes". In this section a range of models that possess some or all of these features is evaluated empirically. For comparative purposes the empirical performance of more traditional models, such as geometric Brownian motion, is also included.

MODELS CONSIDERED

Four types of models were evaluated: Brownian motion, Orstein–Uhlenbeck mean reversion, geometric Brownian motion, and geometric (log) mean reversion. Each model type was evaluated

Table 1 Model property characteristics

	BM	MR	GBM	GMR
Mean reversion		X		X
Seasonal effects	X	X	X	X
Price-dependant volatility			X	X
Occasional price spikes		X		X

with and without jumps, for a total of eight models. The models with jumps incorporate sudden, discontinuous price changes in addition to the continuous changes of the basic diffusion models. Mathematically, the four models without jumps are:

Brownian motion $\quad dP_t = \mu_t dt + \sigma dB_t \quad$ (BM)

Mean reversion $\quad dP_t = \kappa(\alpha_t - P_t)dt + \sigma dB_t \quad$ (MR)

Geometric Brownian motion $\quad dP_t = \mu_t P_t dt + \sigma P_t dB_t \quad$ (GBM)

Geometric mean reversion $\quad dP_t = \kappa\left(\alpha_t + \dfrac{\sigma^2}{2} - \ln P_t\right)P_t dt + \sigma P_t dB_t \quad$ (GMR)

When included, jumps are modelled with a Poisson arrival time, Bernoulli (positive or negative) jump direction, and exponential jump magnitude. Models modified to include jumps are denoted with an additional "J" appended to the model acronyms defined above (GMR thus becomes GMRJ). In the mean-reverting models (ie, MR and GMR), the jump modification produces a "spike" effect as large jumps create a high level of mean reversion. Table 1 summarises the ability of each of the models to capture the four key properties of electricity prices identified above.

EMPIRICAL ANALYSIS

To evaluate the models, their ability to fit price data from the California, Scandinavia, UK, and Victoria markets for the months of April–May, August–September, and November–December 1998 was compared.

The data were deseasonalised before the models were fitted in order to leave the relationship between price and volatility

> **PANEL 2 SUMMARY OF PROPERTIES OF ELECTRICITY PRICES**
>
> The key properties of electricity prices identified in the main text, along with the economic fundamentals that they reflect, are given below.
>
> **Mean reversion**
> Electricity prices fluctuate around values determined by the cost of production and the level of demand. In the short run, mean reversion results from the cyclical, mean-reverting nature of demand, but in the long run it results from bounds imposed by the cost of new generation.
>
> **Seasonal effects**
> The mean price of electricity in individual markets varies by time of day, week and year in response to cyclical fluctuations in demand of the same frequencies. The precise shape and magnitude of the price cycles observed depend on the patterns of economic activity and weather in the region, and on the characteristics and ownership structure of the region's generation assets.
>
> **Price-dependent volatility**
> Like the mean price of electricity, the volatility of electricity varies cyclically with price as fluctuations in demand move different generating assets to the margin. Because higher levels of demand bring assets on the upper, steeper portion of the supply stack to the margin, price volatility increases with price level.
>
> **Occasional price "spikes"**
> Electricity prices exhibit occasional positive price spikes when either key generation of transmission assets suffer an outage or, because of an unusual load condition, demand reaches the limits of available capacity. When the relevant asset is returned to service or demand recedes, prices quickly return to more typical levels. Negative spikes occur when operating costs or constraints limit the ability of generators at or near the margin to curtail production during brief periods of reduced demand.

unchanged in each model. For the BM and MR models, seasonality was removed by subtracting the average hourly or (for the UK and Victoria) half-hourly price changes, and for the GMR and GBM models by subtracting average hourly or half-hourly percentage price changes. No attempt was made to account for day of the week effects in the data.

Using the appropriate deseasonalised data set, the parameters of each model were fitted using the method of maximum likelihood, and the sum of the log-likelihood values were recorded. Larger (less negative) values imply a better fit of the transitional (conditional) distributions to the observed time series. Note that there are twice as many data points for the markets with half-hourly data (the UK and Victoria) than those with hourly prices (California and Scandinavia).[1] The best value for a particular region/season is shown in bold. A quick analysis of the results reported in Table 2 reveals that in every case:

❑ the mean-reverting model outperforms its Brownian motion counterpart;
❑ adding jumps to a model improves its performance; and
❑ the best model without jumps is also the best model with jumps.

SAMPLE PARAMETER VALUES

Table 3 records the annualised versions of the parameter values obtained for the California market for August–September 1998. Notice that in all models the addition of jumps decreases the diffusion term, and in the mean reverting models it also reduces the degree of mean reversion.

LEAST LIKELY EVENTS

While the sum of the likelihood values reported in Table 2 reflects the overall performance of each model on each data series, they do not provide insight into the relative ability of the models to capture the extreme events in the data series. Accordingly, the likelihood values of the least likely price changes of each model are recorded in Table 4. These values range from approximately 1-in-10,000 to 1-in-10^{-21}. In all models without jumps the least likely event actually observed would be expected to occur less than once every 10 million years. This suggests that models without jumps are inappropriate for modelling electricity prices.

LONG-RUN DISTRIBUTIONS AND SIMULATIONS

The log-likelihood values reported above are based on the conditional transition density, so when they are fitted to hourly or

Table 2 Sum of log-likelihood values

Model	BM	MR	GBM	GMR	BJM	MRJ	GBMJ	GMRJ
Scandinavia								
Apr–May	−3829.6	−3819.5	−4179.2	−4166.3	−3442.3	**−3438.1**	−3656.1	−3649.8
Aug–Sep	−4229.3	−4221.8	−4499.0	−4490.5	−3889.2	**−3874.1**	−4062.9	−4056.5
Nov–Dec	−6582.7	−6469.7	−5125.1	−5087.9	−4419.7*	−4406.3*	−3799.3	−3762.7
UK								
Apr–May	−10274.1	−10057.3	−8659.3	−8508.0	−8831.4	−8778.4	−8033.2	**−7973.8**
Aug–Sep	−10122.9	−9785.3	−8448.2	−8219.1	−8083.1	−7980.4	−7426.5	**−7340.2**
Nov–Dec	−11377.0	−11228.8	−9655.1	−9533.1	−10038.4	−9960.3	−9117.4	**−9051.3**
Victoria								
Apr–May	−24601.2	−24296.1	−20546.2	−20436.8	−21651.8	−21566.9*	−19750.5	**−19622.0**
Aug–Sep	−21978.7	−21803.4	−21120.2	−21000.5	−20695.1	−20902.6*	−20848.0	**−20775.5**
Nov–Dec	−30944.0	−30589.2	−21077.4	−21030.9	−22206.0*	−22198.0*	−20518.0	**−20487.6**
California								
Apr–May	−3078.6	−3065.6	−3903.3	−3876.4	−2854.2	**−2844.1**	−3322.3	−3317.9
Aug–Sep	−5191.5	−5170.0	−4105.3	−4084.6	−4555.2	−4554.8	−3836.3	**−3829.6**
Nov–Dec	−4428.1	−4399.1	−3539.2	−3518.8	−3242.4	−3209.8	−3096.2	**−3080.4**

Note: Values marked with an asterisk may be less reliable due to the presence of data points with extremely small likelihood values, which raise issues of numerical precision

Table 3 Annualised parameter values for California, August–September 1998

	μ	α	κ	σ	λ	ψ	γ
BM	−28.6			787			
BMJ	−1753			127	7828	0.520	0.187
MR		36.8	524	800			
MRJ		−24.4	25.9	87.0	8100	0.517	0.191
GBM	−1.22			11.9			
GBMJ	2.80			6.93	1885	0.512	6.22
GMR		3.45	504	12.1			
GMRJ		3.44	221	7.67	1850	0.513	6.19

Table 4 Worst case likelihood values

	BM	MR	GBM	GMR	BMJ	MRJ	GBMJ	GMRJ
Likelihood	$3*10^{-19}$	$5*10^{-20}$	$6*10^{-16}$	$1*10^{-16}$	$4*10^{-7}$	$3*10^{-7}$	$1*10^{-5}$	$1*10^{-5}$

half-hourly prices changes, as they were above, they measure the performance of the models in the short term. Because the level of mean reversion identified (see Table 3) is quite small over a one-hour period, the differences identified between the transitional density distributions of the mean-reversion models and the corresponding Brownian models are quite small. Over longer periods, however, the cumulative effect of this level of mean reversion becomes substantial.

To illustrate this effect Figure 20 shows the distribution over the two-month period (conditioned on the initial value of the period) implied by each model, as well as the actual two-month distribution observed in the California PX prices for August–September 1998. As the figure makes clear, despite their nearly equal average log-likelihood values over the short term, the GBM and GMR models differ substantially in the long-run distributions they imply, with the GMR model's performance clearly superior. The figure also shows the drawback of the GMRJ model relative to the GMR model Specifically, while the log-likelihood values of the two models reported in Tables 2 and 4 highlight the superiority of the GMRJ model in representing extreme events and characterising the transitional error distributions, the long-run distributions in Figure 20

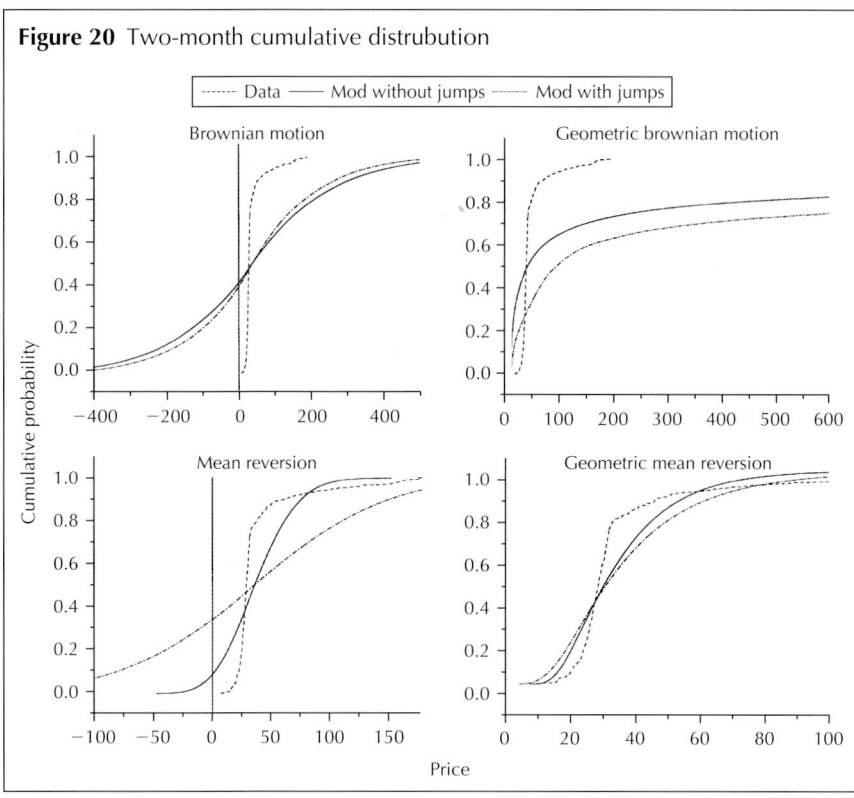

Figure 20 Two-month cumulative distrubution

show that the GMR model better characterises the long-run distribution of prices near their average value. When jumps are included, the amount of diffusion decreases, and as a result so does the degree of mean reversion. Thus, the improved ability of the GMRJ model to characterise short-term price changes comes at the cost of allowing prices to deviate from the average price for longer periods of time.

The relative performance of the models over the longer run can also be seen informally in simulations of each model. In Figure 21 the actual California August–September 1998 data are plotted together with simulated paths from each of the models. Like the graphs of Figure 20, the figure illustrates the wide dispersion of the Brownian models and the much better fit of the GMR and GMRJ models.

Finally, to better illustrate the daily patterns implied by the models, Figure 22 provides simulations of prices over only the first

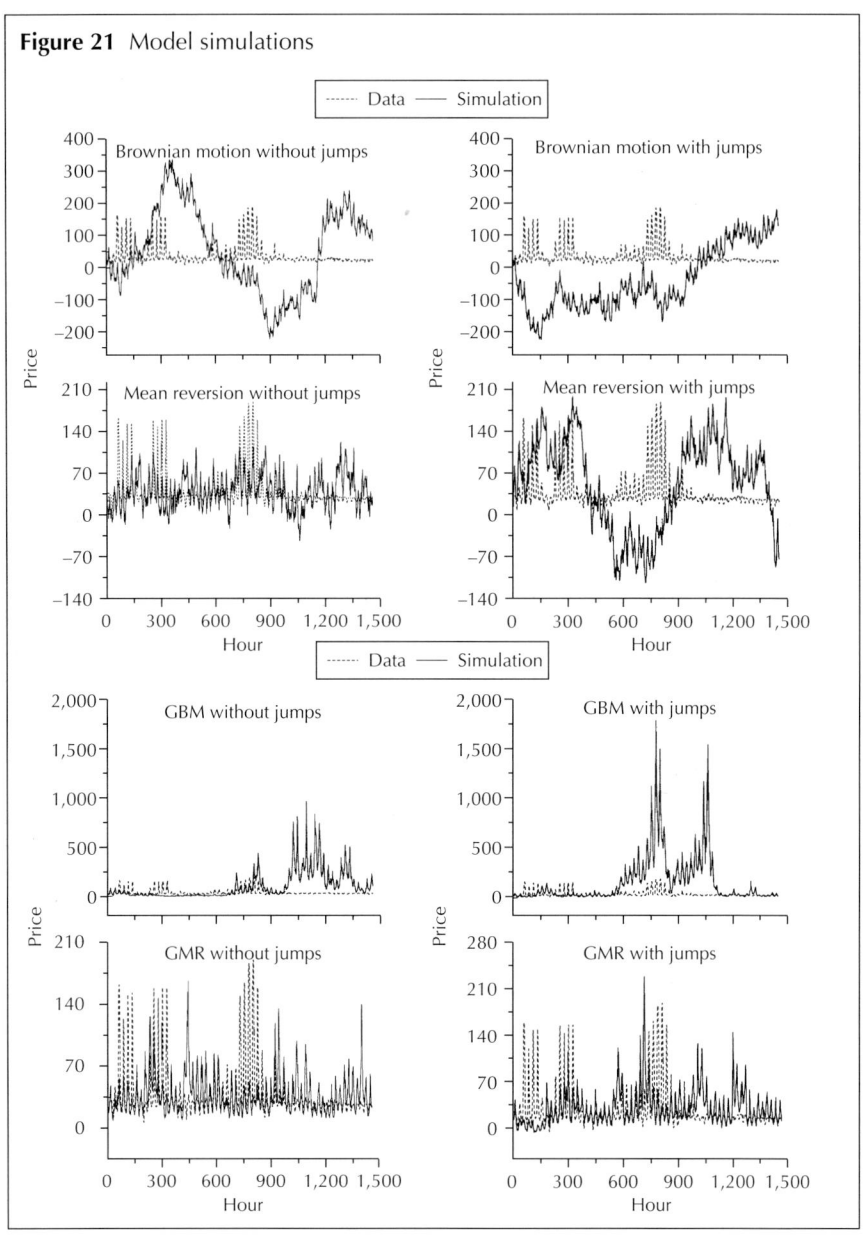

Figure 21 Model simulations

week of the sample period. The figure again illustrates the problems associated with the Brownian models, namely overcorrection of daily price patterns and undercorrection of longer-term price fluctuations. Table 5 summarises this trade off numerically, showing

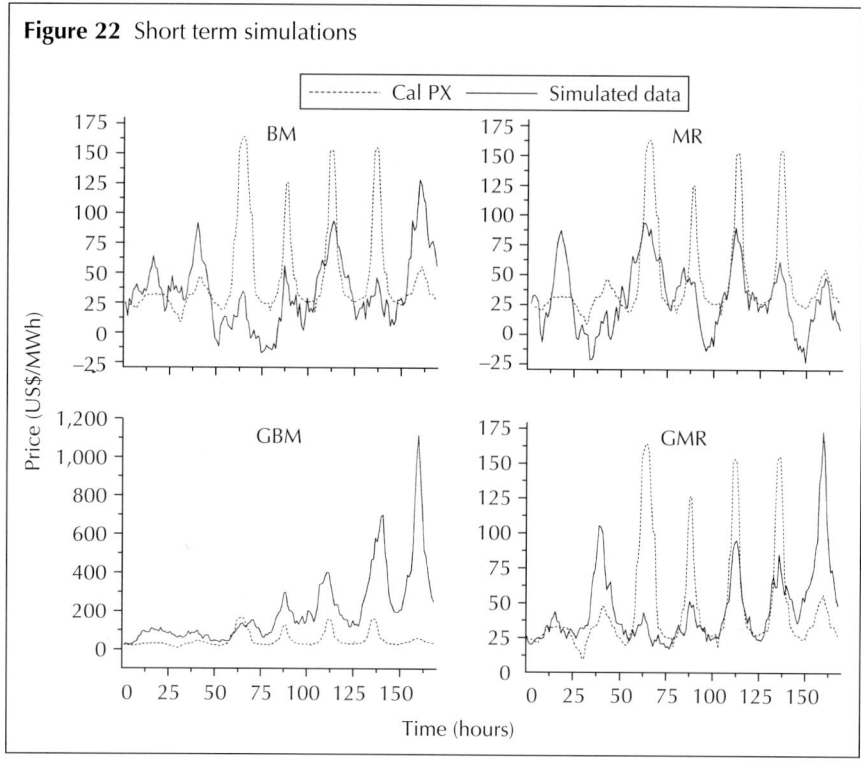

Figure 22 Short term simulations

Table 5 Log-likelihood values for California, August–September 1998

Model	BM	MR	GBM	GMR	BMJ	MRJ	GBMJ	GMRJ
1 Hour	−5191.5	−5170.0	−4104.6	−4085.8	−4555.2	−4554.8	−3836.3	−3829.6
1 Day	−6790.2	−6549.7	−5708.3	−5488.6	−6720.5	−6665.2	−5811.4	−5509.9
1 Week	−8207.9	−6724.7	−7088.8	−5641.9	−8079.1	−7741.8	−7172.1	−5831.4

the log-likelihood for each model and data series for periods of one day and one week along with the values for one hour (for comparative purposes). The GMR and GMRJ models consistently perform the best, with GMR superior for the longer-term predictions. Note, too, that the relative performance of the BM and GBM models (with and without jumps) deteriorates significantly as the period length increases.

PANEL 3 THE FOUNDATION OF A FIRM'S VALUATION AND RISK MANAGEMENT SYSTEMS

Price process models form the foundation of a firm's valuation systems because, in order to place a value on an asset – whether a structured contract, a derivative or a physical asset – the price of the power to which the asset provides a claim must be well understood. Price models play an equally fundamental role in risk management systems because the future power prices on which asset values depend can never be known with certainty.

Due to the foundational role that price models play in valuation and risk management systems, the results that these systems produce can be no more reliable than the price models on which they are based. For example, price models that do not include the hourly and weekly price cycles that characterise electricity prices will clearly misprice contracts or options written on hourly or daily power. Likewise, models that do not include the dramatic price spikes observed in electricity prices will, among other things, substantially undervalue out-of-the-money calls and significantly misrepresent value-at-risk (VAR), as has been made clear, for example, in the many extreme market environments that have been experienced, such as the Midwest of the US during the summer of 1998 and California in 2000.

In addition to an underlying price model, valuation and risk management systems must also include methods for valuing and managing the risk of the assets that the systems are designed to evaluate, such as derivatives, contracts or physical assets. The algorithms that form the basis of these methods are typically complex. To keep this complexity manageable, such algorithms are generally designed around specific price models. Although frequently necessary, this simplification has the unfortunate consequence of making valuation and risk management systems price-model dependent. This fact is extremely important to bear in mind when selecting either a price model or a valuation or risk-management system. The relevant risks are that either:

❑ the valuation and risk management system selected is built on an inappropriate price model, causing its results to be unreliable (as is typically the case for systems originally developed for commodities other than electricity); or
❑ a price model will be chosen for which no adequate valuation and risk management systems are available.

Although there is much work in progress that is attempting to develop valuation and risk management systems tailored to electricity, it should be noted that to do so successfully requires changes beyond simply substituting one price model for another. The most important change required results from the fact that electricity is non-storable, whereas the commodities for which existing valuation and risk management methodologies were developed are not. This is important because it is possible to replicate, or to hedge perfectly, derivatives and other assets that depend on storable commodities by maintaining a physical position in the relevant commodity. As electricity is non-storable, these methods are clearly not applicable, and an entirely different approach to the hedging and replication of electricity based assets is required. To consider how power assets can be replicated, note that an exposure to power prices at a specific future time and location can be created by taking a position in a futures contract for that location, which matures at that time. Thus an electricity-based asset can be replicated by taking positions in the appropriate set of futures contracts. For a complete description of the methodology for doing so when a traded futures contract for the relevant maturity and location exists, see Deng, Johnson and Sogomonian (1998).

When the necessary futures contracts do not exist, other methods are required. An alternative approach is to assume that the underlying factor driving the value of the asset is the market's expectation about the price of electricity at the asset's maturity. The asset can then be valued and hedged using other assets or derivatives with the same maturity date.

While this procedure may be implemented for any price model, to illustrate it consider the model of the spot price of electricity P_t, given by

$$dP_t = \kappa\left(\alpha + \frac{\sigma^2}{2} - \ln P_t\right)P_t dt + \sigma P_t dB_t$$

Let $Z_t^T \equiv E_t[P_T]$ denote the underlying *virtual* asset where t is the current time and T is the maturity date of the derivate. Solving for Z_t^T yields

$$Z_t^T = \exp\left[e^{-\kappa(T-t)}(\ln P_t - \alpha) + \alpha + \frac{\sigma^2}{4\kappa}\left(1 - e^{-2\kappa(T-t)}\right)\right]$$

$$dZ_t^T = \sigma Z_t^T \exp^{-\kappa(T-t)dB_t}$$

Notice that

- Z_t^T is a martingale which converges to the spot price P_T as $t \to T$.
- The time varying volatility of Z_t^T, $\sigma e^{-\kappa(T-t)}$, is due to the mean reversion of the spot price.

Next, to construct a riskless portfolio, let $G(Z, t)$ and $H(Z, t)$ be the prices at time t of European style derivative securities defined on $Z \equiv Z_t^T$ with the dynamics.

$$dG(Z,t) = v_G(Z,t)dt + s_G(Z,t)dB_t$$
$$dH(Z,t) = v_H(Z,t)dt + s_H(Z,t)dB_t$$

To form a riskless portfolio, take $s_H(Z,t)$ units of derivative "G" and $-s_G(Z,t)$ units of derivative "H". Because this portfolio is riskless, it must have a growth rate equal to the risk-free rate r:

$$s_H v_G - s_G v_H = r(s_H G - s_G H)$$

or, equivalently,

$$\frac{v_G - rG}{s_G} = \frac{v_H - rH}{s_H} \equiv \lambda \quad \text{(the market price of risk)}$$

Using standard methods, this equation may be solved to provide expressions for the values of the derivatives G and H.

Applying this methodology to forward prices, the price of a forward given the current spot price, $F_t^T(P_t)$ can be shown to be equal to the expected spot price at time T multiplied by a risk premium.

$$F_t^T(P_t) = e^{\mu_t^T} E_t[P_T] \text{ where } e^{\mu_t^T} = \underbrace{\exp\left[-\frac{\lambda\sigma}{\kappa}\left(1 - e^{-\kappa(T-t)}\right)\right]}_{\text{risk premium}}$$

Additionally, examination of the dynamic behaviour of F_t^T reveals that the forward price follows a geometric Brownian process with time-varying volatility which reflects the mean-reverting nature of electricity prices and drift equal to the market price of risk multiplied by this volatility.

$$d_t F_t^T = \underbrace{\lambda\sigma e^{-\kappa(T-t)}}_{\text{expected return}} F_t^T dt + \underbrace{\sigma e^{-\kappa(T-t)}}_{\text{volatility}} F_t^T dB_t$$

Following similar steps, the formula for a European call can be derived. The resulting formula is equivalent to the Black commodity pricing formula, with the volatility parameter modified to reflect the mean-reverting nature of electricity prices. Details can be found in Barz (1999).

INTRADAY PRICE DYNAMICS

While the non-storability of electricity rules out cost-of-carry-based bounds on the fluctuations of electricity prices, over short time horizons a significant degree of serial correlation in prices is

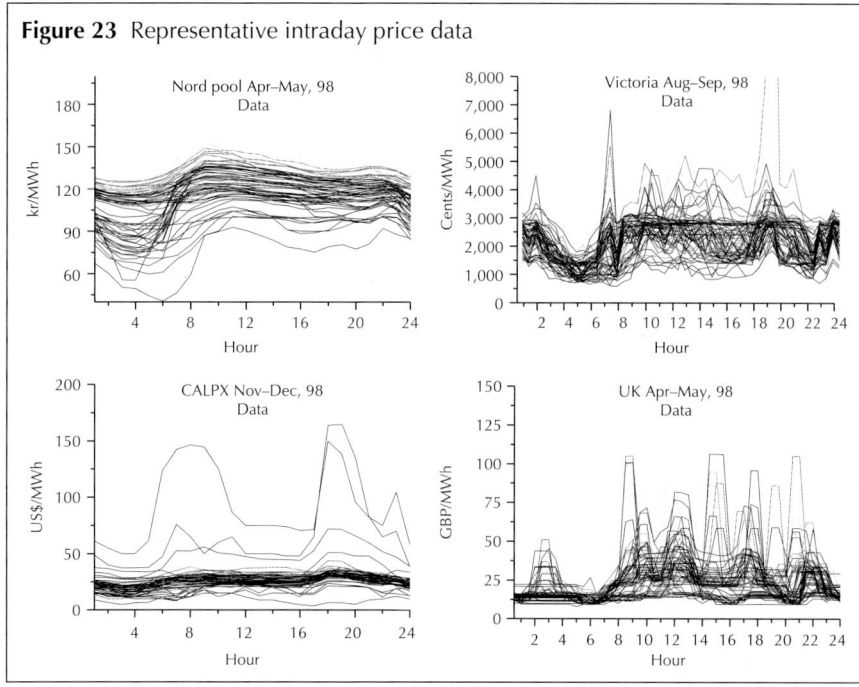

Figure 23 Representative intraday price data

introduced by the serial correlation in demand, and to a lesser extent in supply. The impact of serial correlation in demand and supply is particularly strong over the intraday period, due to the high degree of predictability of intraday demand given knowledge of the day's weather conditions, of local time-of-use patterns and of the availability and operating status of key supply resources.

Serial correlation in intraday prices is clearly visible in the representative intraday price data shown in Figure 23. Each line on the charts in the figure shows the evolution of power prices over the 24 hours of an individual day. As the charts show, while the unconditional distribution of power prices at any given hour of the day is broad, as one would expect given the high unconditional volatility of power prices, the conditional distribution of prices in the same hour given knowledge of actual prices at one or more earlier hours in the day is much narrower.

The conditional distributions in the figure show that days that experience high (low) prices early in the day are much more likely to exhibit high (low) prices later in the day, consistent with the significant intraday serial correlation in demand and supply. Also, note that very little serial correlation persists through the night and early-morning hours of the daily price cycle. This behaviour results from the daily cycle of the dispatch decisions of most generation assets and the limited duration of the serial correlation of demand, and generates a nearly complete "reset" back to the unconditional distribution at the start of each day.

These characteristics of intraday price behaviour present further challenges to electricity price modelling, adding an additional level of complexity to the longer-term volatility and mean-reversion characteristics identified above. To provide an appropriate basis for intraday decisions and risk assessments, such as those required by intraday trading, contract utilisation and asset dispatch decisions, a price model must capture these important characteristics of intraday price dynamics. The remainder of this section presents a generalisation of the top-performing GMR/GMRJ class of models above that enables these intraday dynamics to be captured.

Locally mean-reverting-diverting price process
As illustrated in Figure 23, intraday prices follow relatively smooth paths over the course of the day, consistent with the relative predictability of intraday demand and supply. Specifically, in contrast to the unconditional distributions of hourly prices, conditional distributions of intraday prices:

❏ follow relatively predictable trajectories consistent with the conditional mean prices for prevailing supply-and-demand conditions, which may differ substantially from unconditional mean prices;
❏ exhibit lower volatility, due to the relatively low volatility of intraday demand around its daily time-of-use pattern, and to the lead time constraints on changes in asset operating status; and
❏ reflect the nonlinear transformation imposed on demand by the nonlinear supply stacks of nearly all markets, visible in the divergence of prices from unconditional mean values during periods when demand is moving away from mean levels (that is, ramping towards peak demand periods), and reversion in prices towards unconditional mean values during periods when

demand is reverting towards mean levels (for example, following peak demand periods).

At an intuitive level, this behaviour is consistent with the properties of the GMR and GMRJ models with the following combination of parameter values:

❑ relatively low volatility, primarily driven by the diffusion term (as opposed to the jump terms); and
❑ mean-reverting behaviour during periods when demand generally moves toward mean levels (for example, the end of the day) and "mean-diverting" behaviour during periods when demand generally moves away from mean levels (such as prior to peak demand periods)

This intuitive assessment suggests the following changes in the calibration process for and in the constraints imposed on the parameter values of the GMR/GMRJ model:

❑ remove the constraint on the sign of the mean-reversion parameter, allowing it to take on both positive and negatives values, and by doing so to generate both mean-reverting and mean-diverting behaviour; and
❑ recalibrate the mean-reversion parameter each hour of the day, enabling it to match the type and strength of the mean-reverting or mean-diverting behaviour observed at that time of day.

Mean diversion has generally not been incorporated in the design and calibration of stochastic processes to date, since over sustained periods it allows process values to diverge to both positive and negative infinity. In the electricity context, however, because prices diverge only during certain portions of the intraday period, and strongly mean-revert both over the remaining portions of the intraday period and over longer time periods, this consideration can be eliminated through appropriate calibration of model parameters. In the following, the generalisation of the GMR/GMRJ models to incorporate such time-varying mean reversion and diversion will be referred to as the "geometric locally mean-reverting-diverting" process, both with and without jumps, or "LMRD" and "LMRDJ".[2]

The ability of the LMRD/LMRDJ models to accurately represent the behaviour of intraday electricity prices is illustrated in the charts in Figure 24. For comparison, Figure 25 shows the performance of the GMRJ model for one of the datasets shown in Figure 24. As Figure 25 shows, because the GMRJ model is restricted to mean-reverting behaviour, even with appropriately chosen parameter values it generates excessive intraday volatility, while at the same time failing to generate both the level of high and low prices observed in the data, and sustained sequences of such prices over the course of the day.

Figure 25 also illustrates the economic significance of the performance improvements provided by the LMRD/LMRDJ model. For example, the figure shows that the LMRD/LMRDJ model appropriately values both a call with a strike equal to the upper line shown on the chart, and a put with a strike equal to the lower line on the chart, while the GMB/GMRJ model fails to appropriately value either. Similarly, the excessive dispersion of the central mass of price paths over time that the GMR/GMRJ model generates in an unsuccessful attempt to match the overall volatility of prices results in incorrect valuation of a forward contract with a price equal to the central line imposed on the graph. In contrast, the LMRD/LMRDJ model generates smoother price paths with appropriately sized and timed periods of mean reversion and mean diversion over time, providing a much more accurate assessment of the value and risk of both the options and the forward.

USING DATA ABOUT FUTURE SUPPLY, DEMAND AND MARKET CONDITIONS TO IMPROVE MODEL CALIBRATION AND PERFORMANCE

While spot price data provide the richest and most complete set of data for modelling electricity price dynamics, other data, including data about market fundamentals, such as demand (seasonality in weather and load) and supply (generation and transmission asset availability, fuel prices and supply), and where available forward and option prices, provide additional information that can be used to further improve model performance. This section provides a brief overview of how data of this kind can be used to refine the calibration of the models presented above (rather than used to construct alternative types of models), and the benefits of doing so.

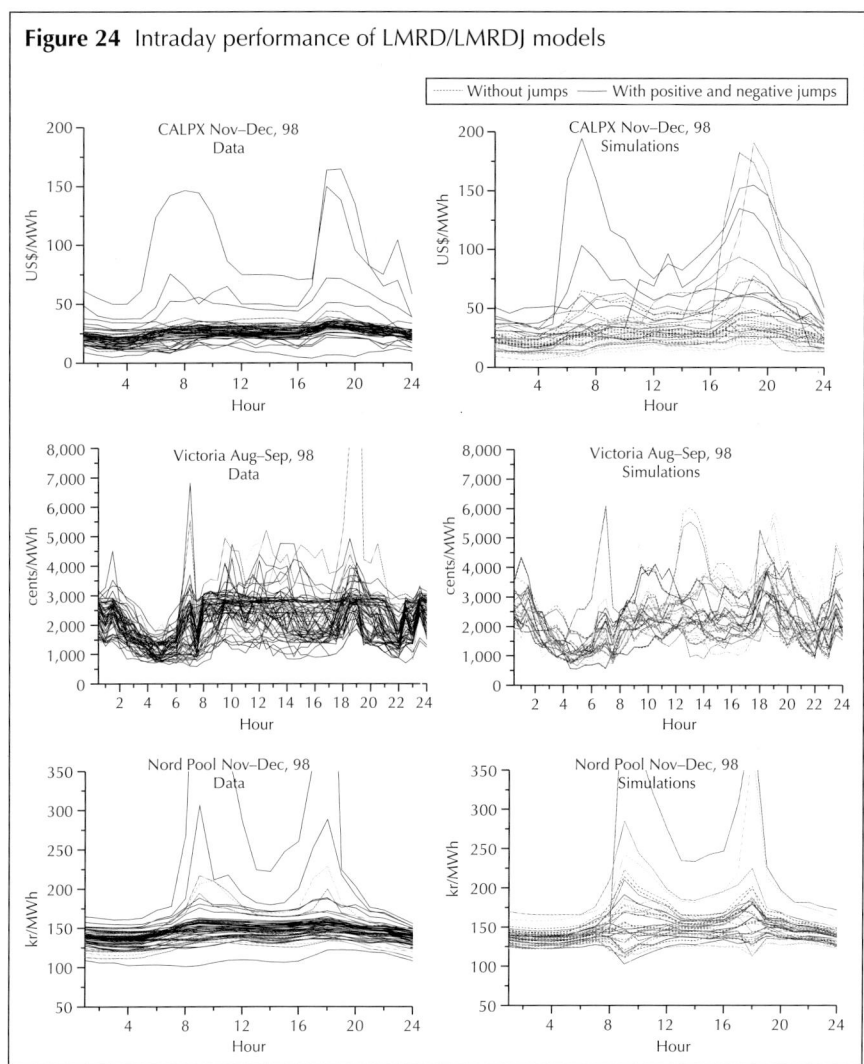

Figure 24 Intraday performance of LMRD/LMRDJ models

Joint distribution between spot price behaviour and supply, demand and forward market information

Information about future supply, demand and market price expectations is relevant to model calibration because it influences, or in the case of market price expectations signals, future price behaviour. Relationships between factors of this kind and future price behaviour allow information about the factors to be used to condition, and

ENERGY MODELLING

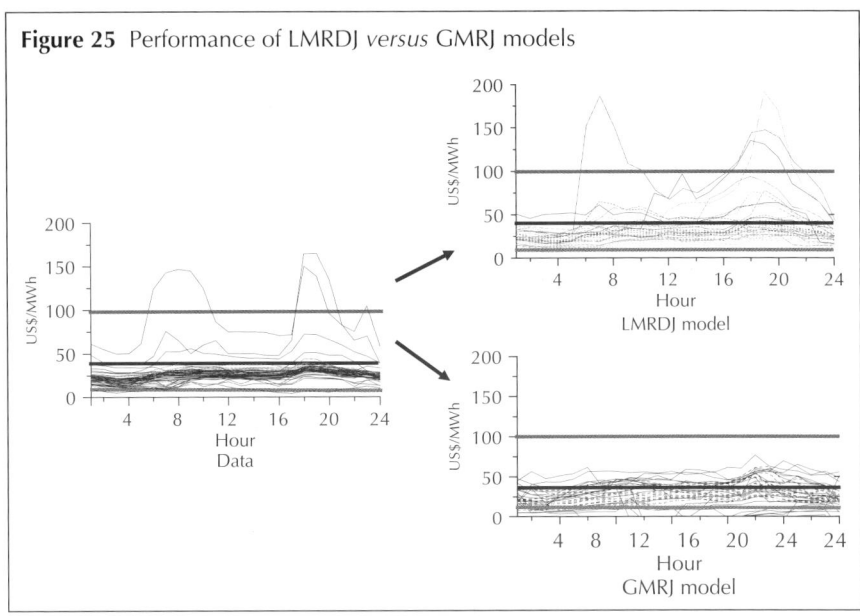

Figure 25 Performance of LMRDJ *versus* GMRJ models

therefore refine, the calibration of price models. Use of the more complete information set contained in the joint distribution of the factors and prices to calibrate the model in turn enables improved model performance.

Seasonality example

As a simple example, the joint distribution between price behaviour and "fundamental" data, which is easiest to analyse and utilise, is the seasonality of electricity prices, as reflected in Figures 1, 3, 4 and 16 and associated analyses and discussion. Seasonality in price behaviour results from seasonality in electricity supply and demand, for example the demand effects of afternoon air conditioning during warm summer months and morning and early-evening heating requirements in cooler winter months, and the supply effects of seasonal fluctuations in the amount of water available for hydroelectric generation, and in prices and supplies of fuels.

In most markets the combined effects of seasonality in supply and in demand results in important differences in price behaviour across seasons, including both the level and the time pattern of the mean and volatility of prices over the course of the day, as illustrated in

Figures 3 and 4. By identifying, analysing and utilising relationships of this kind, for example by calibrating the model separately for each seasonal period using data from comparable seasonal periods in prior years, model performance can be significantly improved.

Important "seasonality" effects can also be observed at the more granular level of "seasonality" by day of the week, for example between weekdays *versus* weekends or holidays. In some markets, differences in price behaviour across individual weekdays may also be material enough to merit independent model calibration, with significant differences most commonly observed on Mondays and Fridays. Day-of-week effects can be captured and incorporated in model performance by calibrating model parameters separately for each category of day identified.

General approach for utilising supply, demand and forward market information in model calibration

Generalising from this example of seasonality, the information content of any factor related to future supply, demand or market conditions can be assessed, and if identified as material utilised to improve model calibration and performance, through the following sequence of steps:

1. *Assess the value of alternative "conditioning" factor(s)*: Assess the information content of a prospective factor or combination of factors by comparing the behaviour of historical spot prices across alternative values of the factor, or combination of factors.
2. *Condition model calibration and performance on the value of factors(s) identified as significant*: For factors, or combinations of factors, determined to have significant information content, calibrate the model separately for each of a representative set of factor values using spot prices from past periods when the relevant factor values prevailed.
3. *Store model parameters from each "conditional calibration", and use the appropriate parameter set when similar values of the "conditioning" factor(s) prevail in the future.*

Thus, rather than being calibrated once using all available data, model parameters are calibrated multiple times using subsets of the aggregate dataset constructed by conditioning on appropriate values of the conditioning factor(s). For example, in the analysis of seasonality above, the overall price dataset may be broken into subsets by month, weekday, weekend, holiday and individual day of the week, as permitted by the available data and identified as significant.

Additional examples

1. Temperature

Figure 26 shows electricity price behaviour over the course of the day in California for days in August and September. The first chart shows all days in August and September, while the second shows days with average temperatures below 70 degrees, and the third shows days with average temperatures above 70 degrees. As the charts show, price behaviour varies substantially as a function of temperature, impacting both the volatility of prices and the time pattern and level of mean prices over the day.

These differences in price behaviour can be captured by calibrating the model separately for each of the conditional datasets. The appropriate parameter set can then be drawn on to evaluate trading, contract, or asset value and risk under alternative temperature scenarios, and to more accurately model price behaviour over the shorter time horizons for which reliable temperature forecasts are available.

A similar process can be followed to analyse and incorporate the impact of other key drivers of supply and demand in model performance, for example:

- load patterns;
- generation asset availability;
- transmission asset availability;
- water supply; and
- fuel prices.

2. Forward price information

Where available, the prices of forward market instruments, such as forwards, futures or options contracts, reflect market expectations about the level and volatility of future prices. While forward prices also reflect net hedging demand, and are therefore not unbiased estimates of future prices, the information they contain can be used to improve model performance by appropriately conditioning the

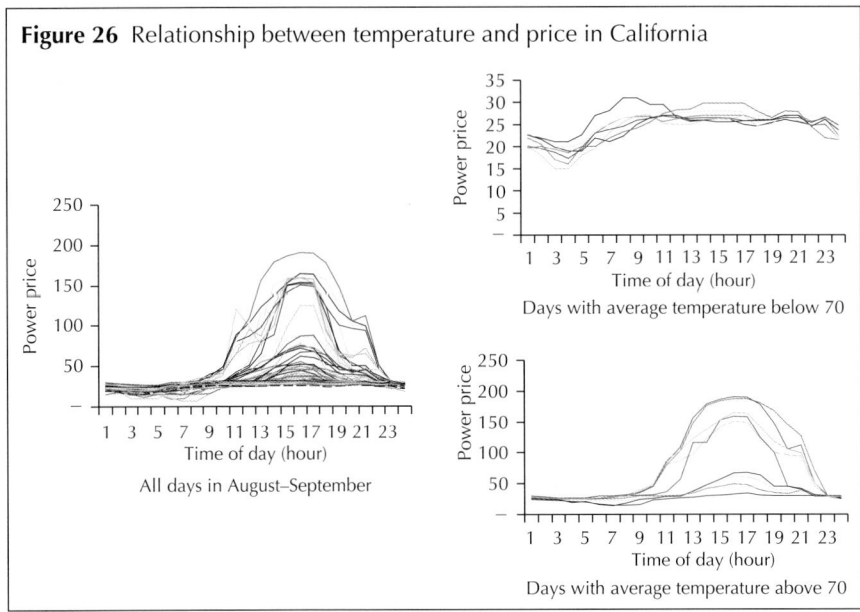

Figure 26 Relationship between temperature and price in California

datasets used for model calibration. By doing so, not just model representation of average price level and overall volatility can be improved, but also of daily patterns of mean hourly prices and volatility characteristics.

For example, as evident in the behaviour of California electricity prices shown in Figure 26, days with different daily average prices levels generally also exhibit different time patterns in mean prices and price volatility. A much more accurate representation of future electricity price behaviour can be generated by drawing on these relationships. As above, this can be done by calibrating a model to the subset of available historical data for a market and season that had mean daily price levels comparable to those signalled by forward prices. The significant information content captured in this way is in contrast lost under the commonly used "parallel shift" method of calibrating to forward prices, where either the data prior to calibration or the model after calibration is shifted to match forward levels.

As an example, the charts in Figure 27 show prices in the PJM market for all days during August and September, as well as for only days with average prices below US$25 and above US$35.

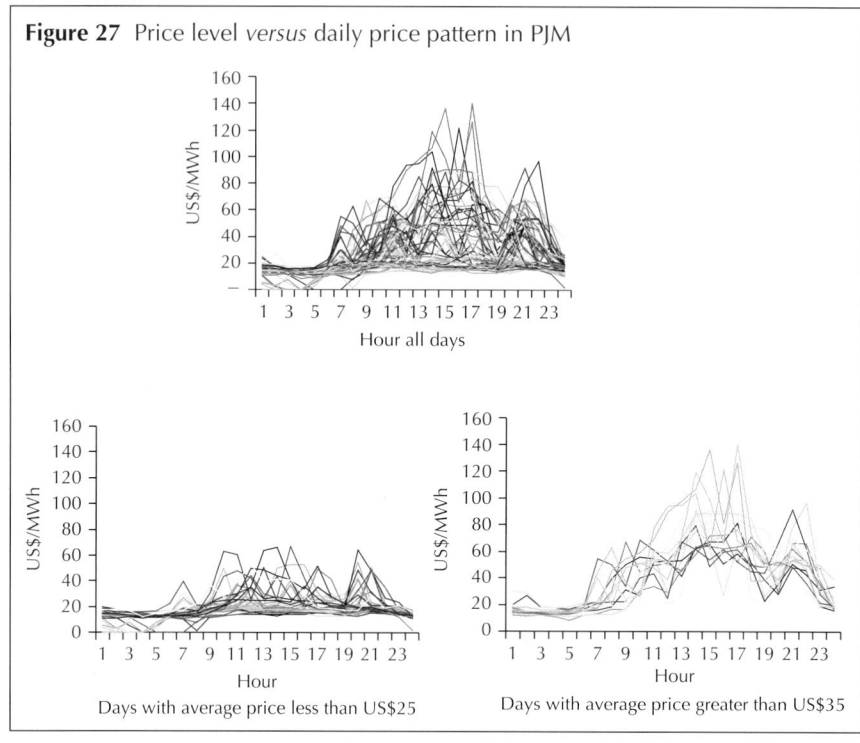

Figure 27 Price level *versus* daily price pattern in PJM

Limitations

As with any data-driven process, the principal limitation on the use of a "conditional calibration" process is the availability of relevant data, for both the conditioning "factors" and for spot prices. Because model calibration under the process is done on the subsets of the overall spot price dataset that prevailed under each of the alternative values of the factors, the specific data constraint is the size of these "conditioned" sets for the factor(s) and factor values of interest. As a result, the completeness and granularity of the conditional calibration that can be achieved is determined by the extent of the available data. Fortunately, in many cases a relatively small number of important factors and their values are enough to differentiate the most important characteristics of price behaviour.

Utilisation of the multiple-parameter sets generated through conditional calibration

As outlined in Step 3 of the general process for utilising supply, demand or market information in the model calibration presented above, after appropriate sets of conditional model parameters have been identified, model performance can be refined by shifting parameter sets over time based on the evolution of the conditioning factors over time. For example, parameters sets for high-temperature periods during summer months, or for specific forward price levels by season, can be utilised when those conditions occur.

With the exception of seasonality, which is deterministic and therefore known in advance, and other potential deterministic factors, such as scheduled maintenance of generation assets, most conditioning factors are stochastic, for example weather, generation asset availability, water supply and fuel price levels. As a result, utilisation of conditional relationships and their associated parameter sets must wait until reliable information about the future factor values is available.

As an example, imagine a trading organisation has identified material relationships between price behaviour and weather conditions, and the availability of key regional generation assets, and that it has generated conditional parameter sets for each relationship. To utilise the relationships identified and associated conditional parameter sets, the organisation monitors evolving weather conditions and asset availability, and uses the most current information available to update the parameter set utilised.

Specifically, assume that weather forecasts are unreliable more than 10 days in the future, and that information about unplanned changes in generation asset status arrives at most one day in advance. Based on this information the organisation models price more than 10 days in the future using "unconditional" parameter sets calibrated using only seasonal and other deterministic factors. Within the 10-day time horizon, conditional parameter sets based on current weather forecasts are utilised, with increasing accuracy and granularity available at shorter forecast horizons. Finally, when a change in the availability status of a key asset occurs, the parameter set(s) used to generated prices is (or are) updated immediately to reflect this change.

Through this process the trading organisation can rapidly and efficiently utilise the most current information available to it about evolving weather conditions and generation asset status to update and refine its price projections. By incorporating weather and asset information in its future price distributions, it ensures the information is appropriately reflected in the value and risk assessments made across the range of its activities and its organisational units.

CONCLUSIONS

As a result of the non-storability of electricity, the interplay between consumer demand and the generation supply stack produces unique and highly volatile electricity price dynamics in every market in which electricity is traded. Because the nature of demand and supply differ from season to season, as well as from region to region, models of the behaviour of electricity prices must be tailored to the specific case of interest.

Though differing in degree between regions and seasons, the interplay of supply and demand results in specific characteristics that are manifest in every case: mean reversion, seasonality, price-dependent volatility and occasional price spikes. Graphical analysis of price data as well as formal analysis of alternative models confirms the economic intuition underlying these characteristics. For a model of electricity prices to be reliable, it should therefore reflect these consistent properties.

Effective models of electricity prices are important due to the central role they play in the structuring, pricing, trading and risk management of physical and financial contracts, and the valuation, risk management and choice of operating policies for generation and transmission assets. The performance benefit identified above – that models tailored to the unique characteristics of electricity prices provide relative to models constructed to represent the behaviour of financial securities or other commodities – makes their utilisation essential to effective electricity valuation and risk assessment.

1. An approximate comparison between regions can be made by dividing the log-likelihood values for the UK and Victoria by two.
2. For further information about the LMRD/LMRDJ models and their calibration methods, see US Patent application #10/621,645, System and Mehtod for Representing and Incorporating Available Information into Uncetainty-Based Forecasts.

REFERENCES

Barz, G., 1999, *Stochastic Financial Models for Electricity Derivatives*, PhD Dissertation, Stanford University.

Clewlow, L. and C. Strickland, 2000, *Energy Derivatives Pricing and Risk Management* (London: Lacima Publications).

Das, S. and Foresi, S., 1996, "Exact solutions for bond and option prices with systematic jump risk", *Review of Derivatives Research*, **1**, pp. 7–24.

Das, S., 1996, "Poisson – Gaussian processes and the bond market", Working Paper, Harvard University.

Deng, S., B. Johnson, and A. Sogomonian, 2001, "Exotic electricity options and the valuation of electricity generation and transmission assets", *Decision Support Systems* 30.

Duffie, D., 1996, *Dynamic Asset Pricing Theory*, (2 edn) (New Jersey: Princeton University Press).

Green, R. and Newberry, D., "Competition in the British electricity spot market", *The Journal of Political Economy*, vol. 100, no 5, pp. 929–53.

Kaminski, V., 1997, "The challenge of pricing and risk managing electricity derivatives", In *The US Power Market: Restructuring and Risk Management* (London: Risk Publications).

Oksendahl, B., 1995, *Stochastic Differential Equations,* (4 edn) (New York: Springer) Verlag.

Pilipovic, D., 1998, *Energy Risk: Valuing and Managing Energy Derivatives* (New York: McGraw-Hill).

Schwartz, E., "The stochastic behaviour of commodity prices", *Journal of Finance*, vol. 52, pp. 923–73.

Schwartz, E. and J. Smith, 2000, "Short-term variations and long-term dynamics in commodity prices", *Management Science*, **46**, pp. 893–911.

Samuelson, P., "Proof that properly anticipated prices fluctuate randomly", *Industrial Management*, vol. 6, pp. 41–9.

Wolak, P., 1998, "Market design and price behaviour in restructured electricity markets: An international comparison", Working Paper, Stanford University.

2

*Fundamentals of Electricity Derivatives**

Alexander Eydeland; Hélyette Geman

Morgan Stanley; University Paris IX Dauphine

Deregulation of electricity is well under way in the United States and is starting in Europe. It represents a multi-billion dollar spot market that is developing very quickly, following the same pattern of evolution that the financial markets took, with growth in a variety of derivative instruments such as forward and futures contracts, plain vanilla and exotic options (Asian options, barrier options and so forth). The main problem associated with the pricing of electricity derivatives is that financial models do not capture the unique features of electricity – in particular its non-storability (except for hydroelectricity) and the difficulties of transmission (such as disruption or access to grids or outages). The aim of this chapter is to investigate possible approaches to the pricing of the most commonly traded electricity options.

DESCRIPTION OF POWER OPTIONS

The first category of options consists of calendar year and monthly physical options. Monthly options roughly follow the specifications of the electricity futures contracts that were introduced on the New York Mercantile Exchange in March 1996. The exercise, at the end of July, of August 1999 call options allows the buyer to receive power (in a given location, defined in the option contract) during all business days (five or six days a week, depending on the specification) of the month of August, 16 hours a day in most cases, from 6 am to 10 pm prevailing time (on peak hours), of a given number

---

*This chapter has been reprinted from the first edition.

of megawatt hours (MWh), at the price k – the strike price of the option. Monthly options are fairly liquid and, as will be discussed below, relatively easy to hedge.

A second category of power options consists of daily options. These options are specified for a given period of time (year, season, and particular month) and can be exercised every day during this period. For example, an owner of a July–August 1999 daily call option can issue, if it so chooses, an advance notice on August 11 to receive a specified volume of electricity on August 12 during the on-peak hours, paying a price k per MWh. Daily options are not very liquid and are difficult to manage. (Although swing and other volumetric options also belong to this category, and the issues discussed below concerning daily options are also relevant to these options, they raise additional constraints and complexities that are beyond the scope of this chapter.)

Lastly, there are hourly options, designed to provide access to power during specified blocks of hours (one, four, eight). The market for these options is currently thin.

In all three of the cases described above, the option payout at expiration is $\max(S_T - k, 0)$, where S_T is the spot price of electricity at the maturity in the defined region.

There were several days in June 1998 when the spot price was above US$2,000 per MWh, up from US$25 a few weeks before (see Figures 1, 2 and 3 for price and volatility data observed in the East Center Area Reliability Coordination Agreement (ECAR) region covering several US midwestern states). Sellers of call options – even deep out-of-the money calls such as $k = 1,000$ – incurred severe losses. In the spring of 1998, these options had been selling for 50 cents per MWh, probably because US$300 per MWh was the highest spot price of power registered during the year 1997.

These extreme spot prices coincided with the heatwave that had struck the midwestern part of the US, together with production and transmission problems. Generally speaking, power prices tend to be remarkably volatile under extreme weather conditions. Prices then become disconnected from the cost of production and may be driven very high by squeezes in the market due to generation shortages or transmission disruptions. Hence, power exhibits exceptional price risk – significantly larger than most other commodities do like currencies, Treasury bonds, grains, metals, or even gas and oil.

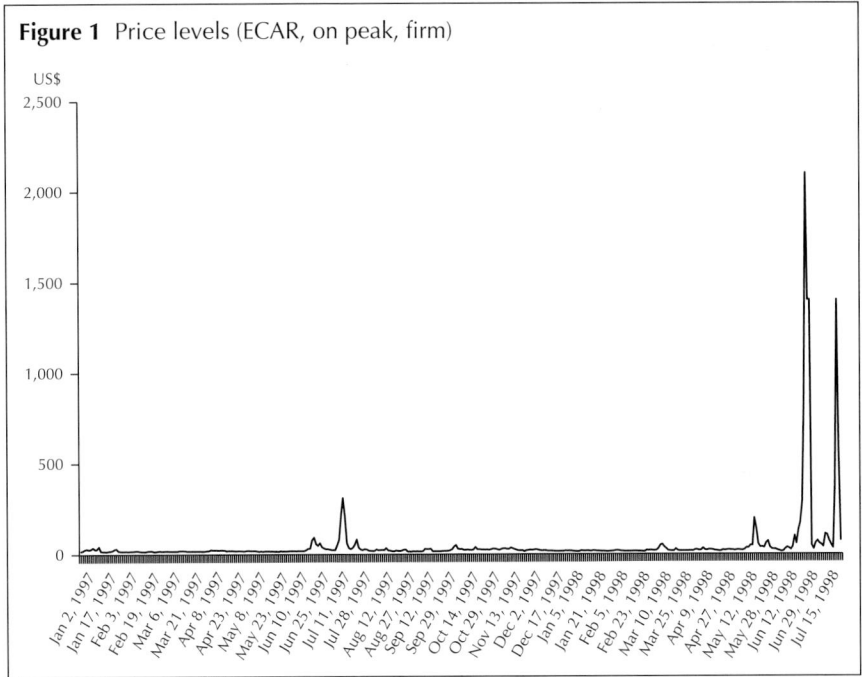

Figure 1 Price levels (ECAR, on peak, firm)

Financially settled power options are gaining popularity. The daily ones exhibit a 10–50% higher volatility than physically settled daily options. In order to avoid complicating issues further, we will restrict our discussion to physically settled options.

As mentioned earlier, the power market possesses some unique features.

- non-storability of electricity, and hence lack of inventories, requires the development of new approaches to study power markets, from both an economic and a financial standpoint.
- US power markets are geographically distinct: there are several geographical regions between which it is either physically impossible or not economical to move power. This explains why new futures contracts are being created to cover these regions: after the COB (California Oregon Border) and PV (Palo Verde, Arizona) contracts introduced in 1996, the Nymex started trading the Cinergy contract (covering the midwestern region) and the Entergy contract (Louisiana region). Another contract on

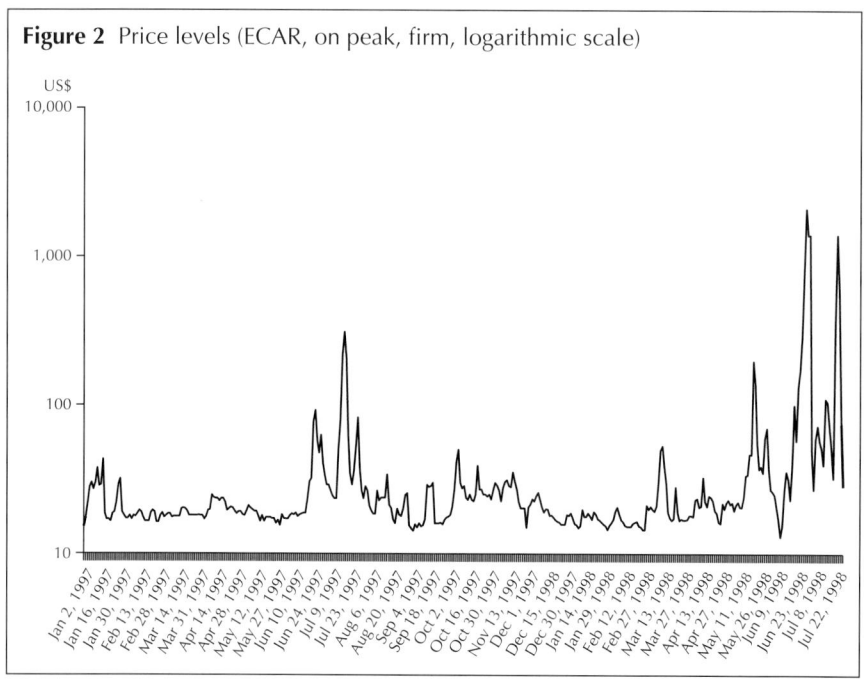

Figure 2 Price levels (ECAR, on peak, firm, logarithmic scale)

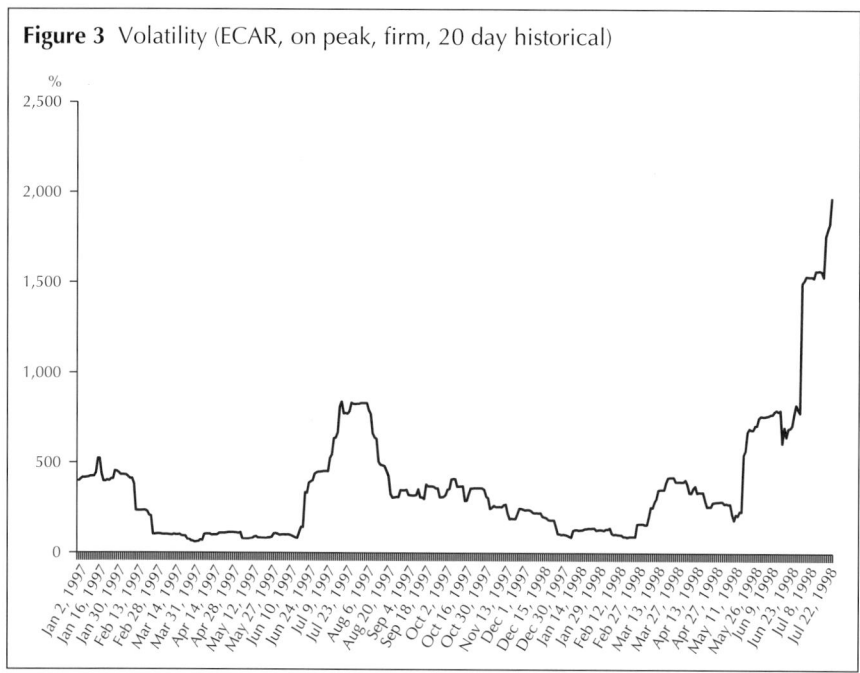

Figure 3 Volatility (ECAR, on peak, firm, 20 day historical)

Pennsylvania, New Jersey and Maryland (PJM) – whose delivery point is the border intersect of Pennsylvania, New Jersey and Maryland – has been recently introduced. Such geographical refinement of contracts is similar to that observed in catastrophic insurance derivatives (see Geman, 1994), first introduced in December 1993 by the Chicago Board of Trade for four regions, and then extended to nine distinct regions in the US.

❑ the market for power options, like the credit market, is not really complete since hedging portfolios do not exist or are at least very difficult to identify, especially for daily options. This incompleteness implies the non-existence of a unique option price, and hence the wide bid–ask spread observed on certain contracts.

In order to introduce a pricing methodology for power options, it is useful first to discuss the valuation methods used for other commodities, particularly energy commodities. In the next section we review the current approach to commodity option valuation for both standard and Asian options (weather derivatives, which are becoming increasingly popular among power traders, most frequently have Asian-type payouts).

POWER VERSUS COMMODITY OPTION PRICING

The economists Kaldor and Working, who studied the theory of storage, introduced the notion of "convenience yield". In the context of commodities, convenience yield captures the benefit of owning a commodity minus the cost of storage. Brennan and Schwartz in their pioneering research (1985) incorporated the convenience yield in the valuation of commodity derivatives and established, in particular, that the relationship prevailing between the spot price $S(t)$ and the future price $F(t, T)$ of a contract of maturity T is given by

$$F(t, T) = S(t)e^{(r-y)(T-t)}$$

where r, the risk-free rate, and y, the convenience yield attached to the commodity, are assumed to be non-stochastic. This remarkable relationship allows one to interpret the convenience yield as a continuous dividend payment made to the owner of the commodity. Hence, under the additional assumption that the price of the underlying commodity is driven by a geometric Brownian motion,

Merton's (1973) formula for options on dividend-paying stocks provides the price of a plain vanilla call option written on a commodity with price S, namely

$$C(t) = S(t)e^{-y(T-t)}N(d_1) - ke^{-r(T-t)}N(d_2)$$

where

$$d_1 = \frac{\ln\left(\frac{S(t)e^{-y(T-t)}}{ke^{-r(T-t)}}\right) + \frac{1}{2}\sigma^2(T-t)}{\sigma\sqrt{T-t}}$$

$$d_2 = d_1 - \sigma\sqrt{T-t}$$

Note that market completeness prevails in the above situation, as we have only one source of uncertainty, represented by the Brownian motion, and one risky asset, namely the underlying commodity, which can be sold, bought or stored to provide the hedging portfolio.

This implies the unicity of the price not only for plain vanilla options but also for exotic options; the latter case only involves solving mathematical technicalities. For instance Asian options, which today represent a huge percentage of the total number of options written on oil or oil spreads (because of the duration of oil extraction and transportation, most indices on oil are defined as arithmetic averages), are also becoming popular for electricity, in particular because of the summer 1998 events. The averaging effect smoothes out the spikes in prices and ensures that the underlying source of risk in the option is the average cost of electricity over a given time period. It is well known that the valuation of Asian options is a difficult problem and several approximations for the call price have been offered in the literature. Geman and Yor (1993) were able to provide the Laplace transform of the exact price of an Asian option using stochastic time changes and Bessel processes. Eydeland and Geman (1995) inverted this Laplace transform and showed the superiority of this approach over Monte Carlo simulations, in particular in terms of hedging accuracy. These results were established under the general assumptions of dividend payments for stocks or convenience yield for commodities.

FUNDAMENTALS OF ELECTRICITY DERIVATIVES

As mentioned above, the main difficulty in the valuation of power options is due to the fact that it is impractical to store electricity. This creates major obstacles for extending the notion of convenience yield to power, as follows.

- by definition, the convenience yield is the difference between two quantities: the positive return from owning the commodity for delivery, and the cost of storage. These two quantities cannot be specified because of the impossibility of storing power.
- the difficulty of storing electricity also leads to the breakdown of the relationship that prevails at equilibrium between spot and future prices on stocks, equity indices, currencies, and so forth. The "no arbitrage" argument, used to establish the cash and carry relationship, is not valid in the case of power because it requires that the underlying instrument be bought at time t and held until the expiration of the futures contract.
- another important consequence of non-storability is that using spot price evolution models for pricing power options is not very helpful. Hedges involving the underlying asset – the famous delta hedging – cannot be implemented as they require power to be bought and stored for a certain period of time.

One way to avoid the problems described above and to extend the hedging strategy explicit in the Black–Scholes–Merton formula to power derivatives is to use forward and future contracts. As we know from the analysis of these contracts in the case of stocks or equity indices, the dividend yield does not appear in the dynamics of forward and futures contracts (regardless of whether interest rates are deterministic or stochastic). Similarly, the dynamics of forward and futures contracts for commodities do not involve the convenience yield. Therefore, when these contracts are used to hedge power options (in particular, monthly or yearly options), the price of the option, which is by definition the price of the hedging portfolio, should not depend on the convenience yield. In other words, even though we fully appreciate the economic interpretation of the convenience yield, we view it as embedded in more relevant state variables for the pricing of power derivatives.

Hence, for a given region and a given maturity T, we need to make an appropriate choice for the dynamics of power futures contracts $F(t, T)$. An example of how the futures prices depend on

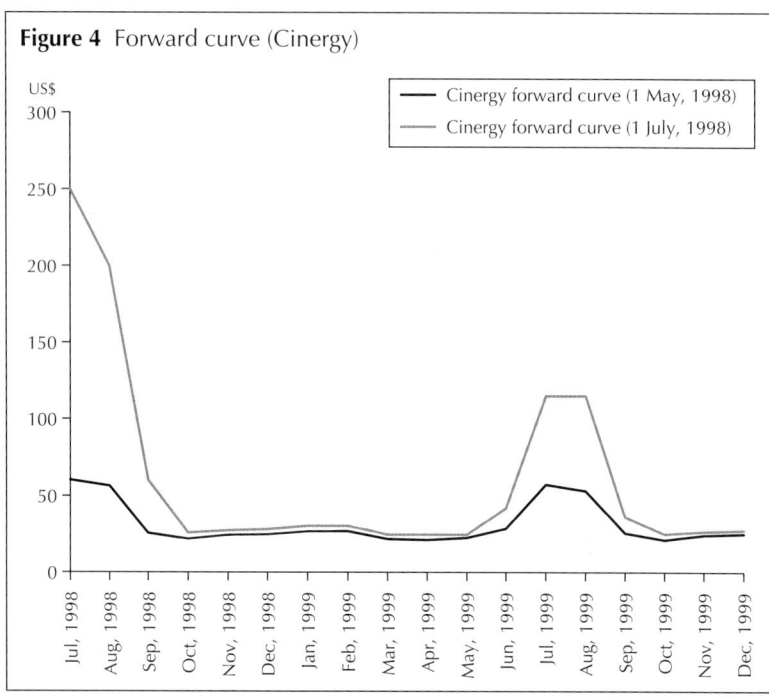

Figure 4 Forward curve (Cinergy)

maturity T can be found in Figure 4. As volatility also varies with time t and maturity T, one has also to specify the forward volatility structure $v(t, T)$, which has, in the context of power, the property of increasing when t goes to T (see Figures 5, 6 and 7). In the next section we will discuss one approach to modelling the evolution of the power forward curve.

A PRODUCTION-BASED APPROACH

We propose to approximate power future prices in the following manner

$$F(t, T) = p_0 + \varphi(w(t, T), L(t, T)) \tag{1}$$

where
p_0 = base load price
$w(t, T)$ = forward price of marginal fuel (gas, oil, etc)
$L(t, T)$ = expected load (or demand) for date T conditional on the information available at time t

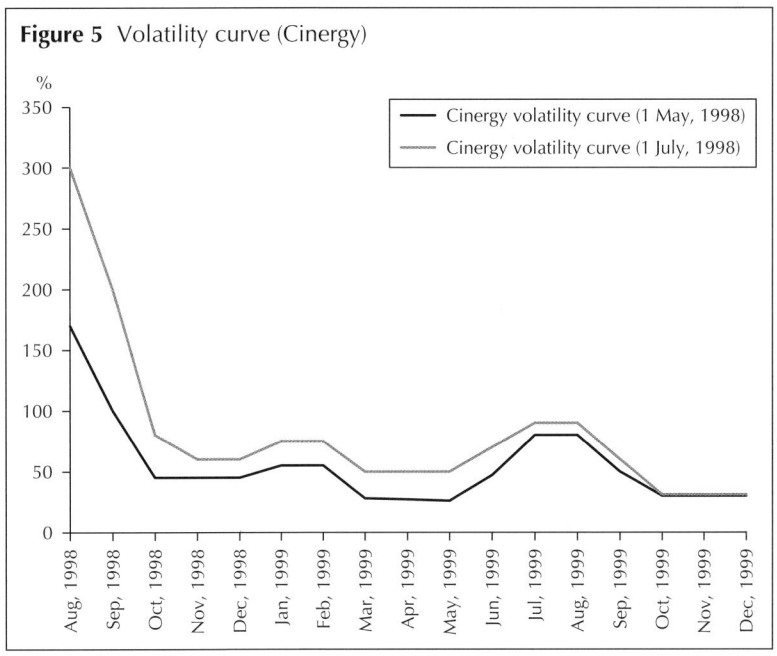

Figure 5 Volatility curve (Cinergy)

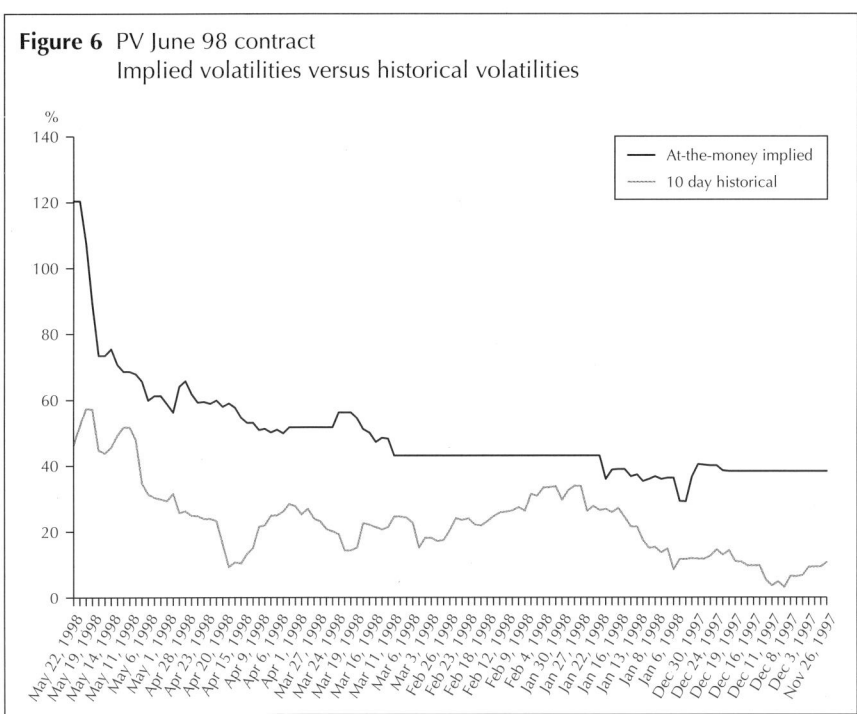

Figure 6 PV June 98 contract
Implied volatilities versus historical volatilities

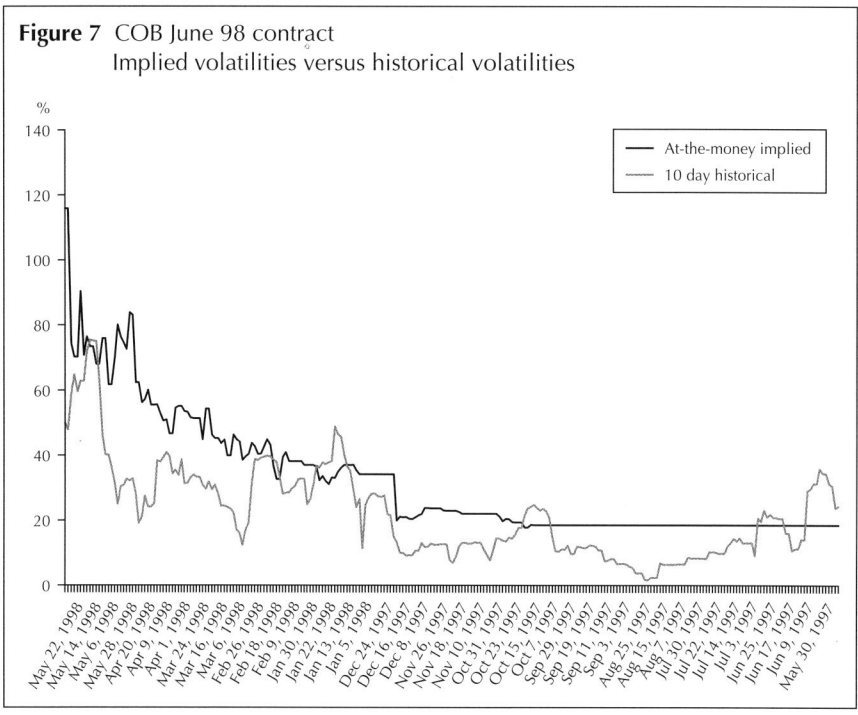

Figure 7 COB June 98 contract
Implied volatilities versus historical volatilities

φ is a "power stack" function, which can either be actual or implied from option prices.

If we assume that φ belongs to a two-parameter family of the type

$$\varphi = w \exp(aL + b) \qquad (2)$$

where a and b are positive constants, we obtain an exponential increase of the cost of generated power with increased demand, which is an adequate approximation to prices observed in the power markets. Moreover, if we assume in a classical manner that the demand, L, is represented by a normal distribution and that the forward fuel price is driven by a geometric Brownian motion, then from (1) we see that the quantity $F(t, T)$ (up to the constant p_0) is also driven by the geometric Brownian motion that has provided us with simple option-pricing formulas for 25 years.

In reality, the power stack function may be more complex than the one proposed in (2). Figures 8 and 9 show that this function should

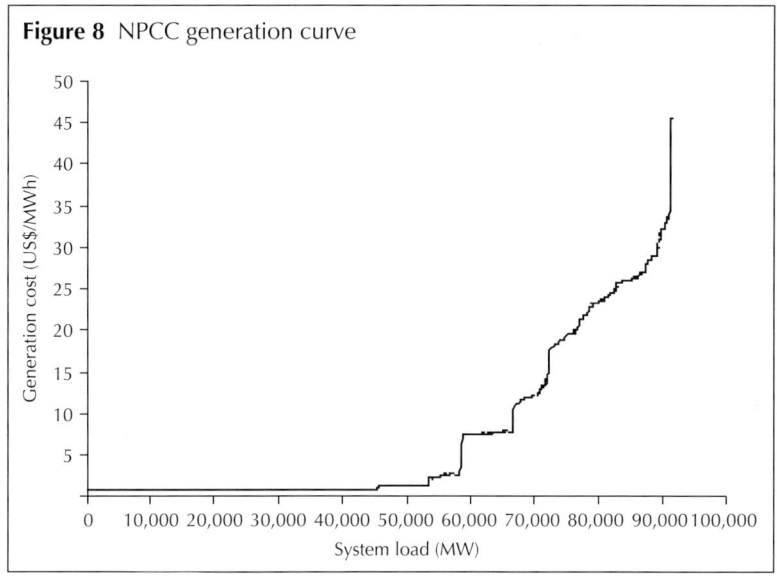

Figure 8 NPCC generation curve

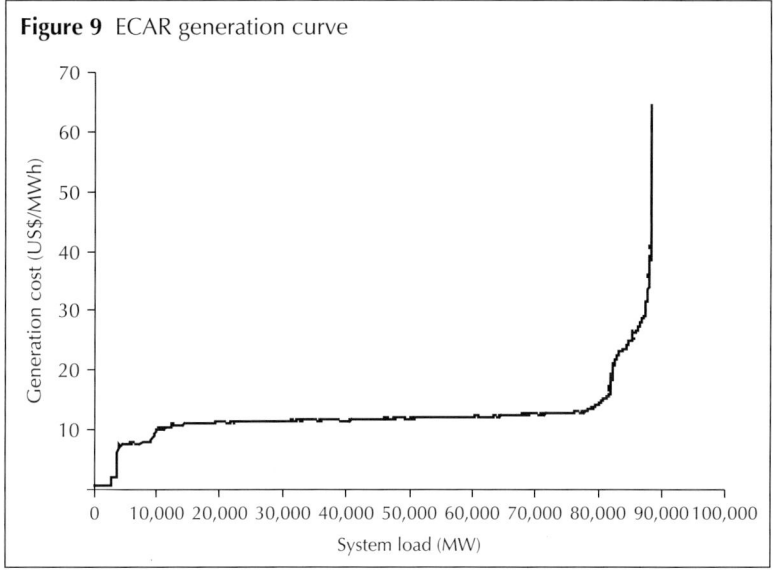

Figure 9 ECAR generation curve

be much steeper than the exponential one at the right-hand end of the graph, where there is a quasi-vertical line for finite values of demand. In this region a small change in load leads to a huge change in price, and this will account for the spikes observed in practice.

Moreover, the probability of higher values is, in fact, greater than in the lognormal approximation we mentioned above and leads to the fat tails clearly exhibited by electricity price return distributions.

To summarise, we note that, in general, we need to model the evolution of fuel prices and demand in order to represent the evolution of the power forward curve, as follows from (1). However, under certain assumptions, such as (2), the evolution of $F(t, T)$ can be modelled using the standard Black–Scholes framework with an appropriately chosen volatility term structure. The hedges generated in this manner will be adequate to manage monthly and calendar yearly options. The situation is quite different for daily options.

THE CASE OF DAILY OPTIONS

If a daily futures market existed, the hedging of daily options would not be different from that of monthly options and the approach described above would be applicable. However, most markets – except perhaps for the Nord Pool – do not have liquid daily forward or futures contracts. We are therefore forced to use an imperfect and sometimes dangerous surrogate for this daily futures contract: the "balance-of-the-month" contract. The balance-of-the-month price is the price of power delivered every day from today until the end of the current month. Going one step further and assuming a strong correlation between this balance-of-the-month price and the spot price, we can allow ourselves to model the spot price evolution in order to derive the option price from the spot price dynamics in a standard way. The balance of the month becomes the traded hedging instrument, rather than the non-storable spot. The main problems that one faces while modelling the spot price dynamics are the difficult issues of matching fat tails of marginal and conditional distributions and the spikes in spot prices. There are a number of techniques for addressing these issues; below, we describe two models that appear to us most relevant.

The first one is a diffusion process with stochastic volatility, namely

$$dS_t = \mu_1(t, S_t) dt + \sigma(t, S_t) dW_t^1$$
$$d\Sigma_t = \mu_2(t, \Sigma_t) dt + y(t, \Sigma_t) dW_t^2 \quad (3)$$

where $\Sigma_t = [\sigma(t)]^2$, $W^1(t)$ and $W^2(t)$ are two Brownian motions with a correlation coefficient $\rho^{(t)}$, and the terms $\mu_1(t, S_t)$ and $\mu_2(t, \Sigma_t)$ may account for some mean reversion either in the spot prices or in the spot-price volatility.

Stochastic volatility is certainly necessary if we want a diffusion representation to be compatible with extreme spikes as well as the leptokurtosity displayed by the distribution of realised power prices. However, stochastic volatility puts us in a situation of incomplete markets since we only have one instrument – spot power (or, rather, its surrogate) – with which to hedge the option. Hence, the valuation formula for the call

$$C(t) = E_Q[\max(S_T - k, 0)e^{-r(T-t)}]$$

where r is the risk-free rate, S_T is the spot price at maturity as defined by (3), and Q is the risk-adjusted probability measure, would require the existence of a volatility-related instrument (for example, very liquid at-the-money option) that could be viewed as a *primitive* security and complete the market.

A second model offers interesting features. As extreme temperatures, and hence, an extreme power demand, happen to coincide with outages in power generation and/or transmission, the spikes in electricity spot prices can be advantageously represented by incorporating jumps in the model. (Geman and Yor, 1997, analyse an example of this type, leading to completeness of the insurance derivatives market.) A classical jump-diffusion model is the one proposed by Merton (1976):

$$dS_t = \mu S_t \, dt + \sigma S_t \, dW_t + US_t \, dN_t$$

where
μ and σ are constant ($\sigma > 0$)
W_t is a Brownian motion representing the randomness in the diffusion part
N_t is a Poisson process whose intensity λ characterises the frequency of occurrence of the jumps
U is a real-valued random variable, for instance normal, which represents the direction and magnitude of the jump.
This model has a number of interesting features. However, the assumption of risk neutrality with respect to the jump component is totally lacking in credibility today (in the power derivatives

market, for instance, the options described earlier are trading, at the time of writing, at 10 times the value at which they traded before the June 1998 spike). Hence, with one tradable risky asset to hedge the sources of randomness represented by W_t, N_t and its random multiplier U, we face an extreme situation of market incompleteness. In some recent popular models for credit derivatives and defaultable bonds, this incompleteness is even more severe since the intensity λ of the jump process is supposed to be stochastic. Events in the credit markets during 1998 demonstrate that this matter should be a serious concern.

Coming back to power derivatives, our view is that currently the safest way to hedge daily power options is to own or lease a power plant. It is known that operating a merchant power plant is financially equivalent to owning a portfolio of daily options between electricity and fuel (spark spread options). Indeed, on any given day one should run a power plant only if the market price is higher than the cost of fuel plus variable operating costs. The net profit from this operating strategy is therefore

$$\pi = \max\left(\text{Price}_{\text{Power}} - \frac{\text{Heat rate}}{1000}\text{Price}_{\text{Fuel}} - \text{Variable costs}, 0\right)$$

where Heat rate is a plant-dependent scaling constant introduced to express power and fuel prices in the same units. (Heat rate is defined as the amount of British thermal units needed to generate 1 kWh of electricity.) The above expression is also the payout of the call option on the spread between power and fuel (spark spread), with variable costs being the strike of this call option. Owning the power plant is therefore financially equivalent to owning a portfolio of stock spread options over the lifetime of the plant.

If the set of daily options we want to analyse matches this portfolio exactly, then its price should equal the value of the plant, which may be obtained from economic fundamentals. (Note that this approach is the reverse of the one used in the theory of *real options*, introduced in corporate finance to value projects and investments. In practice, both viewpoints must be analysed). Of course, in reality, an arbitrary portfolio of daily options will differ from the portfolio of spark spread options representing the power plant, but the residual never explodes, even in a situation where there are extreme prices, and hence can be hedged by classical techniques. For example, the difference between standard daily calls and calls on daily spark spreads in the case of high power prices depends only on the fuel prices, which have comparatively low volatility. Along the same lines, power arbitrage experiences by traders and marketers are not the ones we are used to in the financial markets. They involve arbitraging the real options embedded in the business, such as technology arbitrage (heat rates, fuel switching, response time), transmission/transportation arbitrage, or commodity arbitrage between gas, coal or hydroelectricity.

This chapter has addressed only a few of the numerous issues involved in modelling power prices, but probably some of the most important ones – particularly at a time when intra-day stockmarket volatility tends to resemble power market volatility. In order to have a complete picture, we would need to incorporate the possible discontinuities due to powerplant shutdowns, transmission congestion, changes in environmental policies (in particular regarding emission control) and the development of new technologies to produce electricity.

REFERENCES

Black, F. and M. Scholes, 1973, "The Pricing of Options and Corporate Liabilities", *Journal of Political Economy,* **81**.

Brennan, M. and E. Schwartz, 1985, "Evaluating Natural Resource Investments", *Journal of Business,* **58**.

Eydeland, A. and H. Geman, 1995, "Domino Effect: Inverting the Laplace Transform", *Risk,* (March).

Geman, H., 1994, "Catastrophe Calls", *Risk,* September.

Geman, H. and M. Yor, 1993, "Bessel Processes, Asian Options and Perpetuities", *Mathematical Finance.*

Geman, H. and M. Yor, 1997, "Stochastic Time Changes and Catastrophe Option Pricing", *Insurance: Mathematics and Economics.*

Merton, R., 1973, "Theory of Rational Option Pricing", *Bell Journal of Economics.*

Merton, R., 1976, "Option Pricing and When Underlying Stock Returns are Discontinuous", *Journal of Financial Economics.*

3

*Pricing, Modelling and Managing Physical Power Derivatives**

Corwin Joy

Baylor College of Medicine

As the US electricity market deregulates, it is becoming increasingly obvious that the naïve use of the Black–Scholes pricing model for physical power options leads to severe pricing and hedging errors. In this chapter we introduce a new framework that follows in Black's and Scholes's footsteps but is explicitly tied to the nature of physical electricity. This framework allows us to price physical options with substantially greater accuracy and it opens the door for pricing retail electricity services.

There are three main difficulties when developing a framework for the pricing and management of physical power options. First, the dynamics of electricity markets are inherently different from those of most other commodities due to the lack of physical storage and the strict transmission constraints that exist on the system. Second, as the pace of deregulation increases, existing market dynamics are bound to change rapidly, making histories of regulated prices all but useless. Finally, even the best set of market dynamics is worthless without a practical strategy to manage any options that are sold under such a model. In the electricity market, in particular, careful consideration needs to be given to how to best manage and hedge any derivatives sold in the light of limited liquidity in the physical market. These concerns give rise to the qualified goals summarised in Table 1.

---

*This chapter has been reprinted from the first edition.

> **PANEL 1 IMPLICATIONS OF THE MODEL FOR END USERS**
>
> Consider a typical utility that owns both generation and load obligations. From that utility's point of view, the following three issues are the top risk management concerns:
>
> ❑ how much generation do I own and what is the availability of that generation going forward on a daily basis?
>
> ❑ how much "load", or demand for electricity, do I own (as broken down by residential and industrial customer classes) and what is the variability of that load?
>
> ❑ is there a mismatch between my load and generation? If so, what is my risk and what can my trading group do for me to reduce this risk and/or seize any opportunities that are available to me because of my generation or load position.
>
> In this chapter, we address, in a simplified fashion, each of these items in turn:
>
> ❑ We propose a simple framework to capture generation assets. This approach is rough, but the resulting approximation is fast and forms a practical tool to enable portfolio managers to understand and control their position within the short timeframe required by a trading environment.
>
> ❑ We propose a simple framework for load. Again, this framework is quick and dirty, but the result is a good practical way to understand load variability and customer behaviour by class. This, in turn, promotes better management of this risk within a portfolio context.
>
> ❑ Finally, we tie both system load and generation availability together to form a coherent model for electricity prices at a given location. This physically inspired price dynamic lets us quantify unhedged price and volume risks. It also forms a firm foundation to help manage any mismatch between load and generation via financial trading activities.
>
> Our proposed framework is simple. Probably it is too simple. However, we think that it forms a solid start for utilities that have an increasingly pressing need to begin quantifying both generation and load risks within a rapidly deregulating marketplace.

MARKET DATA

In analysing the structure of the physical power market we started with the Federal Energy Regulatory Commission (FERC) filings data for both load and system lambda (the marginal cost of generation in US$/MWh) from January 1, 1993, to December 31, 1996. The FERC is the US government agency that is charged with

Table 1 Goals in developing frameworks and strategies

Develop a framework for the pricing of physical power options
- ❑ The framework must be financially sound, and should use generally accepted financial methods as much as possible given the special nature of electricity.
- ❑ The framework must not be too heavily conditioned on historical data, since we expect the market dynamics to evolve with deregulation.
- ❑ The framework must reflect the physical nature of power, including the behaviour of the generation stack, and the ability to model plant outages.

Develop strategies to price and manage the risks associated with physical options
- ❑ Examine the strategies available under limited liquidity.
- ❑ Examine full service load deals.
- ❑ Examine generation assets.

Table 2 NERC regions

ASCC	–	Alaskan System Coordination Council
ECAR	–	East Central Area Reliability Coordination Agreement
ERCOT	–	Electric Reliability Council of Texas
MAIN	–	Mid-American Interconnected Network
MACC	–	Mid-Atlantic Area Council
MAPP	–	Mid-Continent Area Power Pool
NPCC	–	Northeast Power Coordinating Council
SERC	–	Southeastern Electric Reliability Council
SPP	–	Southwest Power Pool
WSCC	–	Western Systems Coordinating Council

regulating electric power and natural gas industries at the national level in the US. The data for hourly load and lambda come from the Opri database of prices from 1993–96 (Opri is a division of Resource Data International). These numbers represent system load and marginal cost of generation on a regional basis as reported to the FERC. Each of these reported regions is one of the North American Electric Reliability Council (NERC) regions and represents a regulatory control area. There are 10 NERC which together encompass essentially all the power regions of the contiguous United States, Canada and a small portion of Mexico. The NERC regions are listed in Table 2.

Each NERC region contains several "pools" that represent easily accessible interchange areas where a number of market participants can either deliver or receive power. These pool areas are where most of the deals in the current market are struck. Unfortunately, historical prices at the pool level of detail are hard to obtain or proprietary and so what we have presented here are prices at the regional or NERC level.

There were two primary reasons for using this data set as a starting point. First, this is the only real data we have. Historically, prices have been regulated and the small amount of spot electricity that was traded does not give an indicative benchmark of market prices. Second, with the deregulation of electricity, spot prices are likely to become increasingly volatile as larger amounts are traded on a more active basis. Two things will only change slowly, however:

❑ the supply of available electricity (the fixed capital base of generation and transmission).
❑ the demand for electricity (early pilot studies indicate that demand among most users is relatively inelastic and will respond only slowly to consistent price pressures).

We therefore feel that, by making a careful study of the dynamics of supply and demand in this market, we can develop a framework that will be able to adapt well to changing market conditions.

It is also worth noting that although the data we examine here come from the US power market, the framework we propose is a general one that should be applicable across a wide range of electricity markets by fitting appropriate jump and volatility parameters. For an example of how this framework may be applied to markets outside the US, see Clewlow and Strickland (1999), who apply a similar model to half-hourly electricity prices in New South Wales.

APPROACH

In order to capture the dynamics of the physical power market, and yet retain compliance with widely accepted financial techniques, we have chosen as our starting point the dynamics for the Heath–Jarrow–Morton (HJM) extension of Black's model for options on futures (Cortazar and Schwartz, 1994 provides a description of this method as applied to commodities):

$$\frac{dF(t,T)}{F(t,T)} = b(t,T)\,dW(t)$$

where

$F(t, T)$ = futures price at time t for future expiring at T
$b(t, T)$ = volatility at time t of $F(t, T)$
$W(t)$ = standard Brownian motion.

In examining the system lambda data, some fundamental observations become quickly apparent (Figures 1 to 4).

❑ Current hour prices are strongly conditioned on the price in the previous hour.
❑ Prices have a strong tendency towards mean reversion. Even if lambda starts out high, it has a marked tendency to revert to a more normal level very quickly as the supply shortage passes and the system operators struggle to bring the system back in balance and shut off expensive peaking units as soon as possible.
❑ As the load rises on the system, resources available to meet that load become fewer in number so that the very same event in terms of an outage or a surge in demand has a correspondingly greater impact at a higher system load level than it would at a lower system load level. In financial terms this can be described by saying that as prices rise the level of volatility will also rise. This also makes sense from a physical point of view when we examine a graph of the marginal cost of generation as plotted versus system load. As the system load increases, the marginal cost of generation increases slowly at first, but then jumps sharply as short-term gas-fired or peaking units need to be employed and the system becomes increasingly constrained in terms of power availability.
❑ The graph of system lambda is characterised by patterns that follow regular diurnal cycles but has occasional sharp peaks where lambda surges well beyond these bands. The first possible cause for such jumps would be a sharp change in the supply, as produced by an unscheduled plant or transmission outage. The second possible cause is a jump in demand: this we must also capture. We will address both of these possible causes below.

To apply these properties in our model for electricity, we generalise the HJM framework by adding dependence of volatility on price level and the addition of jumps (see Merton, 1990 for a discussion of general jump diffusion processes). We model the jumps as

ENERGY MODELLING

Figure 1 Graph of system lambda: Current hour versus prior hour at a typical power pool

It can be seen in Figure 1 that as the price in the prior hour rises the volatility of the lambda for the next hour also increases. Figure 2a shows a much weaker correlation, hinting at a reversion effect. This is confirmed by plots of hour-on-hour system lambda for prior hour prices at the 70th percentile (Figure 2b) and the 90th percentile (Figure 2c). Note that the 90th percentile has some expected reversion due to sample bias as the percentiles and next/prior hours are all chosen from the sample data.

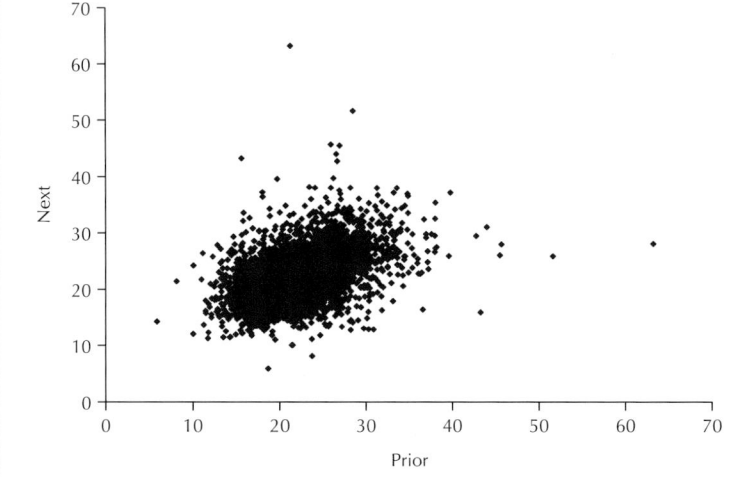

Figure 2a Plot of next-day versus prior-day system lambda at the same hour

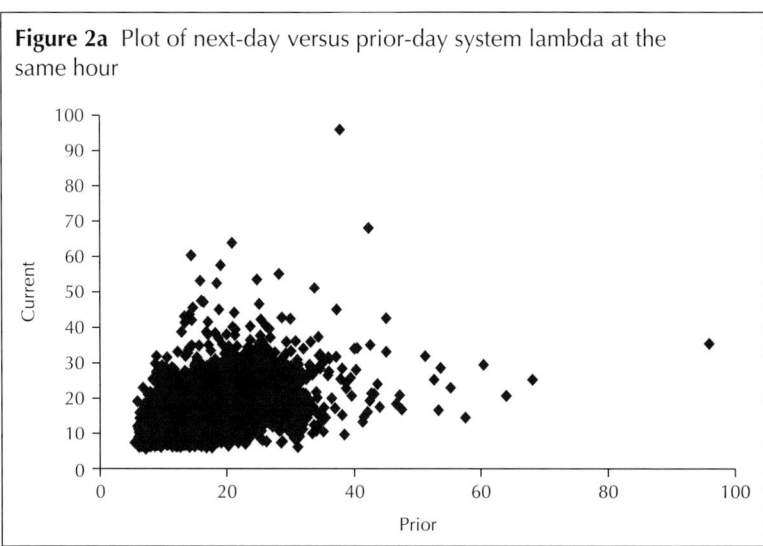

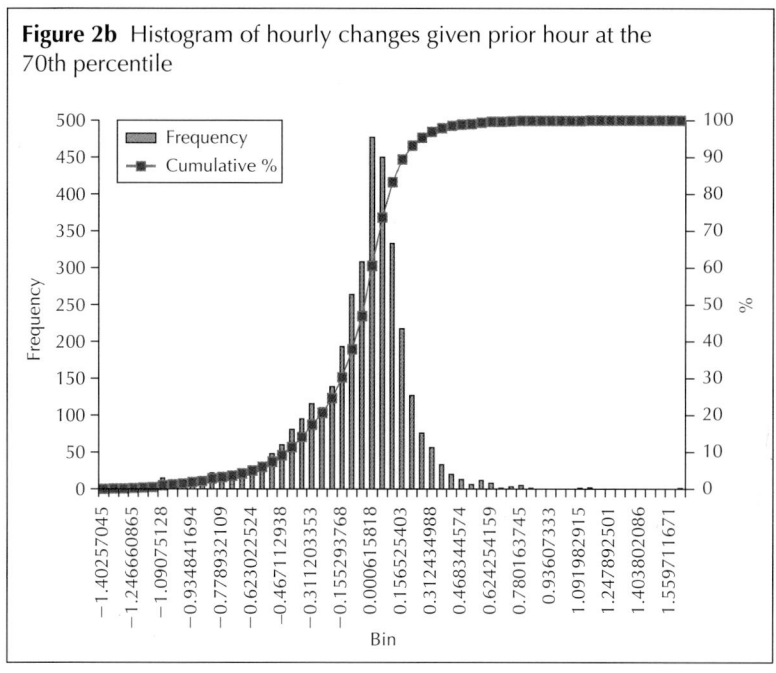

Figure 2b Histogram of hourly changes given prior hour at the 70th percentile

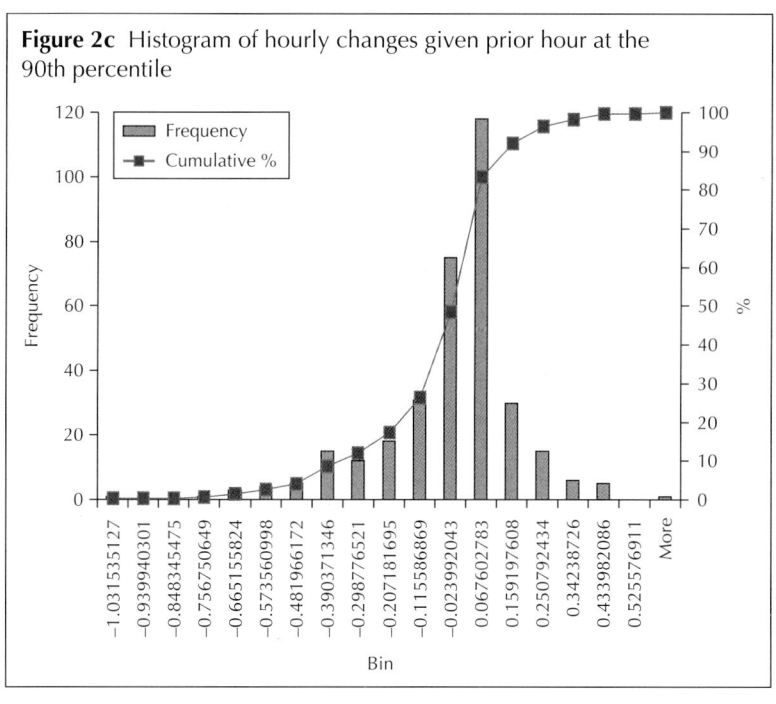

Figure 2c Histogram of hourly changes given prior hour at the 90th percentile

ENERGY MODELLING

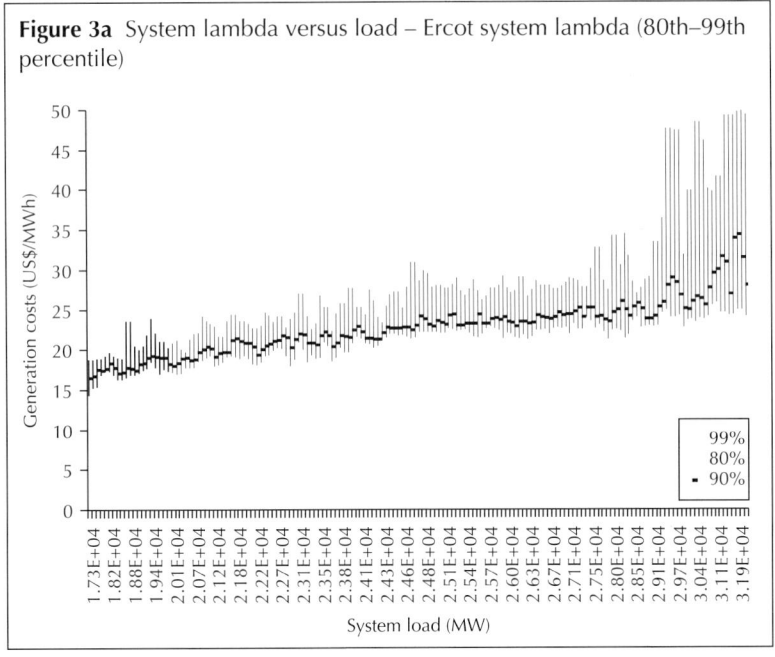

Figure 3a System lambda versus load – Ercot system lambda (80th–99th percentile)

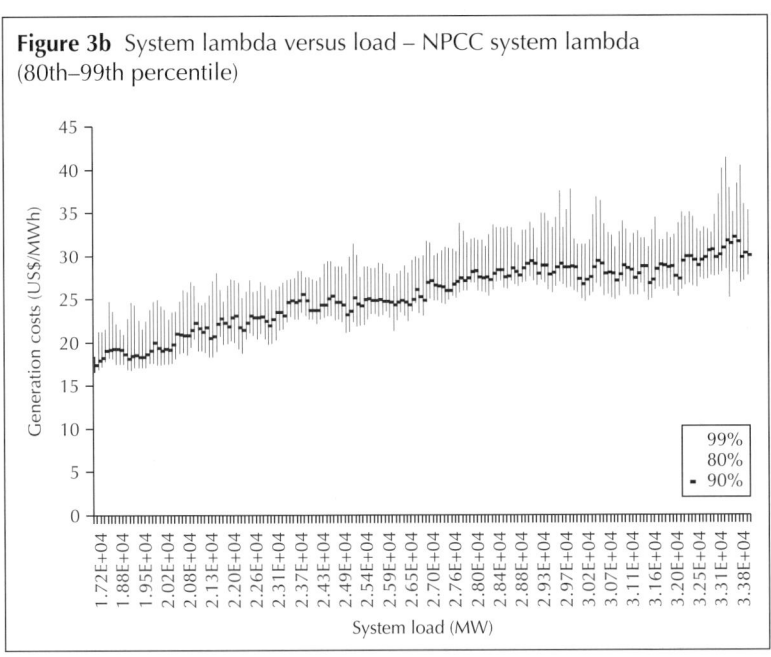

Figure 3b System lambda versus load – NPCC system lambda (80th–99th percentile)

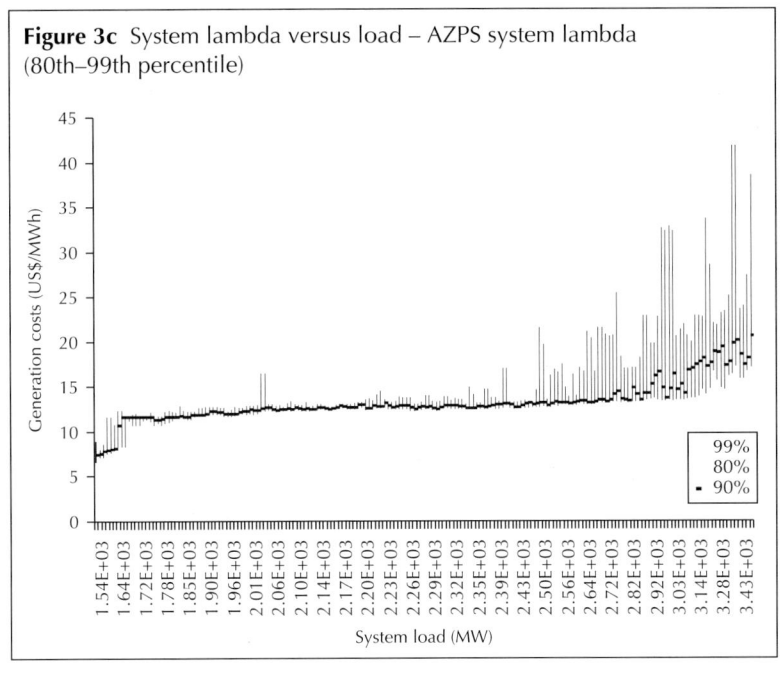

Figure 3c System lambda versus load – AZPS system lambda (80th–99th percentile)

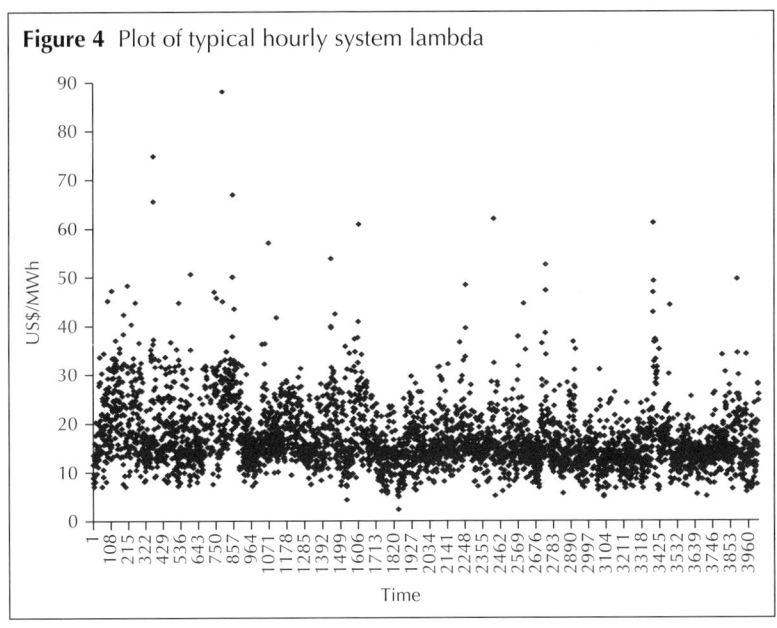

Figure 4 Plot of typical hourly system lambda

Poisson processes because this is the classical distribution used in the literature to model failure events. We also make all jumps of limited duration to reflect their transient nature in the power system. This leads to the generalised dynamic:

$$\frac{dF(t,T)}{F(t,T)} = b(t,T,F)\,dW(t) + \sum dq_i \tag{1}$$

where

$F(t, T)$ = price at time t for future delivery of power at time T
$b(t, T, F)$ = volatility at time t of $F(t, T)$, conditioned on price level $F(t, T)$, and conditional on no arrivals of "jump" information
dq_i = independent Poisson processes, i, representing jumps in system supply or demand.

We associate the following three parameters with dq_i:

λ_i = mean number of jumps per unit time for jump process i
$k_i = E[Y_i - 1]$ where $Y_i - 1$ is the random variable percentage change in the futures price if the Poisson event i occurs and E is the expectation operator over the random variable Y_i
d_i = random variable representing the duration of the jump due to Poisson process i, if a jump occurs.

In other words, equation (1) could be written in the more cumbersome fashion

$\dfrac{dF(t,T)}{F(t,T)} = b(t,T,F)\,dW(t)$ if no Poisson events occur

$\dfrac{dF(t,T)}{F(t,T)} = b(t,T,F)\,dW(t) + (Y_i - 1)$ if one Poisson event, i, occurs

$\dfrac{dF(t,T)}{F(t,T)} = b(t,T,F)\,dW(t) + (Y_i - 1) + (Y_j - 1)$ if two Poisson events, i and j, occur

and so forth.

This equation can be converted to the risk-neutral measure and discretised using the standard first order Euler method as

$$\frac{F(t+\Delta t, T)}{F(t,T)} = \exp\left[(-b^2/2)\Delta t + b(t,T,F)\varepsilon\sqrt{\Delta t}\,\right] Y(n) \tag{2}$$

where
 $\varepsilon = N(0, 1)$ iid r.v,
 $\Delta t =$ simulation time step,
 $Y(n) = 1$ if $n = 0$, $\prod_{\substack{j=1,n \\ i=1,n}} Y(i,j)$ for $n >= 1$ where the $Y(i,j)$

are iid independent Poisson processes distributed with parameter $\lambda_i^* t$ with random duration $d(i,j)$ in the event that a jump occurs. Specifically, if a jump of magnitude $Y(i,j)$ occurs at time t of duration d, then at time $t + d$ we require a jump of magnitude $1/Y(i,j)$ to represent the end of the event.

BENEFITS AND FEATURES OF THE MODEL
This model provides the following benefits:

❏ it extends the Black–Scholes process, familiar to traders and auditors, to electricity markets.
❏ volatility dependence on price level $b(., ., F)$, reflects the real-world constraints of load stack and observed data where price volatility tends to increase as we move up the stack.
❏ volatility dependence on forward maturity date, $b(t, T,.)$ allows us to capture term structure of volatility and ensures that mean reversion takes place through the decay of the volatility curve (Samuelson's hypothesis). This allows us to reflect the long-term mean reversion properties of the system discussed earlier.
❏ the jump process provides for explicit modelling of outage events. This is more realistic from a physical point of view because plant outages happen in discrete steps.

The model thus gives us a more accurate dynamic. A separate process for outages also helps us to price physical options where the payout is tied to an outage event.

ESTIMATION OF PARAMETERS
The many interacting variables mean that estimation of these parameters is somewhat complex.

Volatility
To estimate the term structure of the volatility, b, we proceed in two stages. First, the dependence of b on price, F, is estimated by

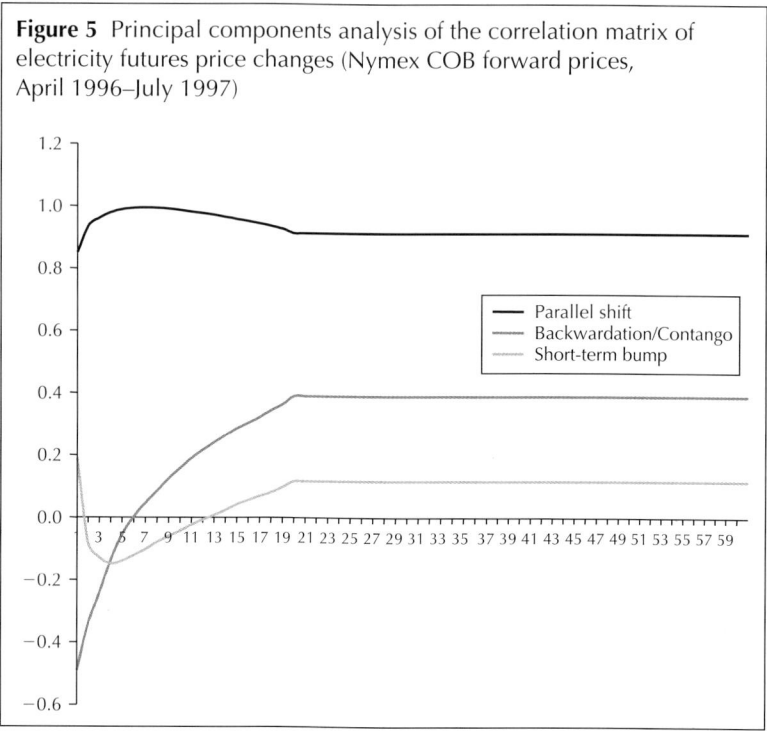

Figure 5 Principal components analysis of the correlation matrix of electricity futures price changes (Nymex COB forward prices, April 1996–July 1997)

comparing the percentage change in volatility versus price level. This gives us a term structure for the volatility smile $b(., ., F)$ that is independent of the absolute level of volatility. Next, $b(t, T)$ is fitted to the observed market implied volatilities by examining the traded prices for options on forward contracts and backing out the implied volatility. As per the standard method for implementing HJM, these volatility functions can be estimated by a principal components analysis of the log historical changes of the forward curve. Figure 5 illustrates some typical results obtained from Nymex COB forward prices between April 1996 and July 1997. The graph shows the eigenvectors corresponding to the three largest eigenvalues obtained from the correlation matrix of forward price changes along with labels indicating the role that these factors play in price curve movements (for example, "parallel shift factor" or "backwardation/contango" factor). In practice, to estimate factors calibrated to current market volatility, we multiply the historical correlation matrix by implied market forward/forward volatilities to create an implied covariance matrix from which we estimate principal components. The resulting factors combine to form the desired overall volatility function

$$b^2(t, T, F) = \sum_i b_i^2(t, T, F)$$

which will closely match the implied forward volatility curve by construction. One key observation that has been made by Schwartz (1997), and Clewlow and Strickland (1999), is that the above approach has a significant advantage over simple mean-reverting models in that it gives a volatility function that flattens out at a non-zero level. To be specific, Schwartz examines a standard one-factor mean-reverting model where spot prices revert towards a long-term price level μ at a rate determined by a positive coefficient α

$$\frac{dS(t)}{S(t)} = \alpha(\mu - \ln(S(t)))\, dt + v\, dW$$

where

$S(t)$ = spot price at time t
μ = long-term price level
α = mean reversion rate
σ = spot volatility of F

When the spot price is above $\exp(\mu)$, the drift of the spot price will be negative, which will tend to bring prices back down toward the long-term level, $\exp(\mu)$. When the spot price is below $\exp(\mu)$, the drift will be positive, tending to bring prices back up towards the long-term level. Schwartz shows that the above model can be equivalently stated as a special case of the generalised HJM model via

$$\frac{dF(t, T)}{F(t, T)} = b(t, T)\, dW(t) \qquad (3)$$

where

$$b(t, T) = \sigma \exp(-\alpha(T - t))$$

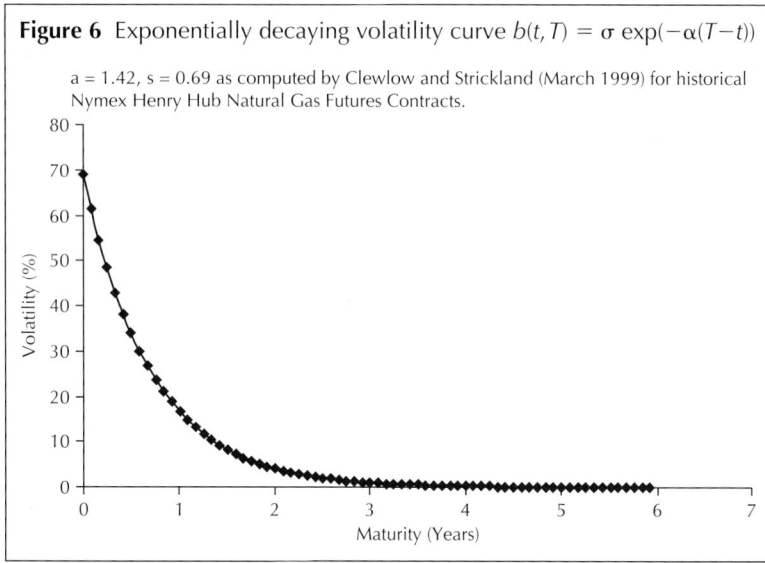

Figure 6 Exponentially decaying volatility curve $b(t, T) = \sigma \exp(-\alpha(T-t))$
a = 1.42, s = 0.69 as computed by Clewlow and Strickland (March 1999) for historical Nymex Henry Hub Natural Gas Futures Contracts.

This forces the forward volatility to decay at an exponential rate, and for long-term contracts forces volatility to be zero (Figure 6). This is an unacceptable property for commodity markets and means that the simple mean-reverting model specified in equation (3) should be rejected and a more sophisticated approach such as equation (1) employed.

Jumps – traditional estimation

Traditionally, the jump parameters for a Poisson jump-diffusion process are estimated by using moment-matching techniques. Following the outline in Ball and Torus (1985) we could proceed as follows. Suppose that $\ln Y \sim N(\mu, \delta^2)$. Then, as the drift for futures prices is zero, we obtain the density for the return on futures prices as

$$p(x) = \sum_{n=0}^{\infty} \frac{e^{-\lambda}\lambda^n}{n!} \phi(x;\ n\mu,\ b^2 + n\delta^2)$$

where

$$\phi(x;\ \mu, v^2) = (2\pi\ v^2)^{-1/2} \exp(-(x-\mu)^2/2v^2)$$

PRICING, MODELLING AND MANAGING PHYSICAL POWER DERIVATIVES

We then obtain sample moments, m_s, from the historical price data via

$$m_s = 1/T \sum_{t=1}^{T} [\Delta Z(t)]^s$$

where $\Delta Z(t)$ is the change in the natural log of the security price during time t, and T is the number of days in the estimation period. Note that because jumps are transient on electricity systems, we must filter the above sample data by excluding down jumps back to a "normal" state that occur after an up jump so as not to double count any jumps that occur. Next, using the sample moments, sample cumulants K_s can be determined as described by Kremer and Roenfeld

$$K_1 = m_1$$
$$K_2 = m_2 - m_1^2$$
$$K_3 = m_3 - 3m_1 m_2 + 2m_1^3$$
$$K_4 = m_4 - 4m_3 m_1 - 3m_2^2 + 12m_2 m_1^2 - 6m_1^4$$

Such estimates are subject to severe numerical round-off problems, however, and so the preferred estimation method is the more careful scheme outlined in Press *et al* (1992).

From these cumulants, we then solve equations for the parameter estimates as described by Press (1967):

$$\mu^4 - \frac{K_3}{K_1} \mu^2 + \frac{3K_4}{2K_1} \mu - \frac{K_3^2}{2K_1^2} = 0$$

$$\lambda = \frac{K_1}{\mu}$$

$$\delta^2 = \frac{K_3 - \mu^2 K_1}{3K_1}$$

$$\sigma^2 = K_2 - \frac{K_1}{\mu} \left(\mu^2 + \frac{K_3 - \mu^2 K_1}{3K_1} \right)$$

In solving for μ, care must be taken since there are usually two possible solutions in the interval of interest. Press shows, however, that the equation for μ has only two real roots of opposite sign. So, we simply choose the root that gives $\lambda > 0$.

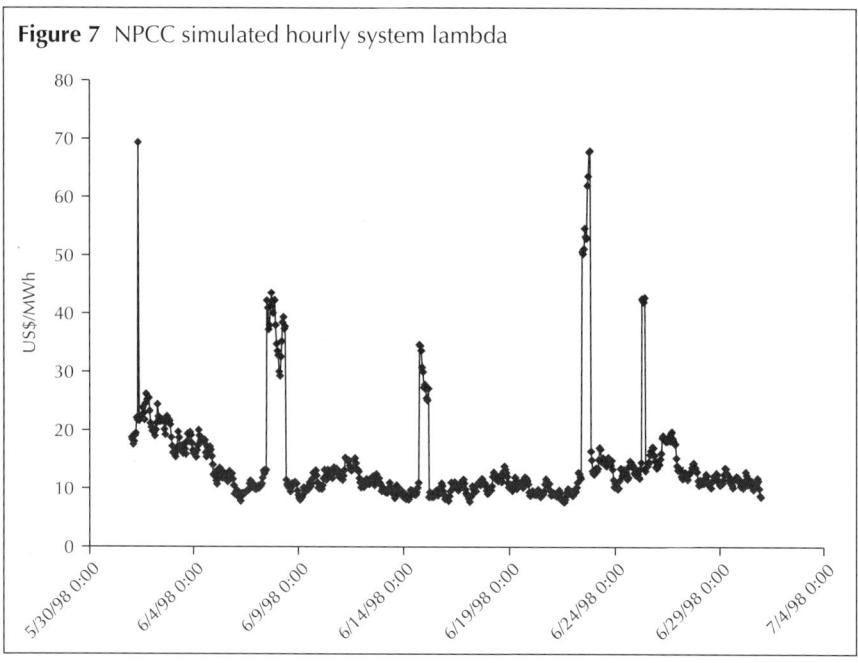

Figure 7 NPCC simulated hourly system lambda

Jumps – estimation for the electricity market

The key observation that makes this model extremely powerful for the physical power market is that, unlike the classical financial case shown above, we can calibrate the jumps with actual physical events on the system. We have hour-by-hour system lambdas, simultaneous data for plant outages, and concurrent system load data, so we can calibrate these events in a rather precise way. Thus, for example, the set of individual plants at a location can be modelled as a log-binomial collection with the impact of each plant's outage on lambda calibrated to its MWh output. This provides a good asymptotic tie-in to the single jump process specified above since, via the central limit theorem, for a large set of log-binomially distributed jump events the distribution converges to $\ln Y \sim N(\mu, \delta^2)$.

Simulated energy prices

In Figure 7 we show the simulated prices that result from estimating the jump and volatility parameters for the Northeast Power Co-ordinating Council (NPCC) NERC region.

HEDGING ISSUES

Even in theory it is impossible to be perfectly hedged against jumps. The reason for this is that no replicating position can protect against such a discontinuity and the best that one can hope for is that one will be able, on average, to cover one's expected cost for these jumps. In practice the problem is even worse because one cannot easily rebalance in the physical electricity market and indeed there are cutoff dates for both pre-schedules and monthly schedules that determine how much power one must block out as one's best estimate for these reserves. There are, however, some concrete steps one can take to estimate the hedging risks that arise from trading at illiquid locations.

No hedging case

The easiest approach is simply to assume that one cannot hedge at a particular location (and therefore that one is not risk-neutral). This leads us to examine the value of the option under the actual distribution of prices

$$\frac{F(t+\Delta t, T)}{F(t, T)} = \exp\left[b(t, T, F)\varepsilon\sqrt{t}\right]Y(n) \qquad (4)$$

We can now compute the payout distribution for a given option and, instead of pricing the resulting security at the expected value, we price the option on an actuarial basis at a particular confidence level. This is shown in Figure 8, where we plot the payout distribution for a particular option and compute a premium such that 80% of the time we make money and the expected loss is manageable the remaining 20% of the time.

Limited hedging case

For most physical locations, it is rarely the case that we have *no* ability to buy or sell power at the points for which we are writing power options. A more typical scenario is that we may have a good ability to trade month-ahead power, a reasonable ability to trade balance-of-month power, some ability to trade week-ahead power and the ability to trade hour- or day-ahead power only with difficulty. How does this limited liquidity in the physical market affect the risk of selling options? Kamal and Derman (1999) have suggested one excellent approach to this problem. They examine

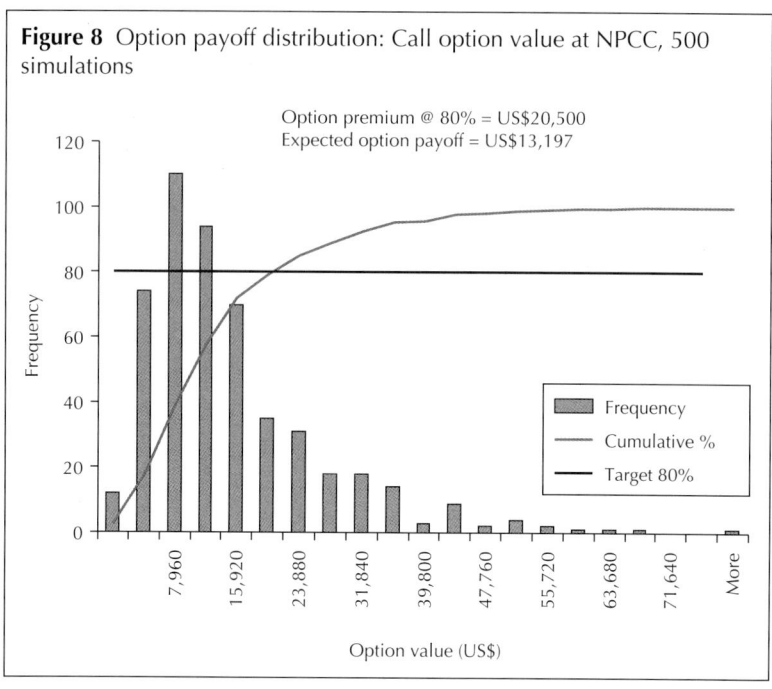

Figure 8 Option payoff distribution: Call option value at NPCC, 500 simulations

what happens when we are able to hedge an option only at discrete time intervals instead of being able to continuously rebalance as assumed by Black and Scholes. They find that, with discrete hedging times (under the Black–Scholes framework):

❏ the average final profit or loss under discrete hedging is zero. Hedging discretely does not bias the outcome in either direction when all other parameters (volatility, interest rates, and dividends) are known.
❏ the final distribution of profit and loss due to replication error resembles a normal distribution.
❏ hedging frequently reduces the standard deviation of the profit and loss at a rate that is proportional to the square root of the hedging frequency.

The standard deviation of the final profit and loss can be approximated as:

$$\sigma_{P\&L} = \sqrt{\frac{\pi}{4}}(\kappa)\frac{\sigma}{\sqrt{N}}$$

where

κ = option vega, evaluated at the initial forward price and trading date

N = number of times the hedger can rebalance over the life of the option (assumed to happen at equally spaced time intervals).

With this approach we can now look at the distribution of profit and loss given a particular hedging frequency. As above, we may combine this with an actuarial approach to calculate the cost of replication error for an option at a given confidence level (Figures 9a and b).

The result shown is the final profit or loss in dollars as calculated via 50,000 simulations for a one month at-the-money put hedged at discrete times to expiry. The initial parameters used were

Stock price = S_o = 100,
Strike price = K = 100
Option time to expiry = T = 1/12 years
Discount rate = r = 5%
Volatility = σ = 20%

The initial Black–Scholes value of the option was 2.512. For each simulation Damal and Derman carried out N rehedging trades spaced evenly in time over the life of the option, using the Black–Scholes hedge ratio (also calculated with r = 5% and σ = 20%).

The upper graph in Figure 9 represents hedging approximately once every business day over the life of the option whereas the lower graph represents hedging four times as frequently.

CUSTOMER BEHAVIOUR AND "SWING" OPTIONS

In the financial world, most of the options that have been developed assume that a particular volume will be taken. In contrast, the gas and electricity markets allow customers to "swing" (vary) the volume that they take. As an example, we might sell a customer a physical power contract to take 100 MWh/h at COB with the right to take up to 20 additional MWh in any hour for the month of June 1998 at the price of US$25/MWh. How much is this option worth? The first questions that arise in this context are as follows:

❑ to what degree is the customer exercising the option economically and to what extent is it the case that the customer just needs the power?

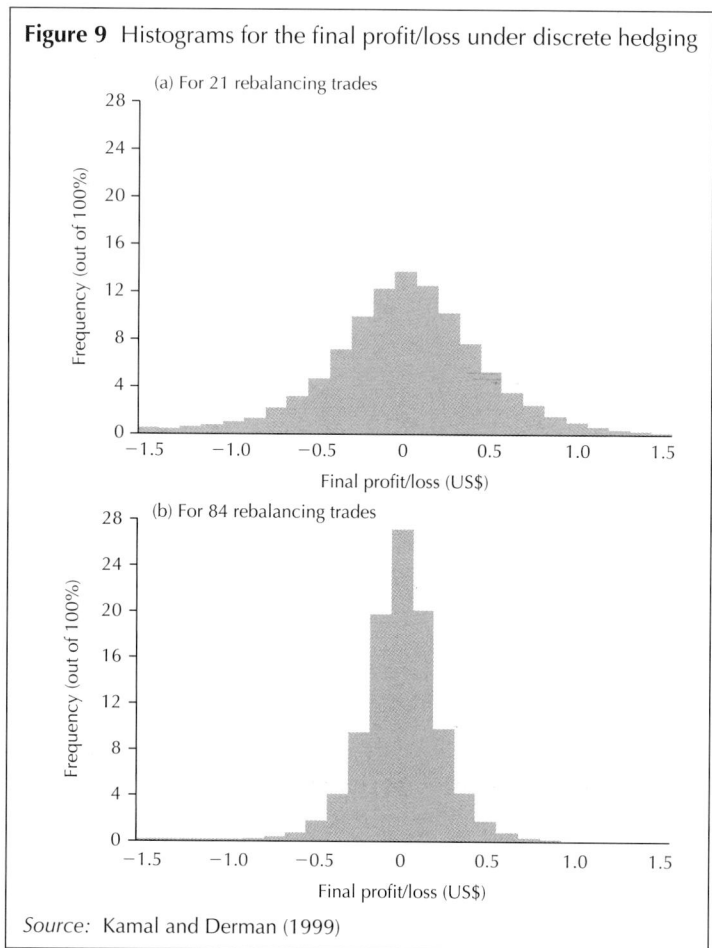

Figure 9 Histograms for the final profit/loss under discrete hedging

(a) For 21 rebalancing trades

(b) For 84 rebalancing trades

Source: Kamal and Derman (1999)

❑ more generally, is the customer's exercise behaviour correlated with price levels?
❑ how elastic is the customer's demand? Would higher price levels or a pass-through of prices affect the customer's behaviour?
❑ how will the American-style features of the above contract affect its value and optimal exercise time?

Most contracts like the one above have additional American exercise features and boundaries that limit either the daily, weekly or monthly swings that can be exercised. As a first attempt to answer the questions, we note that several studies have been carried out

PRICING, MODELLING AND MANAGING PHYSICAL POWER DERIVATIVES

throughout the US by utilities attempting to determine customers' price elasticity. The result of these studies is clear: without strong, consistent and long-term pricing incentives that punish expensive customer behaviour, there is almost no impact of price on the amount of electricity taken by most customer classes (Tiedmann, 1996). Only the largest customers are really price sensitive and even they have a limited ability to shift their load. A simplifying assumption, then, is that customers are inelastic: they do not or cannot exercise these options economically and have a simple correlation of load with price level. Moreover, correlating customer use of power with price levels helps utilities to pick the most profitable customers. As our starting point, we take the famous "quanto" option, the name of which was an abbreviation for "option quantity unknown". The idea behind a quanto option, originally developed for foreign stocks, is that the payout is given by

$$C = \bar{X} \max[S'^* - K', 0] = \max[S'^* \ast \bar{X} - K, 0]$$

where

$\bar{X}$ = a fixed foreign currency conversion rate
S'^* = the final price of the foreign stock, in foreign currency
K' = the strike price in foreign currency
K = the strike price in domestic terms.

For example, a US investor buying a British stock quanto option with a strike of $K' = £20$, a fixed conversion rate of $\bar{X} = 1.50$ (US$/£) and a final stock price of $S'^* = £28$ would receive a payout of

$$C = \bar{X} \max[S'^* - K', 0] = 1.5 \ast (28 - 20)$$

Notice here that two unknowns determine the option writer's exposure. The first unknown is the final value of the stock price. The second unknown is how much foreign currency the option writer will need to convert into domestic currency at the fixed exchange rate $\bar{X}$. Hence the "quantity unknown" is the amount of foreign currency to convert at option payout. The analogy for the energy market is now obvious: think of K' as the strike price for the power swing option, S'^* as the spot price for power in a particular hour, and the "unknown amount of foreign currency" as the unknown amount of power that the user will take.

Under the Black–Scholes-type framework, we can obtain a closed-form formula that gives a price for a single swing option under the assumption that the customer's load is lognormally distributed and correlated with price levels. This leads to the standard quanto formula for the call price as explained by Reiner (1992)

$$C = \bar{X}\left\{S'\left(\frac{r_d}{r_f}\right)^{-t}\exp(-\rho_{s'x}\sigma_{s'}\sigma_x t)N(x_3) - K'r^{-t}N(x_3 - \sigma_{s'}\sqrt{t})\right\}$$

where

$$x_3 = \frac{\log(S'd^{-t}/K'r_f^{-t}) - \rho_{s'x}\sigma_{s'}\sigma_x t}{\sigma_{s'}\sqrt{t}} + \frac{\sigma_{s'}\sqrt{t}}{2}$$

The above formula provides a particularly simple model for pricing retail options where the customer swing is driven by price levels. How well does this type of approach work for electricity? To answer this we return to Figures 3a–c, where we have plotted price versus load at several physical locations. What we see in these graphs is that price is roughly linearly correlated with load except at high load levels where generation outages, transmission outages, or other events can easily lead to a large jump in price. To fit these spikes, we need to include a jump parameter in our model for price movements. If we are willing to estimate load changes as lognormally distributed, then we can apply the following system of equations

$$\frac{dF(t,T)}{F(t,T)} = b_1(t,T,F)dW_1(t) + \sum dq_i$$

$$\frac{dL(t,T)}{L(t,T)} = b_2(t,T,F)dW_2(t)$$

where

$L(t, T)$ = estimated forward load at time t for delivery at time T
$F(t, T)$ = futures price at time t for future expiring at T

and the correlation between $W_1(t)$ and $W_2(t)$ is estimated by correlating price with load *after* trimming out price jumps that occur at high load levels (these jumps are used to fit the parameters for the jump process dq as described earlier). The other significant simplification that results from this approach is that many load options that would normally require an American-style valuation can be priced as European-type assets where exercise is simply a correlated function of price.

GENERATION ASSETS

A key issue for many participants in the energy market is the correct pricing and management of generation assets, as explained in Chapter 9. Although a thorough treatment of generation valuation is well beyond the scope of this paper, we propose the following simple approximation as a useful tool. First, we run a production cost model against a one-week set of prices

$$P = P_0 S$$

where

P_0 = average round-the-clock price for the week
S = hourly electricity shape profile to recover hourly prices from the weekly average.

We compare this with a profile G, that reflects how we would normally operate the plant were we to decide to run it during the week (for example, G would consist of volume multipliers to reflect the fact that we plan to ramp the plant up during the day and back down at night). From this, our production cost model determines a strike level $P_0 = K$ at which it is economically profitable to run the plant for a single week. This forms an exercise strategy for operating the plant that is *a priori* any intra-week price fluctuations.

In practice, we can change our mind during the week as to whether or not to continue running the plant, so this *a priori* exercise strategy forms a lower bound on the American-style operational value. We represent this strategy as a strip of swaptions – see, for example, Table 3.

This forms a sensible approximation of a generation asset as a strip of call swaptions assuming that fuel prices are fixed over the period of interest. It also provides a useful tool for many participants in the market in that this approach is relatively straightforward to implement with existing systems.

Table 3 A *priori* plant operation strategy

Week	Generation profile (G)	Average generation rate in MWh/H	Swaption strike (US$000)
1	G	100	10
2	G	100	11
3	G	100	12
etc.			

REFERENCES

Ball, C. and C. Torous, 1985, "On Jumps in Common Stock Prices and Their Impact on Call Option Pricing", *Journal of Finance*, **40(1)**, March, pp. 155–73.

Clewlow, L. and C. Strickland, 1999, "Power Pricing – Making it Perfect", *EPRM*, February, pp. 26–7.

Clewlow, L. and C. Strickland, 1999, "Valuing Energy Options in a One Factor Model Fitted to Forward Prices", Working Paper, School of Finance and Economics, University of Technology, Sydney, March.

Cortazar, G. and E. S. Schwartz, 1994, "The Valuation of Commodity Contingent Claims", *Journal of Derivatives*, **1(4)**, pp. 27–39.

Reiner, E., 1992, "Quanto Mechanics", *From Black Scholes to Black Holes*, Risk Books, pp. 152.

Kamal, M. and E. Derman, 1999, "Correcting Black–Scholes", *Risk*, January, pp. 82–5.

Kremer, J. and R. Roenfeldt, 1992, "Warrant Pricing: Jump-Diffusion vs. Black–Scholes", *JFQA*, **28(2)**, June, 255–71.

Merton, R., 1990, *Continuous Time Finance*, Blackwell Publishers, pp. 313.

Press, S., 1967, "A Compound Events Model for Security Prices", *Journal of Business*, 40, July, pp. 317–35.

Press, T. and F. Vetterling, 1992, *Numerical Recipes in C*, 2nd ed., Cambridge University Press, pp. 612.

Schwartz, E. S., 1997, "The Stochastic Behavior of Commodity Prices: Implications for Valuation and Hedging", *Journal of Finance*, **52(3)**, pp. 923–73.

Tiedemann, K. H., 1996, "Time-Of-Use Rates, Demand Charges and Residential Peak Energy Demand", *EPRI Conference on Innovative Approaches to Electricity Pricing*, pp. 16–1.

4

*Valuing Power and Weather Derivatives on a Mesh Using Finite Difference Methods**

Craig Pirrong; Martin Jermakyan

Olin School of Business and Washington University;
Vernadun, LLC

Pricing contingent claims on power presents numerous difficulties. The price process for power is highly non-standard, and is not well captured by price process models commonly employed to price interest rate or equity derivatives. Electricity "spot" prices exhibit extreme non-linearities. The volatility of power prices shows extreme variations over relatively short time periods. Furthermore, power prices exhibit substantial mean reversion and seasonality. No reduced form, low-dimension price process model can readily capture these features.

Finally, and perhaps most important, the non-storability of power creates non-hedgeable risks. Thus, preference-free pricing in the style of Black and Scholes is not possible for power. To address these problems, this chapter presents an equilibrium model for pricing power contingent claims (PCCs). This model uses an underlying demand variable (and perhaps a fuel price) as the state variable. The demand variable can be output (referred to as "load" in the power industry) or temperature. The price of power at the maturity of the contingent claim is related to the state variable(s) through a terminal pricing function. This pricing function establishes the payout of the contingent claim, and thus provides one of the boundary conditions required to value the contingent claim.

---

*This chapter has been reprinted from the first edition.

Given a specification of the dynamics of the state variable(s) and the relevant boundary conditions, conventional partial differential equation (PDE) solution methods can be used to value the contingent claim.

There are a variety of techniques for solving PDEs, including finite difference and Monte Carlo approaches. In this chapter we discuss implementation of a finite difference or "mesh"-based approach. We favour this technique for a variety of reasons. Most importantly, it offers great flexibility because it can be used to solve "free boundary" problems that exist when the holder of an option has the right to choose the timing of exercise. As a result, the mesh-based approach can be used to value path-dependent claims such as swings. The approach can also be modified to permit unified valuation and hedging of claims that are sensitive to power price, volumetrics and weather. Finally, it can be used to handle one of the most important problems in power derivatives pricing – the determination of the market price of risk.

This last point deserves emphasis. The risks associated with the demand state variable are non-hedgeable, and so any valuation depends on the market price of risk associated with this variable. Ignoring this feature of power markets will lead to serious pricing errors because the market price of risk may account for a substantial proportion of the forward price for delivery into midwestern power markets in summer months.

One cannot observe the market price of risk directly. It must be estimated. We allow the market price of risk to be a function of time and solve for this function using inverse problem methods and forward prices observed in the marketplace. Given the solution for a risk price function, this function and the fundamental valuation PDE can be used to price any other power contingent claim not used to calibrate the risk price.

We implement this methodology to value power forward prices in Pennsylvania–New Jersey–Maryland (PJM). The results of this analysis are striking. First, given an econometrically estimated terminal pricing function we find that the market price of risk during the summer of 1999 represents a substantial proportion of the quoted forward price of power. In particular, this risk premium represents approximately 24% of the forward price for delivery in July and August. Second, this market price of risk function exhibits large seasonal patterns, peaking in July and August, and reaching troughs near zero in April and October.

These results imply that the market price of risk function is essential to the pricing of power derivatives. Demand and cost fundamentals influence forward and option prices, but the market price of risk is actually quantitatively more important in determining the forward price of power, at least in the current immature state of the unregulated power market. Ignoring this risk premium will have serious effects when attempting to value power contingent claims, including investments in power generation and transmission capacity.

AN EQUILIBRIUM PRICING APPROACH

The traditional approach to pricing derivatives is to write down a stochastic process for the price of the asset or commodity underlying the contingent claim. This approach poses difficulties in the power market because of the extreme non-linearities and seasonalities in the price of power. These features make it impractical to write down a "reduced form" power price process that is tractable and that captures the salient features of power price dynamics.

Linear diffusion models of the type underlying the Black–Scholes model clearly will not do. To address the inherent non-linearities in power prices, some researchers have proposed models including a jump component to power prices. These present other difficulties. For example, a simple jump model like that proposed by Merton is inadequate because, in that model, the effect of a jump is permanent, whereas it is well known that jumps in electricity prices tend to reverse themselves rapidly. Moreover, the traditional jump model implies that prices can either jump up or down, whereas in electricity markets prices jump up and then decline soon after. Barz and Johnson, in this book, incorporate mean reversion and exponentially distributed (and hence positive) jumps to address these difficulties. However, their model presumes that big shocks to power prices dampen out at the same rate as small price moves. This is implausible in some power markets. Estimation in jump-type models also poses difficulties. In particular, a reasonable jump model should allow for seasonality in prices and a jump probability that is also seasonal, with jumps more likely when demand is high than when demand is low. Estimating such a model with the limited time series data available presents extreme challenges.

There are also difficulties in applying jump models to the valuation of volumetric-sensitive claims. For example, a utility that wants to hedge its revenues must model both the price process and the volume process. There must be some linkage between these two processes. Grafting a volume process on top of an already complex price process is problematic.

To address these limitations of traditional derivative pricing approaches in power market valuation, we propose instead an equilibrium approach. In this approach, power prices are a function of two state variables.

The first state variable is a demand variable. To operationalise it, we employ two alternative definitions. The first measure of the demand state is load. The second is temperature. Since load and temperature are so closely related, these interpretations are essentially equivalent. To simplify the discussion we use load as the demand variable in what follows. Later on we discuss how use of weather as the state variable permits unified valuation and hedging of claims that are sensitive to power price, power volume, and weather.

An analysis of the dynamics of load from many markets reveals that this variable has certain important characteristics. Load is seasonal, with peaks in the summer and winter for the US's eastern, midwestern, and southern power markets. Moreover, load for each of the various National Electricity Reliability Council (Nerc) regions is nearly homoskedastic. There is little evidence of Garch-type behaviour in load. Finally, load exhibits strong mean reversion – deviations of load from its seasonally varying mean tend to reverse fairly rapidly.

Referring to load at time t as q_t, the stochastic process for load is:

$$\frac{dq_t}{q_t} = \mu_q(q_t, t)dt + \sigma_q du \qquad (1)$$

The dependence of the drift term $\mu_q(q_t, t)$ on calendar time t reflects the fact that output drift varies systematically both seasonally and within the day. Moreover, the dependence of the drift on q_t allows for mean reversion. One specification that captures these features is:

$$\mu_q(q_t, t) = k[\theta_q(t) - \ln q_t] \qquad (2)$$

In this expression, ln q_t reverts to a time-varying mean $\theta_q(t)$; $\theta_q(t)$ can be specified as a sum of sine terms to reflect seasonal, predictable variations in electricity output. Alternatively, it can be represented as a function of calendar time fitted using non-parametric econometric techniques. σ_q is represented as a constant, but it can depend on q_t and t. Empirical evidence suggests that there is some slight seasonality in the variance of q_t.

The second state variable is a fuel price. For some regions of the US, natural gas is the marginal fuel. In other regions, coal is the marginal fuel. In some regions, natural gas is the marginal fuel at some times and coal is the marginal fuel at others. We abstract from these complications and specify the process for the marginal fuel price. Gas and coal are storable, and there are futures contracts traded on both, so we use a risk-neutralised process for the forward price of the marginal fuel:

$$\frac{dg_t}{g_t} = \sigma_g(g_t, t) \, dz \qquad (3)$$

where g_t is the price of fuel for delivery on date T as of t and dz is a standard Brownian motion.

Given these two state variables, it is well known that the value of any derivative $D(g_t, g_{t,T}, t, T)$ must satisfy the fundamental valuation PDE:

$$rD = \frac{\partial D}{\partial t} + \frac{\partial D}{\partial q_t}[\mu_q(q_t, t) - \lambda(q_t, t)]$$

$$+ 0.5 \frac{\partial^2 D}{\partial q_t^2} q_t^2 \sigma_q^2 + 0.5 \frac{\partial^2 D}{\partial g_t^2} g_t^2 \sigma_g^2 + \frac{\partial^2 D}{\partial q_t \partial g_t} q_t g_t \sigma_g \sigma_q \rho_{qg} \qquad (4)$$

subject to the appropriate boundary conditions, where r is the risk-free interest rate (which is assumed to be constant).

For a forward contract, the relevant PDE is:

$$\frac{\partial F_{t,T}}{\partial t} = \frac{\partial F_{t,T}}{\partial q_t}[\mu_q(q_t, t) - \lambda(q_t, t)]q_t$$

$$+ 0.5 \frac{\partial^2 F_{t,T}}{\partial q_t^2} q_t^2 \sigma_q^2 + 0.5 \frac{\partial^2 F_{t,T}}{\partial g_t^2} \sigma_g^2 g_t^2 + \frac{\partial^2 F_{t,T}}{\partial q_t \partial g_t} q_t g_t \sigma_g \sigma_q \rho_{qg} \qquad (5)$$

where $F_{t,T}$ is the price at t for delivery of one unit of power at $T > t$.

In both (4) and (5), there is a market price of risk function $\lambda(q_t, t)$. The valuation PDE *must* contain a market price of risk because load is not a traded claim and hence load risk is not hedgeable. Accurate valuation of a power contingent claim therefore depends on accurate specification and estimation of the $\lambda(q_t, t)$ function.

Valuation of a PCC also requires specification of terminal boundary conditions that link the state variables (load and fuel price) and power prices at the expiration of a PCC. In most cases, the buyer of a PCC obtains the obligation to purchase a fixed amount of power (for example 25 MW) on each peak hour of the delivery day. Similarly, the seller of a PCC is obligated to deliver a fixed amount of power each peak hour of the delivery day. We specify that the one-day forward price function at time T is given by:

$$F_{T+1,T} = P(q_T, g_T, T) \qquad (6)$$

where F is the forward price, $P(.)$ is a forward price function, and q_T is the load at T.[1] The price function $P(.)$ is increasing and convex in q. It is also a function of calendar time, with higher prices (given load) in spring and autumn months than in summer months because utilities schedule their routine maintenance to coincide with the seasonal demand "shoulders".

This pricing function determines the dynamics of the instantaneous power price. Let us define the instantaneous price as $P^*(q_t, g_t)$. Using Ito's lemma:

$$dP^* = \Phi(q_t, g_t, t)\, dt + P_q \sigma_q q_t\, du_t + P_g \sigma_g g_t\, dw_t \qquad (7)$$

with

$$\Phi(q_t, g_t, t) = P_q \mu_q(q_t, t) q_t + P_g \mu_g(g_t, t) g_t + 0.5 P_{qq} \sigma_q^2 q_t^2$$
$$+ 0.5 P_{gg} \sigma_g^2 g_t^2 + P_{qg} q_t g_t \sigma_q \sigma_g \rho_{qg}$$

where ρ_{qg} is the correlation between q_t and g_t; this correlation may depend on q_t, g_t and t. The volatility of the instantaneous price in this setup is time varying. Specifically, the variance is

$$\sigma_P^2(q_t, g_t, t) = P_q^2 \sigma_q^2 q_t^2 + P_g^2 g_t^2 \sigma_g^2 + 2 P_q P_g g_t g_t \rho_q \sigma_g \qquad (8)$$

This expression is time varying since P is a convex, increasing the function of q. In this case, P_q is increasing with q. Therefore, demand shocks have a bigger impact on the instantaneous price when load is high (when demand is near capacity) than when demand is low. In particular, if the price function becomes nearly vertical when demand approaches capacity, small movements in load can cause extreme movements in the instantaneous price. These non-linearities are a fundamental feature of electricity price dynamics, and explain many salient and well-known features of power price dynamics, most notably the "spikes" in prices when demand approaches capacity and the variability of power price volatility.

SOLVING THE PDE USING A FINITE DIFFERENCE MESH

There are many methods to solve PDEs like (4) and (5). One common approach is to use finite difference methods. This approach has many virtues, including computational efficiency and the ability to value contingent claims with early exercise features. Early exercise features create a "free boundary" that is not readily handled using alternative PDE solution techniques such as integration or Monte Carlo synthesis.

Implementation of the finite difference approach requires the creation of a pricing "mesh" that discretises the state variables (q_t, g_t) and time t. That is, the modeller divides up time into discrete steps, and divides the state variables into discrete points. One then approximates the partial derivatives in equations such as (5) by taking first and second differences across the time and state steps in the mesh. Once this is done, solution of the PDE is transformed into the problem of solving a system of linear equations.

Solution of the PDE requires specification of various boundary conditions. In the contingent claim pricing context, the most important boundary conditions specify what the payoff to the derivative is at its expiration. Once the boundary conditions are known, the solution of the PDE is known at a certain set of points – the time and state steps corresponding to the expiration date of the option, time T. Using these known values, it is possible to solve the discretised versions of (5) at $T - 1$, one time step prior to the expiration date. Given these values, it is possible to solve the PDE at time step $T - 2$, and so on, until one reaches the current date (time step zero).

Working backwards through time in this fashion has an important advantage. If one is pricing a contingent claim that has early exercise features (such as a swing option), one can evaluate the profit of exercising at every point on the mesh. Thus, one can readily solve "free boundary" problems that early exercise features create. The most efficient method of solving such problems is to employ successive over-relaxation (SOR) methods. Some alternative PDE solution techniques (including Monte Carlo or integration) cannot do so.

There are a variety of methods of approximating the partial derivatives in (4) and (5). These include implicit methods, semi-implicit methods (such as Crank–Nicholson), and explicit methods. Choice of technique depends primarily on the nature of the problem to be solved. Among the methods typically employed, semi-implicit methods are preferable because they are more accurate for a given level of computation cost. For example, the well-known binomial and trinomial tree approaches are instances of explicit finite difference techniques. Although relatively easy to implement, the binomial and trinomial methods are less accurate (for given computational cost) than other finite difference methods, including semi-implicit schemes such as Crank–Nicholson. To ensure stability, all explicit methods place constraints on the size of time steps. These constraints frequently require the use of very small time steps to achieve tolerable pricing accuracy. Reducing the size of the time steps increases computational costs. In other words, errors in explicit schemes are of order Δt, whereas errors in implicit or semi-implicit schemes are of the order $(\Delta t)^2$, where Δt is the length of the time step.

Implementation of this technique to value forwards and options requires a specification of the parameters of the stochastic process for the state variables (2) and (3), and determination of the payout function for every time and state step. These, in turn, require:

❏ an estimate of the P function;
❏ parameters of the load and fuel processes;
❏ initial boundary conditions; and
❏ a market price of risk function.

The following sections describe a mesh-based methodology for estimating these, and implement this methodology for a particular pricing region, the PJM market.

DETERMINING THE TERMINAL PRICING FUNCTION

Valuation of a PCC using the equilibrium model requires estimation of the terminal pricing function $P(.)$. There are two basic approaches that one can employ to do so. The first is to assume that the power market is competitive and use data on marginal generation costs as a function of load and fuel prices to determine the terminal power price as a function of these state variables. The second is an econometric approach that does not assume perfect competition.

Use of "generating stack" information poses several difficulties. First, it implicitly assumes that either the power market is competitive and thus prices equal marginal costs, or there is a (relatively) stable spread between power prices and generating costs. Neither assumption is likely to hold for many markets. Environmental Protection Agency (EPA) data on fuel input and power output for power plants that participate in the SO_2 programme permit estimation of marginal costs. For some regions, such as the Cinergy market, the spread between power forward prices (either hourly or daily) and realised marginal generating costs is wide and quite variable. Moreover, the correlation between movements in forward prices and movements in power prices is in the order of 0.45 for into Cinergy. Thus, power prices may bear little relation to generating costs.

Second, the economics of generation are actually quite complex. Start-up and shut-down costs imply that optimal dispatch requires solution of a rather complex dynamic programming problem. This, in turn, implies that the marginal cost of generation at any instant is a function of past loads and operating decisions.

Given the complexities associated with using generating cost information, we use econometric techniques to estimate the terminal price function. Econometric estimates impose no assumptions about competition in the power market. They only assume that there is a stable relation between the state variables and prices. A functional form that captures salient features of the price process is:

$$P(q_t, g_t, t) = g_t^\gamma e^{\alpha q_t^2 + c(t)} \qquad (9)$$

This function implies that prices are a very convex function of load; this convexity is due to the convexity of the exponential function and the fact that the exponent is itself a convex function of load.

This function also allows for a time varying pricing function: $c(t)$ is a deterministic function of time that shifts the pricing function up or down over time.

We estimate this function for the PJM market using data from 1997 and 1998. We find that:

- gas prices explain some of the variation in the price of fuel (γ is statistically significant and positive);
- load drives most variations in prices (α is large and statistically significant); and
- $c(t)$ exhibits substantial seasonal variation: holding load constant, prices are higher in the spring and autumn.

This last result is probably due to the fact that many power plants are idle for maintenance during these periods. This specification does quite well, explaining about 70% of the variation in PJM prices during the sample period.[2]

ESTIMATING THE PARAMETERS OF THE DEMAND PROCESS

The load for any group of utilities exhibits pronounced seasonalities. In eastern and midwestern markets, the load peaks in the summer, has a smaller peak in the winter, and troughs in the spring and autumn. The summer peak corresponds to the peak of the demand for air conditioning, whereas the winter peak corresponds to the peak of the winter heating demand.

To implement the model we need to estimate a $\theta(t)$ function that exhibits these seasonal peaks and valleys. To do so, we approximate the $\theta(t)$ function as a sum of sine terms. This framework allows us to generate a mean load function that has two seasonal peaks of different amplitudes. We seek the function $\theta(t)$ in the form

$$\theta(t) = \alpha_1 \sin(\beta_1 t + \gamma_1) + \alpha_2 \sin(\beta_2 t + \gamma_2)$$

This function will generate a cyclical pattern in load that can exhibit multiple peaks during the year. Defining $x_t = \ln q_t$, it can be shown that under this specification:

$$x(t) = e^{-kt} x_0 + k \int_0^t e^{-k(t-\tau)} \theta(\tau)\, d\tau \qquad (10)$$

The solution of equation (10) can be represented in the following form:

$$x(t) = x_1 e^{-x_2 t} + x_3 \sin(x_i t + x_5) + x_6 \sin(x_7 t + x_8) \qquad (11)$$

The parameters $x_1, x_2, \ldots, x_8$ are obtained as a result of calibration using a minimisation routine.

In our particular problem we are pricing power in PJM, and so we apply this method to load data available from PJM for the 1997–8 period. Combined with the pricing function (9), this load drift function generates seasonality in power prices. Given the convexity in the pricing function, it also implies substantial seasonality in the volatility of power prices.

DETERMINING THE MARKET PRICE OF RISK FUNCTION

As noted earlier, it is essential to incorporate the market price of risk in any power derivative pricing exercise. The market price of risk is inevitably present in any valuation problem due to the fundamental nature of electricity. Moreover, the data make it clear that ignoring the market price of risk is likely to lead to serious pricing errors because it is large.

Data from PJM illustrate this point clearly. If the market price of risk is not zero, the forward price will differ from the expected spot price. Therefore, systematic differences between forward prices and realised spot prices are evidence of a market price of risk. For PJM, on average, there are systematic differences between one-day forward prices and realised spot prices over the 1997–8 period. Over this period, the forward price for peak power delivered on the following day exceeded the average realised peak hourly price of power in PJM on the following day by an average of US$0.92/MWh. Moreover, the median of the difference between the one-day forward price and the realised average peak hourly price on the following day was US$1.36/MWh. This large median indicates that the difference between the forward price and the realised spot price is not due to a few outliers. Furthermore, the forward price exceeded the realised spot price on 311 of the 503 days in the sample. The forward price is also a biased predictor of the next day's realised spot price. The intercept in a regression of the day t average peak spot price against the day $t - 1$ one-day forward price is 8.75 and the slope coefficient is 0.6545. The standard error on the intercept is 1.5, and that on the slope coefficient is 0.047. One can therefore reject the null hypothesis that the intercept is zero and that the slope is significant.

These data make it clear that the forward price is a biased predictor of realised spot prices, even one day hence. Indeed, the bias is large. This indicates that even over a horizon as short as a day there is a risk premium embedded in power forward prices.[3]

The market price of risk is potentially large, so it is imperative to take this into account when valuing derivatives. Unfortunately, the market price of risk is not observable directly. However, it can be inferred from prices of traded instruments. In particular, given a set of quoted forward prices, inverse problem methods can be applied to (5) to generate an implied market price of risk.

At any time there are only a finite number of forward prices quoted in the marketplace. Call the number of available forward quotes Q. There is an arbitrary number of functions λ that could equate the solutions of (5) for these Q quotes. Thus, the problem of determining the market price of risk is ill posed. If a problem is ill posed, small changes in the data input (for example, in the forward quotes) can lead to large changes in the estimates of the λ function. These problems are quite common in a variety of physics and engineering contexts, and methods have been developed to solve them, involving the use of regularisation techniques.[4]

The solution involves choosing a function λ that minimises the sum of squared deviations between the forward prices implied by (5) for a given set of delivery dates and the prices quoted for these dates, subject to some regularisation constraint. The problem is discretised so that (5) can be solved using traditional mesh-based methods.

To make the problem tractable, λ is a function of time only. In the valuation mesh, λ is represented as a vector, where the length of the vector is equal to the number of time steps in the pricing mesh. An initial guess for λ is made. The solution of the regularised minimisation problem generates a new estimate of $\lambda(t)$. Given this new estimate, the regularised minimisation problem is solved again. The process continues until the $\lambda(t)$ function converges.

The regularisation technique in essence penalises overfitting. In the regularised problem, there is a trade-off between the precision with which the forward quotes are fitted and the smoothness of the

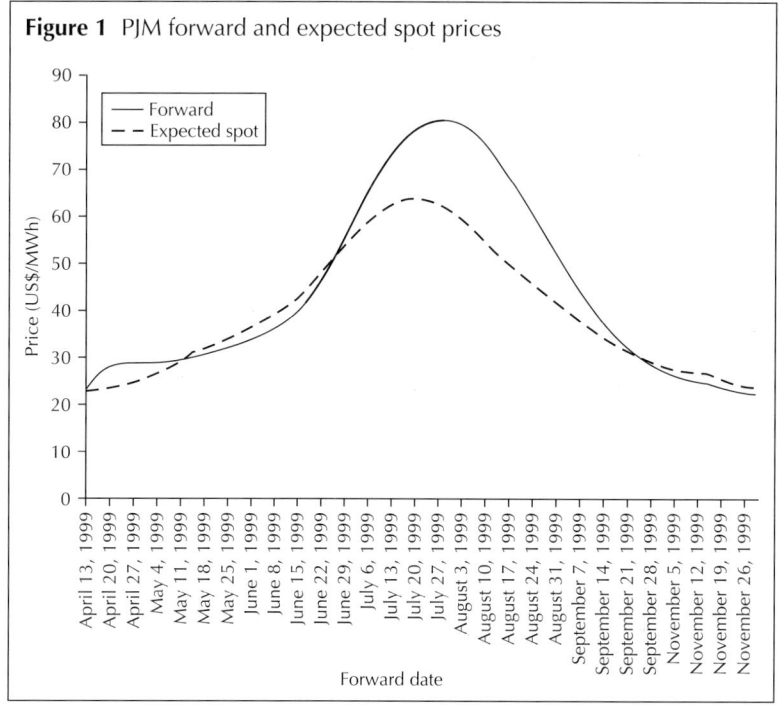

Figure 1 PJM forward and expected spot prices

λ function. Choice of a regularisation parameter determines the smoothness of the resulting fit; the bigger the value of this parameter, the smoother the resulting solution.

Using this method, we evaluated the market price of risk function for PJM prices for April 12, 1999. For this date, we obtained forward price quotes for several delivery dates extending through October, 1999.

The shape of the function generated by the model and the solution of the inverse problem on the mesh implies that the buyer of a July or August forward position must pay a substantial risk premium to the seller. This is best illustrated by Figure 1, which depicts the forward prices for each possible daily delivery date in the February–September 1999 period implied by the model using the λ function jointly implied by the model and the quoted forward prices; and the forward prices implied by the model assuming that $\lambda = 0$. The forward price with $\lambda = 0$ is the expected spot price implied by the dynamics of the load and fuel prices and by the terminal pricing function. Note that the difference between these two functions exhibits large seasonal variations. For delivery dates in April, May, late September, and October, the difference between the two lines is small; this indicates that the market price of risk is small. For July and August delivery dates, there is a large difference between the forward price curve derived from the solution of the inverse problem and the forward curve assuming no market price of risk. It peaks in early August, when the difference between the forward price and the expected spot price is around 32% of the calibrated forward price. The average difference between the forward price and expected price curves is 23.6% of the calibrated forward price during July and August.

These results are consistent with the existence of a "skewness premium" in power prices. During the summer months, in particular, power prices can spike up, generating substantial right skewness in prices. A price spike of this sort can impose a large loss on a seller of power forwards. Thus, the profit distribution of a short power forward position exhibits substantial left skewness. It is well known that those with consistent preferences exhibiting risk aversion dislike left skewness and therefore demand a risk premium to bear it. The results suggest that this premium for bearing skewness risk is large.

These results are extraordinary, and so alternative explanations should be considered. As an example, a so-called "peso problem" may be at work here. That is, the forward prices may incorporate expectations of events that did not occur in the historical data employed to determine the boundary conditions and the probability distribution for load. In this case, we may underestimate the expected spot prices and therefore overestimate the market price of risk. In this context it should be noted that the sample period for the load and price data used to estimate the load process and the pricing function includes 1998, and thus reflects the possibility of very high prices. Thus, unless there is a widespread expectation that 1998 will pale in comparison with 1999, the market price of risk is the most plausible explanation for the findings.

VALUATION OF SWINGS AND OTHER PATH-DEPENDENT CLAIMS

Once a market price of risk has been calibrated using forward prices, the mesh can be used to price any other contingent claim for the same market. For example, a mesh calibrated to the PJM forward price can be used to price daily strike or monthly strike PJM options.

The advantage of a mesh-based approach is its ability to price more complex claims, including path dependent claims such as swing options. Consider a swing with five swing rights exercisable over D days $\hat{t}, t+1, \ldots, t+D$ The strike price is K. For simplicity, there are no quantity choices and no penalty functions. We show here that this swing option can be valued by solving five free boundary problems using the calibrated pricing lattice.

The first step is to solve the value of a Bermudan call that allows exercise on any single day $\hat{t}, t+1, \ldots, t+D$ Designate the value of this call at the i'th load step j'th time step $V_1(i, j)$. Essentially if the option is exercised at i, j, the payout is $P(i, j) - K$ where $P(i, j)$ is the spot price at time step j and quantity i. This option value can be solved using traditional finite difference methods. Note that, as with any $t > \hat{t} + D - 1$, the value of this option is equal to the value of a daily strike option expiring at $\hat{t} + D$. This can be used to determine the option value at day $\hat{t} + D - 1$.

The second step is to solve for the value of an option that can be exercised twice during $[t, t + D]$. Let $V_2(i, j)$ be the value of this call at the i'th load step j'th time step. If the option is exercised at quantity step i and time step j, the exercise proceeds are $P(i, j) - K + V_1(i, j)$. That is, if you exercise this option at i, j, you receive the value of electricity net of the strike payment, $P(i, j) - K$ and the right to exercise one more swing right during the remainder of the option life. This remaining swing right is the Bermudan option defined in the previous paragraph; it has value $V_1(i, j)$. Exercise occurs only if the proceeds from the exercise of the option are greater than the value of the unexercised option $V_2(i, j)$; this value is calculated using the traditional finite difference method assuming no exercise occurs at i, j. The value of the option at i, j is equal to the maximum of the unexercised value and the exercise proceeds. Note that as of any $\hat{t} > \hat{t} + D - 2, V_2(i, j)$ is equal to the value of a portfolio of two strikelets, one maturing on day $\hat{t} + D - 1$, and one maturing on $\hat{t} + D$.

The process should now be apparent. Computationally, the process is as follows.

1. Solve for the Bermudan option value V_1 on the load time mesh. Fill in an array giving the values of the Bermudan option on each

quantity step and for each time step corresponding to a time greater than the initial exercise date t.

2. Solve for the option value V_2 on the load time mesh. Fill in an array giving the values of V_2 on each quantity step and for each time step corresponding to a time greater than the initial exercise date $\hat{t}$.
3. Solve for the option value V_3 on the load time grid. Fill in an array giving the values of V_3 on each quantity step and for each time step corresponding to a time greater than the initial exercise date $\hat{t}$.
4. Proceed in this fashion until you have valued V_5 on each node on the grid corresponding to a time greater than or equal to the initial exercise date $\hat{t}$.
5. The value of the swing for $t<\hat{t}$ is the value of $V_5(i,j)$ calculated at that time step determined using the finite different approach.

In brief, the method involves the sequential solution of five (more generally N) PDEs with free boundaries. This method can obviously be generalised to any number of swing rights N.

Other PDE solution approaches, such as integration or Monte Carlo, cannot readily handle swing features. The mesh-based approach can capture the dynamic decision-making process inherent in optimal exercise of swings because it works from the future back to the present; it shares this feature with classical dynamic programming methods for the solution of dynamic optimisation problems. Monte Carlo or integration both effectively work from the present to the future and therefore cannot readily cope with the dynamic optimisation required to value options with early exercise features.

INTEGRATED VALUATION OF POWER PRICE, VOLUME, AND WEATHER-SENSITIVE CLAIMS

The foregoing uses load as the demand state variable. It is well known that load is largely determined by weather, especially by temperature. It is therefore possible to recast the model using weather as the state variable. This allows unified hedging of derivatives or assets with values that depend on power prices, loads, or weather, or all three.

Formally, define w_t as the value of the weather variable as of time t. The weather variable follows an Ito process:

$$dw_t = \alpha_w(W_t, t)\,dt + \sigma_w(W_t, t)\,dz \qquad (12)$$

where dz_t is a standard Brownian motion.

Whereas in our original analysis power spot prices depended on load, we now specify that they depend on this weather variable and calendar time:

$$P_t = f(w_t, t) \qquad (13)$$

Moreover, power output at t depends on w_t:

$$q_t = h(w_t, t) \qquad (14)$$

Consider the value of any power derivative with a payout that depends on the spot price of power at $T > t$. The value of this derivative at t can be written as $V(w_t, t, T)$. The value of a derivative with a positive value at t (such as an option) must satisfy the partial differential equation:

$$rV = \frac{\partial V}{\partial t} + \frac{\partial V}{\partial W_t}[\alpha_w - \lambda(W_t, t)] + 0.5\frac{\partial^2 V}{\partial W_t^2}\sigma_W^2 \qquad (15)$$

The term $\lambda(w_t, t)$ is a market price of w_t risk. The expression must include a market price of risk because w_t is not a traded asset, and thus is not a hedgeable risk.

For a forward contract, the PDE is:

$$0 = \frac{\partial F}{\partial t} + \frac{\partial F}{\partial W_t}[\alpha_W - \lambda(W_t, t)] + 0.5\frac{\partial^2 F}{\partial W_t^2}\sigma_W^2 \qquad (16)$$

This PDE must be solved subject to boundary conditions. For a daily strike call option on a forward contract with strike price K, the payout is $V(w_T, T, T) = \max[f(w_T, T) - K, 0]$. For a put option with strike price K, the boundary condition at expiration is $V(w_T, T, T) = \max[K - f(w_T, T), 0]$. For a forward contract, the boundary condition is $F(w_T, T, T) = f(w_T, T)$.

Weather derivatives can be evaluated in an identical framework. Specifically, let $W(w_t, t, T)$ be the price of a weather contingent claim with a payout that depends on w_t that expires at T. This contingent claim must satisfy the following PDE:

$$rW = \frac{\partial W}{\partial t} + \frac{\partial W}{\partial W_t}[\alpha_W - \lambda(W_t, t)] + 0.5 \frac{\partial^2 W}{\partial W_t^2} \sigma_W^2 \qquad (17)$$

subject to the appropriate boundary conditions implied by the payout function for the weather derivative.

The solutions to the PDEs for a power contingent claim and a weather contingent claim generate information necessary to construct a hedge position. Specifically, consider a weather derivative with $\Delta_w = \partial W(w_t, t, T)/\partial w_t$ and a power derivative with $\Delta_v = \partial V(w_t, t, T)/\partial w_t$. A firm holding a long position in the power derivative can hedge this position against w_t risk by trading $-\Delta v/\Delta_w$ units of the weather derivative.

Mesh-based methods are appropriate for valuation of weather derivatives. In particular, these methods and inverse problem techniques can be used to determine a market price of risk function $\lambda(w_t, t)$.

Moreover, mesh-based techniques are especially useful in valuing weather derivatives due to the path-dependent nature of the latter; payouts to the most common weather derivatives depend on the average temperatures across many days, rather than the temperature on a single day. For example, consider an option with a payout that depends on the average temperature over some time period $[t_0, t_N]$. This average temperature is given by the integral:

$$I = \frac{1}{t_0 - t_N} \int_{t_0}^{t_N} W(\tau) \, d\tau \qquad (18)$$

It is possible to show that the average temperature option must satisfy the following PDE (see Wilmott *et al*, 1993):

$$rW(W_t, I, t) = \frac{\partial W}{\partial t} + \frac{\partial W}{\partial W_t}[\alpha_W - \lambda(W_t, t)] + 0.5 \frac{\partial^2 W}{\partial W_t^2} \sigma_W^2$$
$$+ \delta(t) W_t \frac{\partial W}{\partial I} \qquad (19)$$

where $\delta(t) = 1$ for $t \in [t_0, t_N]$, and $\delta(t) = 0$ otherwise. Capturing the state dependence therefore requires increasing the dimensionality of the problem. However, although this makes the problem more computationally expensive, it is still quite tractable and can be solved using traditional approaches. Now the pricing mesh must have an I dimension in addition to the w and t dimensions. The solution must also satisfy certain "jump" conditions.

Practical implementation of the weather-based approach must address two issues. First, it is necessary to specify the drift process $\alpha_w(w_t, t)$. This function should exhibit mean reversion and a seasonally varying mean. Historical data can be used to estimate this mean function and speed of mean reversion.

Alternatively, it may be possible to use weather forecasts to specify the mean temperature to which temperature reverts. Advanced forecasting models typically have many dimensions. For example, meteorologists may use air and water temperature information from many locations around the world to forecast temperatures in Chicago for the next cooling season. The forecast mean temperature changes as these temperatures change. Thus, modern weather forecasting models condition their estimate of the drift in w_t not just on some deterministic function derived from historical data and the most recent observation of w_t, but on this information plus other variables x_t, y_t, and z_t (or more). Although this higher dimensionality can in theory be incorporated into the mesh-based approach, this rapidly becomes computationally intractable as dimensions are added. Use of a high-dimension forecasting model would probably require use of other PDE solution approaches, such as Monte Carlo. It is an open question whether the additional accuracy that *may* accompany the use of a higher-dimension weather forecasting model justifies the additional computational cost.

Second, estimation of a market price of risk function is likely to prove important in weather markets. Indeed, the issues of estimating a market price of risk and the use of a richer forecasting setup are linked. It may prove intractable to estimate a market price of risk function in the context of a richer forecasting setup involving many state variables. This is particularly true in as much as each state variable has its own state price. As a result, dimensionality may overwhelm the number of quoted derivatives available to calculate risk prices. Given the potential importance of a market price of risk in determining weather derivatives, this may prove too high a price to pay to achieve the benefits of increased forecast accuracy.

SUMMARY

In sum, valuation of power claims can be performed in a mesh-based framework. Moreover, mesh-based methods have the potential to allow integrated valuation and risk management of power derivatives, weather derivatives, and claims with payouts that depend on volume. Mesh methods are well developed. Moreover, they are very flexible and offer a variety of advantages in a power-pricing context. First, they can handle free boundary problems that arise in the valuation of claims such as swings. Second, they can value path-dependent claims. Most common weather derivatives include some path dependence. Moreover, payouts that depend on power prices or volumes over some period of time also exhibit path dependence. Unlike other methods, the mesh approach can handle both path dependence and free boundaries, and can do so in a computationally efficient manner.

An equilibrium model of power derivatives pricing attempts to exploit the economic fundamentals of the power market to price power contingent claims. An important aspect of the approach is its emphasis on the market price of power output (or weather) risk. The demand state variable in the model – load or the weather – is not traded and hence not hedgeable. Therefore, power derivatives must embed a market price of risk. Moreover, we find that this market price of risk can represent a large proportion of power forward prices for some months in some markets. Therefore, ignoring this component of power prices can lead to serious valuation errors.[5] Mesh-based methods for estimation of the market price of risk implicit in quoted forward and option prices have already been developed and tested. This is another potential virtue of finite difference valuation methods in power markets.

1 It is possible to create a richer price process by assuming that $F_{T+1}, T = P(q, g, T) + \varepsilon_T$ where ε_T is a white noise price process. The addition of the ε term can introduce further "spikiness" to the power price process.
2 We have also included the quantity of nuclear capacity off-line as an explanatory variable (in the exponential term in (11)). The coefficient on nuclear outages is positive and statistically significant, but inclusion of this variable only increases R^2 from about 0.7 to about 0.73. For simplicity, we exclude this variable from our subsequent analysis.

3 This finding is remarkable given the difficulty of detecting risk premiums in other commodities. Economists since Telser (1958) have used increasingly sophisticated methods to attempt to find risk premiums in commodities, with mixed results. The findings in power represent a pronounced case of "normal backwardation" posited by Keynes.
4 See Tikhonov and Arsenin (1977).
5 It should be noted that, due to the non-storability of electricity, it will be necessary to incorporate a market price of risk into more traditional valuation approaches based on a specification of the spot power-price process. Thus, the market price of risk is not an artefact of the particular equilibrium model we discuss; it is an integral feature of power derivatives pricing.

REFERENCES

Telser, L. G., 1958, "Futures Trading and the Storage of Cotton and Wheat", *Journal of Political Economy*, **66**, pp. 233–55.

Wilmott, P., J. Dewynne, and S. Howison, 1993, *Option Pricing: Mathematical Models and Computation*, Oxford University Press, Oxford.

Merton, R. C., 1976, "Option Pricing When Underlying Stock Returns are Discontinuous", *Journal of Financial Economics*, **3**, March, pp. 125–44.

Tikhinov, A. and V. Arsenin, 1977, *Solution of Ill-posed Problems*, Halsted Press, New York.

5

*Modelling Energy Prices and Derivatives using Monte Carlo Methods**

John Putney
National Power plc

Monte Carlo simulation is a powerful technique for determining the outcome of processes that depend on uncertain behaviour or events. Monte Carlo techniques for valuing derivatives and calculating portfolio value at risk (VAR) have been around for a number of years and were initially developed to support trading in financial markets. As energy markets have evolved, the techniques have been adapted to address the particular difficulties presented by energy products and the specific behaviour of energy prices. This chapter discusses these difficulties and shows how Monte Carlo simulation may be applied to tackle them. It focuses on the practical application of the method to price commonly traded energy options – primarily in natural gas and electricity markets – but also briefly considers its use in calculating VAR.

THE MONTE CARLO SIMULATION TECHNIQUE

Monte Carlo simulation is often applied in conjunction with forecasting models to take account of uncertainties in key input parameters. If these uncertainties can be represented by probability distributions, Monte Carlo sampling may be used to generate multiple input scenarios to which the model can be applied to produce a distribution of forecasts. The mean of this distribution is the expectation of the quantity being forecast, and its standard deviation provides a measure of the uncertainty in the forecast. An example where this

---
*This chapter has been reprinted from the first edition.

may be useful in the UK electricity industry is in forecasting outturn pool prices using computer models to emulate the plant-scheduling and price-setting processes. Uncertainties in system demand, plant availability and generator bidding may be incorporated in the forecasts by embedding these models in a Monte Carlo simulation.

The Monte Carlo approach is also applicable for modelling processes that are stochastic in nature, as opposed to deterministic processes with uncertain inputs. A process especially relevant to energy trading and risk management is the behaviour of energy prices over time, in both spot and forward markets. Here Monte Carlo simulation may be used to generate distributions of possible price paths and, in particular, model the performance of derivative contracts written against these prices. Two areas where the method is frequently employed are in pricing exotic options and in calculating the VAR for a portfolio of contracts. Although Monte Carlo simulation is not the only method in these areas, it offers a lot of flexibility in handling complexities in derivative contracts and underlying price processes. These complexities often make analytical methods impossible to derive and cannot always be accommodated by other numerical techniques.

The main barrier to using the method is the often significant amount of computing time needed to obtain a satisfactory solution. There is also a perception that the intellectual investment required to apply the method is too high. However, although it is difficult to implement the method in a spreadsheet without undertaking some macro programming, the mathematics necessary to specify a Monte Carlo algorithm is more accessible than that required to derive analytical pricing formulas, or to work through the matrix algebra behind analytical VAR methods. Furthermore, once a Monte Carlo framework has been set up, it can accommodate new derivative products with only minimal changes. In contrast, analytical pricing formulas are only applicable to specific derivatives and have to be derived for each new product.

OVERVIEW OF MONTE CARLO METHOD FOR PRICING ENERGY OPTIONS

Monte Carlo simulation offers a versatile numerical method for valuing energy options when exact formulas are not available or when there is a need to verify analytical approximations. It is simpler to implement than tree and finite difference methods and is more applicable than these methods when the option payout depends on the history of the underlying prices. It also becomes relatively more efficient as the number of underlying variables increases.

The key attraction of the Monte Carlo method, however, is its flexibility in handling varied and complex option-pricing problems. Once the basic framework of the method has been set up, it is relatively straightforward to "slot in" different payout functions, consider alternative evolutions of the underlying prices, and accommodate market rates and volatilities that are time-, maturity- and even price-dependent. Even when analytical pricing formulas exist, it often requires considerable mathematical intellect and effort to derive them (or even determine whether they exist), and the Monte Carlo method may prove a more effective approach.

The main drawback of the method is its difficulty in treating early exercise decisions in American and Bermuda-type options. This is an area of current research and a number of advances have been made (see Tilley, 1993; Barraquand and Martineau, 1995, for example). Moreover, although various techniques are available for speeding up Monte Carlo simulations (see Panel 4), the computation time necessary to achieve a desired level of accuracy can be high. However, as the dramatic increases in computer processing power seen in recent years are unlikely to abate, and opportunities are taken to exploit parallel processing, computation times for Monte Carlo calculations are likely to become less of an issue.

The following sections illustrate the various features of the method with examples specific to energy markets. The chapter begins by describing the general Monte Carlo methodology for pricing options and its numerical implementation. Later the particular problems posed by energy derivatives and how the method can be applied to address them are also considered. Key features of the approach are then demonstrated by showing how it can be employed to value products commonly traded in energy markets – Asian options, spark-spread options, swap options and swing options – including its ability to accommodate different price-behaviour models. Finally, the convergence and accuracy of the method are discussed.

The intention is to show that although Monte Carlo simulation may never be appropriate for real-time option pricing, it provides a powerful technique for valuing new and complex products, and may be the method of choice when accurate pricing of an energy derivative is essential.

MONTE CARLO METHODOLOGY

The Monte Carlo approach for valuing options was first introduced by Boyle (1977). The basic idea is to use appropriately distributed random numbers to simulate possible price paths for the instruments underlying the option. The simulations are repeated many times, and for each price path the option payout is calculated and discounted to the present. If the process ensures that there are no arbitrage opportunities between purchasing the option or its underlying instruments, the expected (average) discounted payout represents an estimate of the value of the option. For many situations, this no-arbitrage condition can be achieved by performing the simulations in a risk-neutral world.

To illustrate the principles involved, consider a European-style option on a single underlying variable whose price, S, is lognormally distributed (geometric Brownian motion):

$$\frac{dS}{S} = \mu\, dt + \sigma\, dz \tag{1}$$

where μ is the drift rate, σ is the volatility and dz is a random variable drawn from a normal distribution with mean 0 and variance dt. In the absence of arbitrage opportunities, if a risk-free portfolio can be set up consisting of a position in the option and a position in the underlying instrument, the return on the portfolio must be equal to the risk-free interest rate r. This is the basis of the Black–Scholes analysis (Black and Scholes, 1973), which shows that the option value, f, satisfies the differential equation

$$\frac{\partial f}{\partial t} + rS\frac{\partial f}{\partial S} + \frac{1}{2}\sigma^2 S^2 \frac{\partial^2 f}{\partial S^2} = rf \tag{2}$$

and, as a consequence, is independent of the risk preferences of investors. The key point to note is that this is true for any option written on an underlying price that is lognormally distributed. Different types of option are accommodated by specifying appropriate boundary and initial conditions on this equation, and there are only certain conditions for which the equation can be solved analytically (vanilla European calls and puts, for example).

However, even when an analytical solution does not exist and we are forced to adopt a numerical approach, the absence of risk preferences from the option value still holds.

This result implies that, in the Black–Scholes world, risk preferences can be ignored when valuing an option and, in particular, the simple assumption that all investors are risk neutral can be made. The no-arbitrage condition may therefore be built into the Monte Carlo process by setting μ equal to the expected growth rate of the underlying in a risk-neutral world, and discounting all cashflows at the risk-free rate. Thus, $\mu = r$ for an option on a non-dividend-paying stock; $\mu = r - q$ for a stock paying a continuous dividend yield q; and $\mu = 0$ for an option on a future.

In the case of commodities that are held primarily for consumption rather than investment, the expected growth rate in a risk-neutral world is more difficult to specify. In particular, there may be benefits from holding the physical commodity that are not available from holding a derivative – such as the ability to profit from temporary local shortages or to keep a production process running. If these benefits can be quantified as a continuously compounded convenience yield, y, the risk-neutral growth rate may be expressed as $r - (y - u)$, where u denotes the storage costs. Alternatively, the risk-neutral growth rate may be set to $m - \lambda\sigma$, where m is the expected growth rate of the commodity in the real world and λ is the market cost of risk.

NUMERICAL IMPLEMENTATION

For a small time step Δt, equation (1) may be written as

$$\frac{S^{(n+1)} - S^{(n)}}{S^{(n)}} = \mu \Delta t + \sigma \varepsilon \sqrt{\Delta t} \tag{3}$$

where ε is a normally distributed random number with a mean of 0 and a variance of 1 (standardised normal variate). Various routines for generating values of ε can be found in Press et al (1992), but a reasonable approximation is

$$\varepsilon = \sum_{i=1}^{12} R_i - 6 \tag{4}$$

where R_i ($i = 1, \ldots, 12$) are independent uniform random numbers between 0 and 1 – for which most programming and spreadsheet macro languages provide inbuilt routines.

A single Monte Carlo simulation run involves applying equation (3) to generate a sample price path $S^{(0)}, S^{(1)}, \ldots, S^{(N)}$, where $S^{(N)}$ is the price when the option is exercised. For European options, these prices provide all the information necessary to determine the discounted payout for the sample. In the case of a vanilla call option with strike price K, the payout P is simply given by

$$P = \max(S^{(N)} - K, 0) \tag{5}$$

but for more exotic options, such as barrier and Asian options, the whole price path may need to be considered. Repeating the process over M sample paths allows the option price to be estimated as

$$f = e^{-rT} \frac{1}{M} \sum_{j=1}^{M} P_j \tag{6}$$

where T is the time to expiry and, for simplicity, we have assumed that r is constant or uncorrelated with P.

In the particular case where μ and σ are constant, the stochastic differential equation (1) can be solved analytically to obtain the price at any time $t > 0$:

$$S(t) = S_0 \exp\left[\left(\mu - \frac{\sigma^2}{2}\right)t + \sigma \varepsilon \sqrt{t}\right] \tag{7}$$

where S_0 is the initial price. Hence, for options whose payout depends only on the price at the time of exercise, $S^{(N)}$, there is no need to generate the full price path, and each Monte Carlo simulation run can be reduced to a single step by applying equation (7) with $t = T$. Of course, in most option pricing problems for which Monte Carlo simulation is appropriate it will be necessary to divide the time to expiry into multiple steps. However, in such cases, equation (7) may still be used at each step to provide a more accurate estimate of the price evolution.

The option hedge parameters, or "Greeks", can also be calculated using Monte Carlo simulation. To calculate delta, for example, which represents the rate of change in option value with respect to the underlying price, $\partial f/\partial S$, Monte Carlo simulation is first applied in the normal way to value the option. A small increase, ΔS, is then made to the underlying price and the option is revalued, using the same random number sequence, to determine the consequent change in option value. An estimate for delta may then be obtained from:

$$\Delta \approx \frac{f(S+\Delta S)-f(S)}{\Delta S} \qquad (8)$$

The same approach can be applied to calculate the other first-order Greeks – namely vega, theta and rho – and gamma can be determined using the approximation

$$\Gamma = \frac{f(S+\Delta S)-2f(S)+f(S-\Delta S)}{\Delta S^2} \qquad (9)$$

ENERGY OPTIONS

When the asset underlying an option is an energy commodity a number of difficulties enter the pricing process. Firstly, as discussed above, because energy commodities are held primarily for consumption rather than investment, the convenience yield or cost of risk must be introduced to take the lognormal price process into a risk-neutral world. More worrying, however, is the complete lack of any empirical evidence to show that spot energy prices actually follow a lognormal price process. In fact, such prices normally exhibit strong mean-reverting tendencies and are often affected by random jumps.

Fortunately, most energy commodities have actively traded futures or forward markets in which prices tend to follow a lognormal process. Many energy options are now written on an underlying forward or futures contract and can therefore be valued in a risk-neutral world by setting the drift rate, μ, equal to zero. Moreover, options whose payout depends on the spot energy price at only one time, T, can also be priced by assuming that the underlying variable is the forward price maturing at time T – ie, F_T – as this equals the spot price at time T (or more precisely $F_T(t) = S(t)$ when $t = T$).

The difficulties associated with a non-lognormal spot-price process do not disappear completely, however, by valuing options

PANEL 1 EQUIVALENCE AND CONVERGENCE OF OPTION PRICING METHODS

For European futures options, the standard Monte Carlo and binomial tree methods produce results that converge to the option value given by the Black (1976) analytic formula. This is because the methods are all based on the same underlying (lognormal) price process and the same arbitrage-free pricing condition. In the binomial tree approach, convergence is achieved by reducing the time step used to construct the tree, and in the Monte Carlo approach convergence is achieved by increasing the number of sample runs. (Recall that for such options we only need to generate the terminal price distribution, which can be obtained in a single time step using equation (7).)

The following figures illustrate how the Monte Carlo and binomial tree methods converge to Black's solution for an at-the-money European futures option expiring in one year's time, when the price volatility is 25% and the risk-free rate is 6%. For a given number of

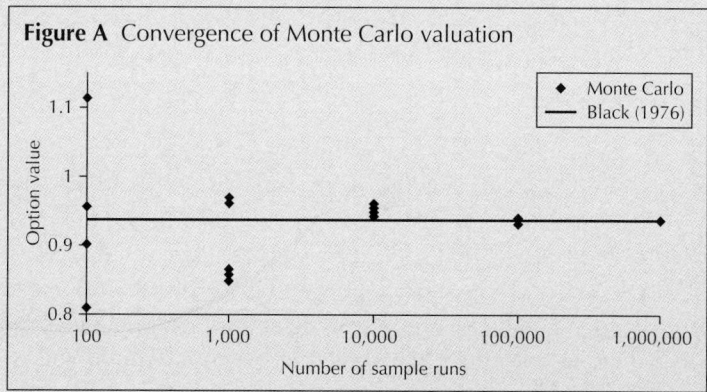

Figure A Convergence of Monte Carlo valuation

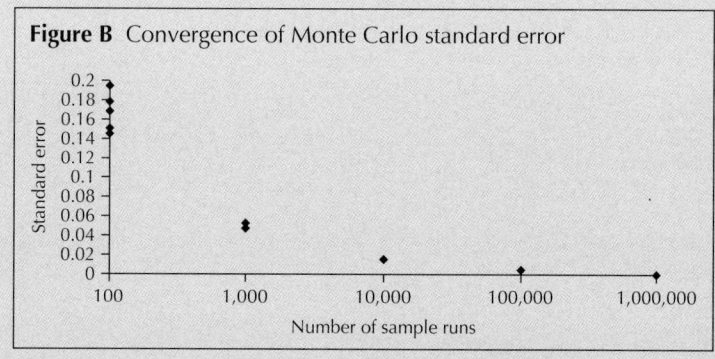

Figure B Convergence of Monte Carlo standard error

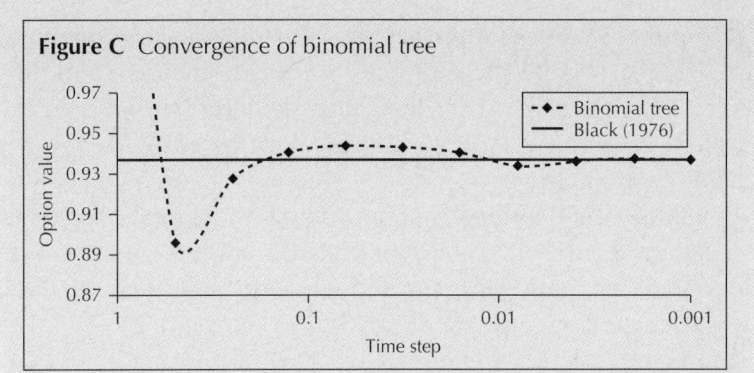

Figure C Convergence of binomial tree

sample runs, the Monte Carlo valuation depends on the initial random seed, but the spread in values reduces as the number of runs increases and acceptable results are obtained with around 10,000 sample runs. The standard error in the option value behaves in a similar way.

With the binomial tree method, good results are obtained with a time step of 0.2 – ie, by dividing the year into five time steps.

against forward prices. In particular, a consequence of mean reversion in the spot price is that the volatility of the forward price exhibits strong term effects – it tends to increase in an exponential fashion as the time to maturity reduces. This means that the volatility controlling the lognormal price process cannot be assumed to be constant throughout the life of the option, and a distinction has to be made between the volatility estimated from historic forward prices and that implied from options. Not only do these represent different forecast views of volatility (past behaviour and market implied) but they also describe quite different quantities. The former represents the instantaneous volatility of the forward price as seen now (σ_i) and the latter is the average volatility of the forward price between now and the time when it matures (σ_A):

$$\sigma_A^2 = \frac{1}{T}\int_0^T \sigma_i^2(t)\,dt \qquad (10)$$

Clearly, it is the average volatility that must be used when pricing the option using a method that assumes that volatility is constant, although the instantaneous volatility is still required for value-at-risk calculations. The ideal approach is to obtain market option prices and derive an implied volatility appropriate to the term and duration of the forward price. In reality, however, such prices are relatively sparse and other techniques must be introduced in the volatility estimation.

A common and promising approach is to construct a parameterised form of the instantaneous volatility curve from theoretical models of the underlying price behaviour, and then calibrate the parameters against the available market data. The simplest example is

$$\sigma_i(\tau) = \sigma_\infty + (\sigma_c - \sigma_\infty)e^{-\kappa\tau} \qquad (11)$$

where $\tau = T - t$ is the tenor, $\sigma_\infty = \sigma_i(\infty)$ is the infinite volatility, $\sigma_c = \sigma_i(0)$ is the cash volatility, and κ is the volatility decay.

When pricing an option using a constant volatility model, the parameterised instantaneous-volatility curve must be integrated to determine the average volatility between pricing and exercise. For the above curve the result is

$$\sigma_A^2(\tau) = \frac{1}{\tau}\left[\sigma_\infty^2 \tau - \frac{1}{2\kappa}\{\sigma_i^2(\tau) + 2\sigma_\infty[\sigma_i(\tau) - \sigma_c] - \sigma_c^2\}\right] \qquad (12)$$

In practice, equation (11) does not provide a sufficient description of the instantaneous-volatility behaviour and more complex parameterisations are required. Often seasonal factors have to be introduced and these can make the integration of the curve difficult. An advantage of the Monte Carlo approach is that this integration can be avoided by performing the simulation using the time-dependent instantaneous volatility curve directly (although, if the average volatility is known, this can still be applied as a constant volatility). Moreover, once the process is set up, it is very easy to explore different instantaneous volatility parameterisations.

Note that, for options whose payout depends on the path followed by the underlying price, it will not in general be valid to use the average volatility in a constant-volatility price model. Here Monte Carlo simulation may provide the only approach.

ASIAN OPTIONS

Asian options are path-dependent options whose payout depends on the average price of the underlying instrument during all or at least part of the life of the option – such as the last few days of trading. They are popular for thinly traded assets such as energy commodities as they ameliorate price distortions that may arise due to lack of depth in the market and provide protection against price manipulations near maturity. Many producers and consumers of energy products are also interested in hedging average fuel costs and achieving average product prices. Asian options therefore fit their risk profiles better than European options and are generally cheaper. They are also easier to hedge with market contracts and, if the averaging period is sufficiently long, have reducing gamma risk as they approach expiry (unlike at-the-money European options).

Asian options generally fall into two categories: average price and average strike. An average-price call option has a payout defined by $\max[\text{avg}(F) - K, 0]$, where K is the strike price and $\text{avg}(F)$ is the average price of the underlying instrument over the specified period. An average-strike call option on the other hand pays out $\max[F - \text{avg}(F), 0]$ at expiry, where F is the final price, which guarantees that the average price paid for the instrument is not greater than the final price.

The vast majority of Asian options are defined in terms of arithmetic averages of the underlying prices, rather than geometric averages. As the arithmetic average of a set of lognormal distributions does not have tractable analytical properties, exact analytical pricing formulas do not exist for Asian options. A number of good approximations have been derived (see Haug, 1998; Vorst, 1996, for example), although these do not account for volatility term effects, which can be important for energy options and are not always applicable when the underlying instrument is a forward contract. Tree and finite-difference methods are also very difficult to apply to Asian options.

Monte Carlo simulation, however, although slower than an analytical approximation, is very flexible at valuing Asian options. It is simply a matter of calculating the appropriate average price and consequent payout for each price path (see Panel 2). The method can easily handle different averaging periods, including weighted

PANEL 2 EXCEL MACRO TO PRICE ASIAN CALL OPTION USING MONTE CARLO

Many analysts are reluctant to use Monte Carlo simulation to price options because of the need to use a programming language to implement the method. However, the level of programming expertise required is fairly basic, and the amount of code involved is relatively small.

This is illustrated by the following Visual Basic macro to price an Asian call futures option, which may be attached to a button in an Excel spreadsheet. The macro begins by extracting the option details, market data and simulation parameters from cells in the spreadsheet (named *ranges*), and determines the simulation time step from the time to expiry and number of steps required. The Monte Carlo simulation then consists of two loops: an outer loop over the number of sample runs and an inner loop to generate the price path for the run and accumulate the average price. At the end of the inner loop the payout for the sample run is determined, and at the end of the outer loop the option price is calculated as the discounted average payout over all sample runs. The result is then written to a cell in the spreadsheet.

The macro employs two user-defined functions: normal rand(), which returns a random number drawn from a univariate standardised normal distribution; and max(*a, b*), which simply returns the maximum of *a* and *b*.

In the example considered, the averaging period covers the whole life of the option, but the macro can easily be extended to handle a shorter averaging period. It may also be adapted to price the option after the averaging period has started, if the average price to date is specified.

```
Sub monte_carlo()

'Get input data from spreadsheet
price = Range ("price") 'current price of underlying
vol = Range ("vol") 'volatility of underlying
strike = Range ("strike") 'option strike price
time_expiry = Range ("time_expiry") 'time to expiry
interest = Range ("interest") 'risk free interest rate
num_steps = Range ("num_steps") 'number of time steps
num_runs = Range ("num_runs") 'number of simulation runs

'Perform Monte Carlo Simulations
dt = time expiry / num_steps
sqr_dt = Sqr(dt)
total_payout = 0
  For i = 1 To num_runs
  p1 = price
  average_price = 0
  For j = 1 To num_steps
  eps = normal_rand()
```

```
    p2 = p1 * (1 + interest * dt + vol * eps * sqr_dt)
    average_price = average_price + 0.5 * (p1 + p2)
    p1 = p2
    Next j
    average_price = average_price / num_steps
    total_payout = total_payout + max(average_price – strike, 0)
    Next i
average_payout = total_payout / num_runs
option_price = average_payout * Exp(–interest * time_expiry)
'Write result to spreadsheet
Range ("option price") = option_price
End Sub
```

averages, accommodate instantaneous volatility curves to account for volatility term effects, and price the option after the averaging period has started. In all cases, a full set of Greeks can also be calculated.

If an Asian option is written against the average spot price rather than an average forward price, the issues surrounding the assumption of a lognormal spot-price process must be faced. One approach is to apply Monte Carlo simulation to evolve the forward price at expiry, $F_{t,T}$, and introduce an appropriate relationship to recover the spot price at each time step, S_t. Unfortunately, the modelling necessary to achieve this is not straightforward and is difficult to calibrate – the usual cost-of-carry relationship, for example, namely

$$F_{t,T} = S_t \exp[(r - y + u)\tau] \qquad (13)$$

is actually based on a lognormal spot-price process (and constant convenience yield).

ALTERNATIVE SPOT-PRICE MODELS

A better approach when applying Monte Carlo simulation to value Asian and other energy options whose payout depends on the path followed by the spot price is to simulate the spot price evolution directly using appropriate models of the underlying processes. The Monte Carlo approach is very flexible is this respect – changing the underlying price processes merely involves introducing discretised forms of the stochastic price equations and generating random variates from the associated probability distributions. Moreover, these distributions need not have a closed-form analytical expression – empirical distributions may be used – and any parameter in the price model may be sampled from a distribution.

One spot price model commonly used for energy markets is the log-of-price mean reversion model:

$$\frac{dS}{S} = \alpha[\beta - \ln(S)]\,dt + \sigma\,dz \tag{14}$$

or

$$dx = \alpha[b - x]\,dt + \sigma\,dz \tag{15}$$

where $x = \ln(S)$, α is the rate of mean reversion and $b \equiv \beta - \sigma^2/(2\alpha)$ is the long-term equilibrium value of x. The model has the same form as the Vasicek (1977) mean-reverting interest-rate model and has the attraction that it guarantees non-negative prices. It may be applied in a Monte Carlo simulation using the discrete approximation

$$x^{(n+1)} - x^{(n)} = \alpha[b - x^{(n)}]\Delta t + \sigma\varepsilon\sqrt{\Delta t} \tag{16}$$

Another model that has received much attention is one in which random "jumps" in price are superimposed upon a lognormal price process. This "jump diffusion" model (Merton, 1976) takes the form

$$\frac{dS}{S} = (\mu - \gamma k)\,dt + \sigma\,dz + dq \tag{17}$$

where dq denotes a Poisson process independent of dz that generates the jumps, γ is the rate at which jumps arrive and k is the average proportional jump size. When a jump event occurs the price jumps from S to SY (ie, $E(Y) = 1 + k$), where Y is a non-negative random variable that is generally assumed to follow a lognormal distribution. Monte Carlo simulations involving this model therefore require random variables to be drawn from three distributions.

An example of a spot-price model that has been specifically developed for energy markets and appears to fit historic price data in a number of oil, gas and electricity markets is the Pilipovic (1998) model. The model assumes a two-factor representation of price behaviour in which a spot price stripped of seasonal effects mean-reverts to a lognormally distributed long-term equilibrium price, L (the price when supply and demand are in balance), such that:

$$dS_t = \alpha(L_t - S_t)\,dt + S_t \sigma\,dz_t \tag{18}$$

$$dL_t = \eta L_t\,dt + L_t \xi\,dw_t \tag{19}$$

where η and ξ are the drift and volatility of the long-term equilibrium price, and dw is a normally distributed random variable with mean 0 and variance dt, independent of dz.

Introducing arbitrage-free arguments and convenience yield considerations enables a differential equation to be derived for the forward price evolution, which, when solved with the spot and equilibrium price equations, leads to the following closed-form solution for the forward price curve

$$F_{t,T} = (S_t - hL_t)e^{-(\alpha + \lambda \xi)\tau} + hL_t e^{(\eta - \lambda \xi)\tau} \tag{20}$$

where $h = \alpha / (\alpha + \eta) \approx 1$ and λ is the cost of risk. By taking expectations, analytical expressions can also be derived for the forward volatility and within-curve correlation.

The beauty of the Pilipovic model is that, once the parameters have been calibrated against the market, the whole forward curve can be evolved from a Monte Carlo simulation of the spot and long-term equilibrium prices, along with the volatility and correlation structure (if needed). Because of the two-factor nature of the model, the forward price curve also accommodates different dynamic behaviours in the near and far terms – for example, contango in the front end and backwardation in the back end, or vice versa. Seasonal effects may be incorporated in the spot and/or forward prices after they have been evolved by applying factors built from period functions about the peaks – for instance, the presence of a single annual peak in forward prices may be represented by multiplying $F_{t,T}$ by

$$1+\frac{1}{2}A\cos[2\pi(T-t_A)] \qquad (21)$$

where t_A is the time of the peak and $A/2$ is its amplitude.

When adopting a spot-price model other than lognormal in order to price an option using Monte Carlo simulation, some means must be found of building the no-arbitrage condition into the process, because the risk-neutral assumption cannot in general be applied. The usual approach is to adjust the market rates used in the model to match market forward prices. The Pilipovic model, however, already incorporates no-arbitrage and convenience yield conditions between its spot and forward price models, and may therefore be used in Monte Carlo pricing by calibrating these models to the markets.

SPARK-SPREAD OPTIONS

Monte Carlo simulation can easily be extended to price options on more than one underlying asset. An example appropriate to the power industry is the spark-spread option, which is the option to exchange the cost of generating a unit of electricity for the market price for this unit. It is particularly relevant to gas-fired stations operating as merchant plants in traded markets. Such plants are exposed to price volatility in both the gas and electricity markets, but have the option to switch off when the gas price is greater that equivalent electricity price. Modelling these plants as a series of spark-spread options provides a means of assessing their true market value. Alternatively, if the plant is backed up by fuel supply and power offtake contracts, it provides the option of buying electricity and selling gas when the gas price exceeds the electricity price. This embedded optionality can be valued by the option to exchange electricity for gas (the reverse spark-spread option).

Spark-spread options are also sold as financial tolling deals by plant owners as a means of hedging the full value of their plant. They are also useful for setting up internal markets between a generator's power production and power trading businesses.

If the spark-spread option is written as a simple option to exchange gas for electricity, it may be valued using an extension to the Black–Scholes model obtained by Margrabe (1978):

$$c = [F_e N(d_1) - F_g N(d_2)]e^{-rT} \qquad (22)$$

where

$$d_1 = \frac{\ln(F_e/F_g) + 0.5\sigma^2 T}{\sigma\sqrt{T}}$$

$$d_2 = d_1 - \sigma\sqrt{T}$$

and

$$\sigma^2 = \frac{1}{T}\int_0^T [\sigma_e^2 + \sigma_g^2 - 2\rho\sigma_e\sigma_g]\,dt \qquad (23)$$

is the average spread volatility. The parameters in these equations are defined as:

- F_e forward price of electricity;
- F_g forward price of gas per unit of electricity produced, which depends on the plant heat rate;
- σ_e instantaneous volatility of forward electricity price;
- σ_g instantaneous volatility of forward gas price;
- ρ instantaneous correlation between forward gas and electricity price returns;
- T time to delivery;
- r risk-free interest rate;
- $N(x)$ cumulative probability function for a standardised normal variable.

In many cases, however, it is necessary to include the effect of plant operational characteristics in the specification of a spark-spread option, and the above formulas are not valid. For example, to account for the presence of non-fuel marginal costs K, the spark-spread option becomes an option to buy the difference (spread) between the electricity and gas prices, $F_e - F_g$, at a strike price K. As this spread can be negative, the standard Black–Scholes model cannot be applied because the lognormal distribution only produces positive values. If K is small, its effect may be approximated in the Margrabe formula by including it in the gas price, scaling the gas volatility by

$$\sigma_g^* = \sigma_g \left[\frac{F_g}{F_g + K} \right] \tag{24}$$

and ignoring the effect on correlation. Another approach is to assume that spread itself is normally distributed, which allows a closed-form solution to be obtained but implies an underlying price behaviour that differs from that represented by lognormally distributed gas and electricity prices when volatilities are greater than about 20%.

The Monte Carlo method provides a means of valuing complex spark-spread options without the need to introduce approximations to obtain a particular solution. In the above example, each Monte Carlo simulation run involves generating sample forward-price paths for both gas and electricity:

$$F_g^{(n+1)} = F_g^{(n)} \exp\left(-\frac{1}{2}\sigma_g^2 \Delta t + \sigma_g \varepsilon_1 \sqrt{\Delta t} \right)$$

$$F_e^{(n+1)} = F_e^{(n)} \exp\left(-\frac{1}{2}\sigma_e^2 \Delta t + \sigma_e \varepsilon_2 \sqrt{\Delta t} \right) \tag{25}$$

where ε_1 and ε_2 are a pair of random variables from a standardised bivariate normal distribution – standardised normal variates correlated by ρ. These variates can be obtained by applying equation (4) to generate a pair of independent standardised normal variates, ϕ_1 and ϕ_2, and then applying the relationships

$$\varepsilon_1 = \phi_1$$
$$\varepsilon_2 = \rho\phi_1 + \phi_2\sqrt{1-\rho^2} \tag{26}$$

At the end of each simulation run, the spark-spread payout is calculated from

$$P = \max[F_e^{(N)} - F_g^{(N)} - K, 0] \tag{27}$$

and the option is valued in the normal way – as the discounted average payout over all sample runs.

This process for valuing spark-spread options provides all the flexibility offered by the Monte Carlo technique and, in particular, can easily accommodate parameterised volatility and correlation curves. It can also be extended to value other spread options such as regional spreads and even Asian spreads.

SWAP OPTIONS

The above approach can be generalised to value options whose payout depends on N (ie, many) underlying variables. All that is required is a method to take account of the correlation between these variables when simulating their evolution, which means that N correlated random variables must be generated at each time step. The most efficient method is to use Cholesky's decomposition (Press et al, 1992) to factorise the $N \times N$ correlation matrix, C, to the form

$$C = LL^T \qquad (28)$$

where L is a lower triangular matrix. Then, if ϕ is a vector of N independent standardised normal variates,

$$\varepsilon = L\phi \qquad (29)$$

is a vector of N correlated standardised normal variates. Unfortunately, this factorisation is only possible if the correlation matrix is positive semi-definite. If this is not the case, more complex decomposition techniques – such as eigenvalue decomposition – must be applied. If the correlation matrix has a term structure, the decomposition must also be performed at each time step.

Interestingly, the Monte Carlo method tends to be more efficient than other numerical methods for valuing options as the number of underlying assets increases. This is because the time taken to perform a Monte Carlo simulation increases approximately linearly with the number of underlying variables, whereas the time required by tree and finite-difference techniques can increase exponentially. In general, when the number of underlying variables is three or more, the Monte Carlo method is the more efficient.

An example of an energy option whose value may depend on several underlying instruments is a swap option (or swaption). This is an option to buy or sell a swap – a contract to buy or sell a specified volume of an energy commodity at a fixed "swap price" over a predefined "swap period" in the future – which is normally exercised shortly before the swap period starts. Call swap options are popular with energy consumers who need the assurance of a maximum fixed purchase price, but who feel that prices may drop before the option is exercised – in which case they can leave the option unexercised and buy a swap at the lower price. A call swap option is also cheaper than a cap covering the swap period, as there is two-way risk on the swap after the option is exercised.

Unless a single forward swap price is quoted in the market, pricing a swap option can be difficult because it is necessary to model the behaviour of the entire forward curve over the swap period, not just a point on this curve. For example, to price a one-year swap option when market information/models have been derived at a monthly level, the correlation along the curve must be respected when modelling the process followed by the average swap price

$$F^S = \frac{1}{12} \sum_{i=1}^{12} F_i \qquad (30)$$

(ignoring differences in the number of days in each month). With Monte Carlo simulation, this can be achieved by treating each monthly forward price, F_i, as an underlying variable and applying the method described above to simulate their evolution:

$$F_i^{(n+1)} = F_i^{(n)} \exp\left(-\frac{1}{2}\sigma_i^2 \Delta t + \sigma_i \varepsilon_i \sqrt{\Delta t}\right), \qquad i = 1, 2, \ldots, 12 \qquad (31)$$

where the ε_i are correlated random variates. The value of swap option may then be estimated as the discounted average payout.

An alternative (and analogous) approach to using Cholesky or eigenvalue decomposition to generate N correlated random variates at each time step is to apply a principal component or factor analysis to find a set of orthogonal factors that capture the stochastic behaviour of the forward curve over the swap period. These factors may be viewed as perturbations to the curve at increasing frequency – for example, parallel shift, tilt, bow and flex – and allow the swap price evolution to be simulated as

$$F_i^{(n+1)} = F_i^{(n)} \exp\left(-\frac{1}{2}\sum_{j=1}^{12} \omega_{ij}^2 \Delta t + \sum_{j=1}^{12} \omega_{ij} \phi_j \sqrt{\Delta t}\right), \quad i=1,2,\ldots,12 \quad (32)$$

where ϕ_j are a set of *independent* standardised normal variates, and the same set is used for each monthly forward price, F_i. The volatility coefficients, ω_{ij}, are a product of the principal component/factor analysis and are related to the eigenvalues and eigenvectors of the covariance/correlation matrix.

Evidently, this approach on its own is not necessarily any more efficient than the decomposition approach, particularly if Cholesky's method can be used. However, the analysis normally reveals that the behaviour of the forward curve can largely be explained by a few key factors. Using these factors to approximate the swap price evolution allows the number of underlying variables in the Monte Carlo simulation to be significantly reduced.

SWING OPTIONS

Many contracts involving the sale and purchase of energy commodities – particularly gas, electricity and coal – include elements that provide some flexibility in the timing and quantity of delivery. Such contracts have arisen from the uncertainty in the amounts of electricity and gas that residential and business sectors consume, and are often called swing options as they give holders the right, but not the obligation, to swing up or down the amount of energy they take. This flexibility, however, is often tempered by constraints on the minimum and maximum takes over certain periods (daily and annual totals) and on the number of times the swing rights can be exercised. Swing options therefore have an implicit dependence through time: the exercise of an option on one day reduces the flexibility to take energy in the future, potentially leaving the consumer short of energy or forced to take it when not required. They are also "American" in nature, as the exercise dates are not predetermined and can be selected by the holder depending on market conditions.

The various types of energy swing options that exist generally fall into two distinct categories: price-driven and demand-driven.

Price-driven swing options normally occur when both counterparties can buy and sell the underlying energy commodity in the market. This means that the holder of the option is able to extract the maximum economic value of the embedded swing. In contrast, demand-driven swing options are present in contracts in which one of the counterparties can only take or withhold from taking delivery of the energy commodity. Typically, this would be a counterparty whose business is set up to purchase and physically consume the commodity, but not to sell and deliver it. The counterparty is therefore likely to exercise the swing rights more on energy need than on energy price – sometimes referred to as non-ruthless rather than ruthless exercise.

Although energy swing contracts have been around for many years, it is only recently that their optionality content has been fully understood or even appreciated. Consequently, this embedded optionality was often mispriced or neglected in the past, and utilities that held swing rights did not always have the strategies to exercise them optimally in the marketplace. Now that the importance of swing options has been recognised, significant efforts are being devoted to developing techniques to value and operate them. The coupled, constrained and American nature of these options, and the possibility of non-ruthless exercise, make this a very challenging task.

An example of a simple swing option is an annual forward contract to deliver a specified volume, V, of gas each day at a fixed price K, in which the consumer has the right to vary the daily amount taken between V_{min} and V_{max}, but must meet minimum and maximum limits on the total volume taken over the year. Such "take-or-pay" arrangements are often found in contracts to supply gas to power stations, to provide fixed-price protection with the flexibility to tailor the amount taken to match power station demand. More complex variants can include limits on the number of times the volume taken can swing up or down from V, and a strike price that is linked to a floating index.

One approach to valuing price-driven swing options is to attempt to decompose them into components that can be valued analytically. For example, the above simple take-or-pay contract may be represented as:

❑ 365 daily forward contracts to buy volume V at price K;
❑ N daily European call options to buy $V_{max} - V$ at K;

❑ M daily European put options to sell $V - V_{\min}$ at K;
❑ a set of options to switch between these options.

Estimates of N and M may be calculated from the annual take limits by valuing the daily call and put options, ranking these values and using the delta to determine the expected takes. If a "value" can be assigned to the options to switch, the total value of the contract can be estimated from the various components. Such semi-analytical swing-pricing methods can be very efficient, but can never in general capture all the embedded optionality in the contract (they will generally give a lower bound on the true value). They are also more difficult to construct as the complexity of the contracts increases.

An alternative approach to valuing swing options is to use a binomial tree to evolve the spot price underlying the contract. For gas and electricity swing contracts, the spot price process assumed should ideally incorporate mean-reversion and seasonal dependent parameters, and should match the forward price and volatility term structures seen in the market. When applying backward induction on the tree to price the option, the number of exercises remaining needs to be considered at each node. This requires a multi-layered extension to the tree, sometimes referred to as a binomial forest (Jaillet, Ronn and Tompaidis, 1998a; 1998b). These forests probably represent the state of the art in valuing swing options. However, the modelling of a realistic spot-price process and the need to include must-take constraints with penalties, which make the option component path dependent, can cause implementation difficulties. The computer memory required to hold the multiple layers can also be significant when the number of exercise rights is high.

Monte Carlo simulation is not a competitor to the semi-analytical and binomial forest methods for valuing swing options, but it offers a somewhat different approach to the problem. The principal advantage of the method is that it does not place restrictions on the underlying price processes that can be modelled and it can accommodate complexities in the contract specification without significant implementation difficulties. It also provides a framework for incorporating non-ruthless exercise rules, which are necessary to value demand-driven contracts. The key drawback of the method, of course, is its difficulty in accounting for the American elements of the swing options, and hence an appropriate early-exercise rule must be provided. This in itself need not be a disadvantage, as many swing options are operated in practice using such exercise rules, particularly if there are plant or production constraints that may inhibit the full value of the optionality from being realised.

Monte Carlo simulation coupled with exercise rules, although generally underestimating the theoretical value of swing, therefore provides a means of determining the value that can be extracted from the contracts under actual operating and trading conditions. Conversely, it can be applied to determine appropriate and optimal early-exercise rules for operating swing contracts.

An example of a daily exercise rule that may be used with the simple take-or-pay contract outlined above is one that maximises the value of the planned gas-take over the remainder of the contract year against the latest forward curve, subject to the daily and annual volume constraints. In effect, this amounts to valuing the remaining swing contract without day-ahead volume optionality (zero volatility) and may be achieved in practice by

❑ calculating the value of the daily fixed-price forward contracts to the end of the year and ranking the days in order of value;
❑ taking V_{min} on each day;
❑ taking an additional $V_{max} - V_{min}$ on days in decreasing order of value until the annual minimum take limit has been reached (taking account of volume already taken);
❑ taking an additional $V_{max} - V_{min}$ on days in decreasing order of value until this value becomes negative or the maximum take limit has been reached.

A swing contract operated with this exercise rule may be valued using Monte Carlo simulation by adopting the Pilipovic model to describe the underlying price behaviour. Each sample run in the simulation involves applying equations (18) and (19) to generate the daily spot and long equilibrium price paths over the contract life. For each day, equations (20) and (21) are then used to construct the forward price curve, and the exercise rule is invoked to determine (and accumulate) the daily take. At the end of the run, the total value of the gas taken is calculated against the spot prices. Once all sample runs have been completed, the average of these values provides an estimate of the value of the swing contract.

PANEL 3 MONTE CARLO VAR

The bulk of this chapter has focused on the ability of Monte Carlo simulation to model stochastic price behaviour when valuing complex energy derivatives. This capability of the method, however, can also be used to measure the risk associated with trading energy derivatives by facilitating the calculation of their value-at-risk (VAR).

Value-at-risk has become a popular method of describing risk as it captures the interactions between the various sources of risk in a portfolio of assets and produces a single summary measure of the total risk. It is defined as the maximum expected loss for the portfolio under normal market conditions, over a given time horizon (unwind period) and to a specified confidence interval. The method takes account of diversification and hedging effects and considers adverse price movements that are consistent with market volatilities and correlations.

The first step in calculating VAR is to identify the factors that present significant risk and, if necessary, to map the underlying instruments onto these "risk factors". For portfolios of energy derivatives, these factors would typically be segments of the forward price curves – for example, monthly prices in the short term, quarterly in the medium term and annual in the long term – but could also include exchange and interest rates or other market parameters. The portfolio is then "marked-to-market" by valuing each contract in the portfolio against the latest forward prices and market rates. For option contracts, this requires an appropriate option-pricing model.

The Monte Carlo VAR method consists of performing multiple simulations of the change – or return – in the portfolio mark-to-market over the VAR horizon. Each simulation run involves generating a scenario for the changes in the values of the risk factors. Thus, for example, if risk factor i is a lognormally distributed forward price, F_i, and the VAR horizon, Δt, is small, one scenario for its value at the end of the horizon is

$$F_i^{\Delta t} = F_i \exp\left(-\frac{1}{2}\sigma_i^2 \Delta t + \sigma_i \varepsilon_i \sqrt{\Delta t}\right)$$

where σ_i is the instantaneous volatility and ε_i is drawn from a set of correlated standardised normal variates. The latter can be obtained from a set of independent standardised normal variates, by applying Cholesky's decomposition to the risk-factor correlation matrix in a similar way to the method described in the text for valuing swap options.

For each scenario, the portfolio is marked-to-market and the return is calculated. At the end of the simulation runs, the distribution of portfolio returns is constructed and the VAR corresponding to the desired confidence interval is read off from the appropriate percentile – for example, the 95% VAR is the portfolio return for which 5% of the simulations produced a worse return.

The Monte Carlo VAR approach is often compared with the delta-normal method, which offers a simpler, analytical calculation of VAR. This is made possible by representing the change in value of each contract over the VAR horizon, as a linear function of the changes in the risk factors values. Thus, for a derivative with value f written on an underlying forward price F

$$f^{\Delta t} - f = \Delta[F^{\Delta t} - F]$$

where Δ is the delta of the derivative. Then, if price returns are normally distributed, it follows that derivative and thus portfolio returns will also be normally distributed, and the VAR can be obtained from the portfolio variance. Further, the latter can be calculated directly by pre- and post-multiplying the risk factor covariance matrix with a vector of cashflows derived from the risk factor volatilities.

The delta-normal VAR method works well for energy portfolios dominated by forward and futures contracts, without significant option components. Figure A shows the distribution of 1-day returns for a portfolio of US electricity forward contracts. The contracts represented a six-year exposure to peak and off-peak power prices in the Cinergy Pool, and the distribution was obtained by collapsing this exposure onto 12 risk factors and performing 20,000 Monte Carlo simulations. The distribution shown is essentially normal and, as a consequence, the delta-normal and Monte Carlo VAR estimates agree to within 0.5%.

The risks associated with option contracts may not, however, be captured adequately by the delta-normal method, as their value and delta are non-linear functions of underlying price. This is particularly true for at-the-money options close to expiry, where the change in

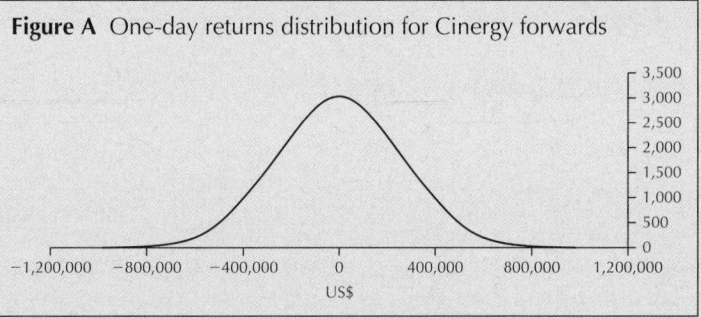

Figure A One-day returns distribution for Cinergy forwards

delta with price (gamma) can be significant. Moreover, portfolios of options can exhibit quite complex return behaviour as the individual non-linearities interact. An example is the sale of a call and a put – a so-called short straddle. At-the-money, this has a delta close to zero

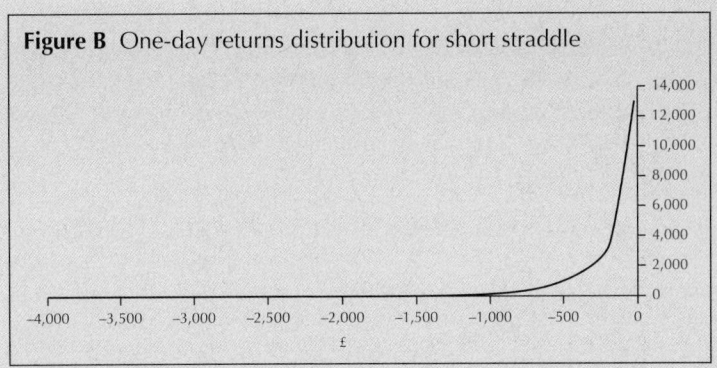

Figure B One-day returns distribution for short straddle

and thus the delta-normal VAR estimate is very low. However, either an increase or decrease in underlying price will cause delta to change to reflect the fact that the expected loss for the straddle will increase – as an increase in price will result in a payout from the call, and a decrease in price will result in a payout from the put. The Monte Carlo VAR approach can capture this effect but the delta-normal method misses it completely.

The effect is illustrated in Figure B where we show the one-day returns distribution for a 50 MW one-month short straddle expiring in three months time, generated using 20,000 Monte Carlo simulations with a 20% option (implied) volatility and a 10% instantaneous volatility. The payout behaviour of the straddle is reflected by a distribution that is far from normal and completely skewed towards a loss. For a 95% confidence interval, the delta-normal method underestimates VAR by a factor of two for a one-day horizon and a factor of five for a 10-day horizon. This increase in error demonstrates that, for non-linear contracts, VAR is not proportional to the square root of the VAR horizon, which the delta-normal method assumes.

Another advantage of the Monte Carlo VAR approach over the delta-normal method is that because it produces the full distribution of returns, as opposed to just a single VAR estimate, it provides a richer picture of portfolio risk – for example skewness – and allows other risk measures to be presented. By dividing the VAR horizon into steps, the method can also account for changes in market parameters over the horizon and changes in the portfolio contents due to contracts expiring. It is therefore particularly appropriate for long VAR horizons, which may be necessary when trading in illiquid markets or measuring the risks associated with generation assets. The method can also accommodate different underlying price-behaviour models, including non-normal models.

The principal drawback of the method is that it can require a large amount of computing time and is thus not very suitable for real-time VAR or incremental VAR analyses. Fortunately, many of the methods described in Panel 4 for speeding up Monte Carlo simulations for

> option pricing can also be applied to Monte Carlo VAR calculations. Clearly, it would not, in general, be practical to embed a Monte Carlo option valuation inside a Monte Carlo VAR calculation. However, if a suitable pricing model is not available, the change in option value over the VAR horizon may be represented by a high order approximation incorporating other Greeks, for example
>
> $$f^{\Delta t} - f = \Delta(F^{\Delta t} - F) + \frac{1}{2}\Gamma(F^{\Delta t} - F)^2 - \theta\Delta t$$
>
> where Δ, Γ and θ may be estimated using Monte Carlo simulation before the VAR calculation commences.

Since the Pilipovic model may be used to evolve the complete forward price vector and associated forward volatility and within-curve correlation, the above method can be used to evaluate a wide range of early exercise rules. These include rules employing semi-analytical valuations and rules based on binomial forests applied to an approximation to the swing problem – for example less frequent exercise decisions (say monthly) – and a simpler representation of the spot-price process. The method may also be used to develop appropriate exercise rules, either manually by judicious trial and error, or automatically by embedding it in an algorithm such as genetic programming (Koza, 1992) to search for an optimum combination of market parameters and conditions.

Interestingly, in markets where the forward price curve is dominated by seasonal rather than term effects, such as the UK gas market, Monte Carlo simulation of a take-or-pay contract using the standard Pilipovic model may not show significant extra value over pricing the contract against a static forward curve. This is because seasonal differences in price swamp variations driven by the spot and equilibrium price processes, and the seasonal profile assumed in the model is deterministic. However, this deficiency can be removed in a Monte Carlo framework by modelling the time of the seasonal peak (t_A) as a stochastic variable. The ability of the exercise rule to exploit a shift in the seasonal peak will then be captured in the valuation calculation.

Demand-driven swing contracts can also be valued using Monte Carlo simulation by extending the exercise rule to incorporate the demand requirements of the holder. The latter could take the form of a functional relationship between the quantity demanded and the underlying price, or a stochastic demand model correlated to the price process. Again, Monte Carlo simulation is very adaptable to the inclusion of such models.

CONVERGENCE AND ACCURACY OF MONTE CARLO SIMULATION

The number of sample runs required to value options using Monte Carlo simulation clearly depends on the level of accuracy required. Fortunately, a measure of the error in the estimated option value may be obtained by calculating the standard deviation of discounted payouts from the sample runs. If this quantity is denoted by s and M runs are performed, the standard error in the estimated option value, f^* (the mean of the discounted payouts) may be approximated by

$$\frac{s}{\sqrt{M}} \qquad (33)$$

and a 95% confidence interval for the true value of the option, f, is given by

$$f^* - \frac{1.96\,s}{\sqrt{M}} < f < f^* + \frac{1.96\,s}{\sqrt{M}} \qquad (34)$$

The accuracy of the Monte Carlo method also depends on the quality of the pseudo-random-number generator – a poor random-number generator can lead to slow convergence or even incorrect results. Unfortunately, many system-supplied random number routines are fairly basic and should be treated with caution – most implementations of the ANSI C rand() function, for example, fall into this category. Various methods for testing the adequacy of random-number generators are described by Chaplin (1993). If these tests fail, techniques for improving system-supplied random-number routines and complete stand-alone random-number generators are given by Press et al (1992). These authors also discuss methods for sampling from different probability distributions.

PANEL 4 SPEEDING UP MONTE CARLO SIMULATIONS

Monte Carlo simulation provides an extremely flexible numerical method for pricing options, but it is generally slower than tree and finite-difference techniques unless the number of underlying variables is three or more. The standard error equation (33) shows that the accuracy of the method is inversely proportional to the square root of the number of sample runs. This means that to reduce the error in an estimated option value by a factor of 10, the number of sample runs must be increased by a factor of 100.

Several techniques have therefore been developed for reducing the number of Monte Carlo simulations necessary to achieve a desired level of accuracy. Such techniques are generally referred to as "variance-reduction procedures" as they attempt to reduce the standard deviation of the simulated payouts. The most common and simplest procedures are the antithetic variable and control variable techniques. Others include importance sampling, stratified sampling, moment matching and the use of low-discrepancy sequences.

A good review of the above variance reduction procedures is given by Hull (1997). Below we summarise the main features of each one.

Antithetic variable technique
In this technique, each simulation run involves calculating two values for the option payout. The first is calculated in the normal way; the second is calculated by changing the sign of all samples from the standard normal distribution when producing the simulated price path(s). The option payout for the simulation run is then calculated as the average of these two values. The method works well because when one value is above the true value, the other tends to be below it, and vice versa.

Control variate technique
This technique involves finding an option similar to the option being value for which an analytical solution is available. Both options are then valued by Monte Carlo simulation using exactly the same random number sequence and time steps. If f_A^* and f_B^* denote the Monte Carlo estimates of the value of the option under consideration, A, and the similar option, B, and f_B is the known true value of B, an improved estimate of A is obtained from

$$f_A = f_A^* - f_B^* + f_B$$

which, in effect, assumes that the error in using Monte Carlo to price the two options is the same.

Importance sampling
This technique attempts to restrict the sampling of random variates to those price paths that lead to a non-zero payout from the option – as

zero payout paths contribute very little to the value of the option and are therefore a waste of computation time. The method requires the probability, k, of the underlying price being "in-the-money" at expiry to be known analytically. Then, if F is the unconditional probability distribution for the underlying price, the simulation is performed by sampling from F/k rather than F – the probability distribution of the underlying price conditional on it being in-the-money at expiry. The estimated option value is the average discounted payoff multiplied by k.

Stratified sampling
Stratified sampling, of which Latin hypercube sampling is an example, involves partitioning the underlying probability distributions into distinct intervals, or strata, and sampling from each stratum according to its probability. If the number of intervals is sufficiently large, a "representative" value can be derived for each one and used whenever sampling from the interval. The approach can lead to a significant reduction in the sampling space.

Moment matching
Moment matching involves adjusting the random samples used in the simulation so that the first, second and possibly higher moments are correct – they match those of the theoretical distribution from which the samples are drawn. Thus, when sampling from a standardised normal distribution, the samples would be adjusted to have a mean of zero and a standard deviation of unity. The method saves computation time but requires sufficient memory to store every sample until the end of the simulation. It is often used in conjunction with the antithetic variable technique, which automatically matches all odd moments.

Low-discrepancy sequences
A low-discrepancy sequence is a deterministic sequence of samples from a probability distribution that has the property that the samples are dispersed evenly throughout the sampling region. Moreover, each new addition to the sequence preserves this property, so that no matter when we stop adding samples, the existing samples are evenly distributed. These sequences generally provide better convergence than traditional random sequences, with standard errors typically proportional to $1/M$ rather than $1/\sqrt{M}$. An example of the use of low-discrepancy sequences in valuing an Asian option is given by Boyle (1996).

Finally, apart from the moment-matching technique, all Monte Carlo simulation runs are independent of each other. The method is thus very amenable to parallel processing – performing the simulations on K processors will result in a direct speed-up by a factor of K.

REFERENCES

Barraquand, J. and D. Martineau, 1995, "Numerical Valuation of High Dimension Multivariate American Securities", *Journal of Financial and Quantitative Analysis*, **30(3)**, pp. 383–405.

Black, F., 1976, "The Pricing of Commodity Contracts", *Journal of Financial Economics*, **3**, pp. 167–79.

Black, F. and M. Scholes, 1973, "The Pricing of Options and Other Corporate Liabilities", *Journal of Political Economy*, **81**, pp. 637–59.

Boyle, P. P., 1977, "Options: a Monte Carlo Approach", *Journal of Financial Economics*, **4**, pp. 323–38.

Boyle, P. P., 1996, "Valuation of Exotic Options Using the Monte Carlo Method", in I. Nelken (ed.) *The Handbook of Exotic Options*, pp. 316–26 (New York: Irwin Professional Publishing).

Chaplin, G., 1993, "Not So Random", *Risk*, **6**, pp. 56–7.

Haug, E. G., 1998, *The Complete Guide to Option Pricing Formulas* (New York: McGraw-Hill).

Hull, J. C., 1997, *Options, Futures and Other Derivatives*, 3rd edition (Englewood Cliffs: Prentice-Hall International).

Jaillet, P., E. Ronn, and S. Tompaidis, 1998a, "The Quest for Valuation", *Energy and Power Risk Management*, **3(3)**, pp. 14–16.

Jaillet, P., E. Ronn, and S. Tompaidis, 1998b, "A Ruthless Business", *Energy and Power Risk Management*, **3(4)**, pp. 28–9.

Koza, J. R., 1992, *Genetic Programming: On the Programming of Computers by Means of Natural Selection* (Cambridge MA: MIT Press).

Margrabe, W., 1978, "The Value of an Option to Exchange One Asset for Another", *Journal of Finance*, **33**, pp. 177–86.

Merton, R. C., 1976, "Option Pricing when Underlying Stock Returns are Discontinuous", *Journal of Financial Economics*, **3**, pp. 125–44.

Pilipovic, D., 1998, *Energy Risk: Valuing and Managing Energy Derivatives* (New York: McGraw-Hill).

Press, W., B. P. Flannery, S. A. Teukolsky, and W. T. Vetterling, 1992, *Numerical Recipes in C*, 2nd edition (New York NY: Cambridge University Press).

Tilley, J. A., 1993, "Valuing American Options in a Path Simulation Model", *Transactions of the Society of Actuaries*, **XLV**, pp. 499–520.

Vasicek, O., 1977, "An Equilibrium Characterization of the Term Structure", *Journal of Financial Economics*, **5**, pp. 177–88.

Vorst, C. F., 1996, "Averaging Options", in I. Nelken (ed.) *The Handbook of Exotic Options* (New York: Irwin Professional Publishing).

6

Fundamental Analysis of Power Price Modelling

Roman Kosecki
MAK Energy Consultants

Power markets have undergone significant changes since their inception. They have achieved certain maturity, both as far as liquidity is concerned and in the variety of structures traded. There has been a steady influx of new participants, most notably large financial institutions and, most recently, hedge funds. Simultaneously, the generation assets ownership (either outright plant ownership or contractually via tolls) by the new entrants is increasing. Market design, as exemplified in PJM[1] with location marginal pricing (LMP), has sufficient robustness and regulatory clarity in most of the regions. In addition, we witness greater unification of the structure of all market designs, which facilitates a more general approach to risk management problems. In what follows, we describe the major ingredients of asset valuation in today's markets.

The general problems facing an active participant in the energy markets can be broadly grouped into three independent areas:

❏ unit characteristics;
❏ market structure; and
❏ contractual energy obligations.

Depending on the profile of an energy firm, more weight is given to the different components. A pure trading company, for instance a hedge fund, is mostly concerned with the market structure, while an independent power producer considering building a new power plant or a bank entering into load following deals has to consider the interplay of all the elements.

GENERATION ASSETS AND LOAD CHARACTERISTICS

Electricity is an instantaneous commodity – its generation and delivery must be balanced in real time, rather than produced and stored for future use. From an operational point of view, power production involves a myriad engineering problems, significant high-tech and human resources and of course the actual generation of energy. The simplest economic trigger indicating the profitability of the power plant at a given point in time is a spark spread S

$$S = P - HR \times F - OC - R \qquad (1)$$

where P is the price of power to be delivered, HR (heat rate) is the efficiency of the plant expressed in MMBtu/MWh, F is the price of fuel and OC is operational costs. The additional constraints, such as availability of transmission, connection to the grid, ramp-up time or global cumulative factors such as start-up considerations, maintenance schedule or emissions, are denoted by R. Daily fluctuations of fuel prices impact the generation cost of electricity, $HR \times F + OC + R$, but with S positive it is economical to run the power plant. The actual power price in the market, P, is the result of equilibrium between the regional demand, driven by weather and industry-based loads, and the available capacity. We analyse the weather impact first.

The impact of the weather

For simplicity, we concentrate on temperatures, but it is clear that factors such as humidity and wind chill influence the amounts of electricity used by consumers. It is generally accepted that the average daily load depends on the deviation of temperature from a set level, T_0, deemed "comfortable". Usually T_0 is set at 65°F. The deviation from T_0 is measured in terms of heating or cooling degree days (HDD and CDD), which are given by the following:

$$HDD(T) = \text{Max}(T_0 - T, 0)$$
$$CDD(T) = \text{Max}(T - T_0, 0)$$

where T is the actual temperature for the day. This realised temperature could be the high, the low, or the weighted average for that day. The choice of T depends on the type of average load that we are trying to explain. One can use HDD with the high for the day as an

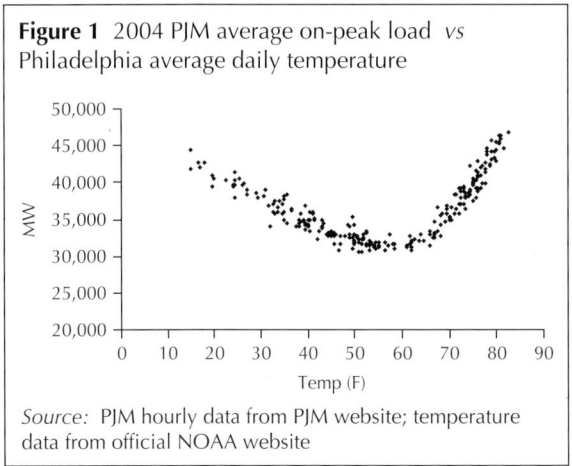

Figure 1 2004 PJM average on-peak load vs Philadelphia average daily temperature

Source: PJM hourly data from PJM website; temperature data from official NOAA website

input variable in the case of summer on-peak load and CDD with the daily lows for the winter on-peak load. The data suggest that the generic average load L_{Ave}, can be modelled as

$$L_{Ave}(T) = f(HDD, CDD) \qquad (2)$$

where f is location specific function with a single minimum at T_0, and is increasing it its arguments. A possible generic choice is an asymmetric "parabola"

$$L_{Ave}(T) = A + B|T - T_0|^2$$

with the coefficient B depending on the sign of $T - T_0$.

Figure 1 represents the on-peak load/temperature relationship in PJM in 2004. As the industrial base and population size in PJM are stable, there is not much difference in load composition in the last years. In general, the economic and demographic changes result in load charts of similar (parabolic) shapes but with parallel shift accounting for the growth. It is worth noting that the graph in Figure 1 is based on the realised temperature and the realised load, while a load forecast would use forecast temperatures as the inputs. The random error so introduced tends to be systemic rather than occasional, and, until the perfect forecast will be available, any quick-response units will command a premium.

Figure 2 represents a typical forecasting error (standard deviation of a rolling 10-day period) in terms of the lead time. By comparison,

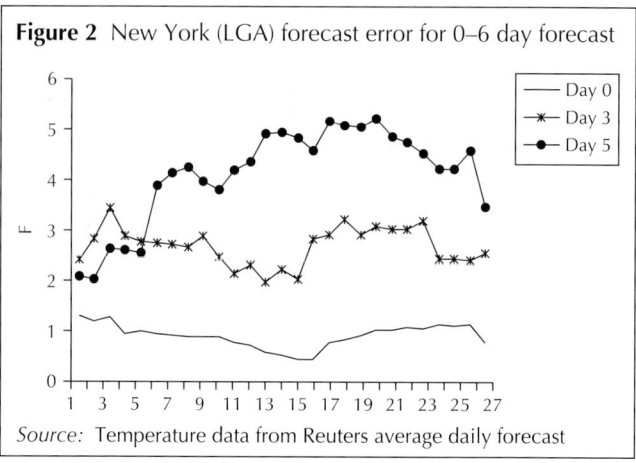

Figure 2 New York (LGA) forecast error for 0–6 day forecast

Source: Temperature data from Reuters average daily forecast

a statistical summer daily forecast based on the last 30 years has a standard deviation of about 6 degrees.

On any given day, the universe of units available to system operators forms a generation stack. The stack is a sequence of generators, ordered with respect to marginal cost; in most cases it is representative of the order in which the unit is bid into the market. A market price may diverge from the cost of the marginal unit depending on the bidding strategy, transmission constraints or reliability considerations. This generation stack concept is a useful but idealised model to understand the price fluctuations of energy.

Figure 3 depicts the marginal price, based on the generation stack in PJM. If we denote the total available capacity in the region by L^{Cap} and define a marginal price as the price of the last unit necessary to run to serve the load, then, with unit A_i having

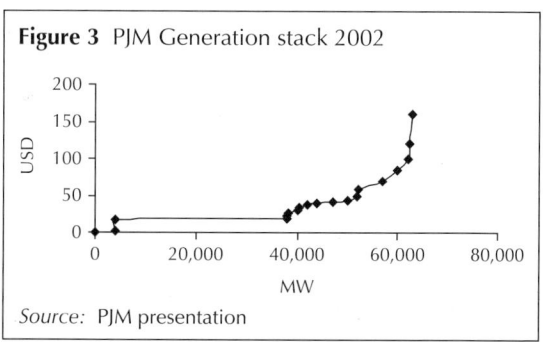

Figure 3 PJM Generation stack 2002

Source: PJM presentation

available capacity C_i and the price per MWh p_i, the marginal price is obtained by:

$$p(L) = \min\nolimits_{\Sigma C > L} (\max p_i)$$

where the minimum is taken over all ensembles with total capacity greater than L.

The particular structure of the stack varies greatly with the region due to both geography and unit operating history, but there are also some common features. The stack tends to reflect level prices, with the large-base-load nuclear, hydro and coal units, then tends to rise with gas-fired, combined-cycle intermediate units, and approaches a perpendicular limiting capacity asymptotes with high heat-rate gas-, oil- (or even wood-) fuelled small units. On a daily basis, the generation stack undergoes changes due to outages, scheduled maintenance, fuel price change etc. Let us assume that the price per MWh for a unit A does not differ significantly from the cost of running it (as is the case in capacity-rich regions in absence of constraints). It follows that the volatility of the marginal prices is a function of fuel volatility and the marginal heat rate or, as is the case in most regions, natural gas volatility and load.

One can model the daily marginal price as:

$$p_{\text{MARGINAL}} = p_{\text{MARGINAL FUEL}} \times S(L) \qquad (3)$$

where S is a non-decreasing (stack) function with $S(L^{\text{Cap}}) = \infty$. Together with (2) we can write a simplified model equation as

$$p_{\text{MARGINAL}} = p_{\text{MARGINAL FUEL}} \times S(f(\text{HDD}(T), \text{CDD}(T))) \qquad (4)$$

Despite the non-linear and discontinuous dependence of the price components on load and fuel type the daily spread S and the consequent run/no run decisions are relatively day-ahead predictable and stable. The non-linearity plays a crucial role in modelling the forward value of the unit.

Most of the models use correlation between power and fuel futures prices (lognormal returns) to account for the dynamic implied in (1), the equation for unit profitability, by (4), the marginal price. This has to be contrasted with the fact that, on any given day, the correlation between spot power prices and specific

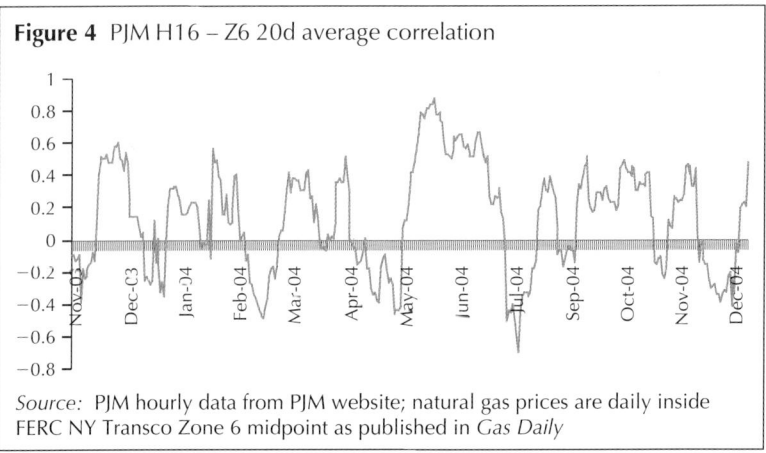

Figure 4 PJM H16 – Z6 20d average correlation

Source: PJM hourly data from PJM website; natural gas prices are daily inside FERC NY Transco Zone 6 midpoint as published in *Gas Daily*

fuel prices is either very high or non-existent. Given that there is a certain range of loads where the marginal units are gas driven, and operate at heat rates of a similar efficiency, there are certain seasons when the correlation between natural gas and firm power prices is stable and high. However, even if load levels and the fuel prices remain the same, the correlation might be very low as the available capacity changes.

In Figure 4, the 20-day rolling correlation exhibits the commonly seen pattern of high and low regimes. It might be added that, in the particular time frame depicted in Figure 4, crude oil prices were at the record levels (US$55/bbl) and SO_2 emission allowances had moved from US$200 to US$700, significantly impacting the spark spread values for the coal plants. In fact, the high emission costs were partly responsible for moving the base-load price on the generation stack up by several US dollars, as was the coal arbitrage between Europe and the US in the second half of 2004.

Hourly Load Distribution

The load distribution undergoes hourly changes. It is driven, as described above (2), by the weather components, but also by a normal course of daily human activity. The factors include seasonality, geographical location, holidays, school closings and sports events. The standard statistical model is a mixture of historical data and functional dependence on temperature. Figure 5 is a typical winter load shape with morning and evening peaks. The relatively low

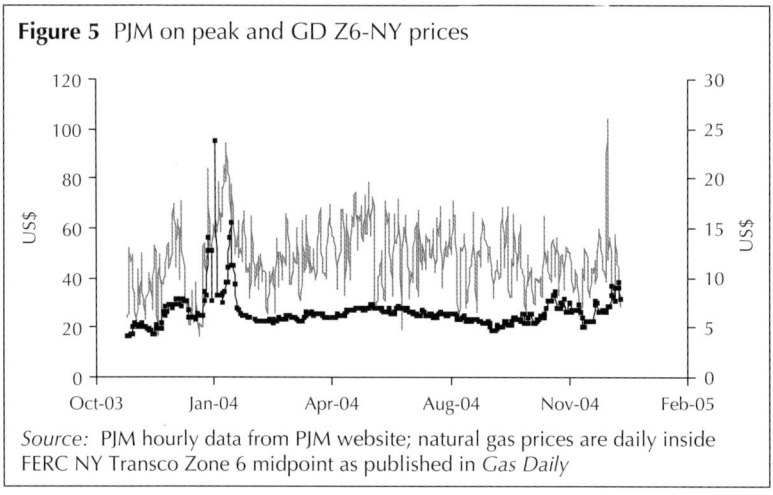

Figure 5 PJM on peak and GD Z6-NY prices

Source: PJM hourly data from PJM website; natural gas prices are daily inside FERC NY Transco Zone 6 midpoint as published in *Gas Daily*

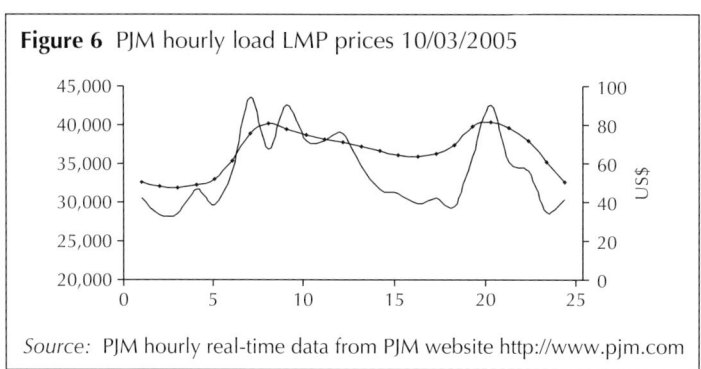

Figure 6 PJM hourly load LMP prices 10/03/2005

Source: PJM hourly real-time data from PJM website http://www.pjm.com

prices in the middle of the day can be ascribed to the need to run some units continuously (to avoid shutdown and startup costs) to capture the second price peak.

UNIT REPLACEMENTS AND WEATHER

Daily operations of a generation unit are only marginally impacted by the longevity of the unit. There are engineering limits on the number of startups, there is seasonal scheduled maintenance, but any other impacts are of a non-transparent, usually cumulative nature. The typical examples are unscheduled maintenance (wear and tear on the unit), emission limits and new technological solutions not available to the older units. One can obtain

Table 1 Spark spread values 03/03/05

			Heat rates		
Marginal heat rate		7,000	8,000	10,000	12,000
Mass Hub	8,330	US$10.49	US$2.60	(US$13.18)	(US$28.96)
NY Zone J	12,300	US$42.50	US$34.50	US$18.50	US$2.50
PJM West	8,340	US$10.42	US$2.68	(US$12.80)	(US$28.28)
ERCOT	7,572	US$3.65	(US$2.74)	(US$15.53)	(US$28.32)
SP15	8,943	US$11.95	US$5.80	(US$6.50)	(US$18.80)

Source: Gas Daily Spark Spread Table. The operational life of the plant is therefore affected by the new generation coming into the region, or, in general, capacity considerations. The PJM construction timetable puts the new generation in 2007 at about 4,000 MW, while the NEPOOL for the same period is about 400 MW

outage insurance for forced outages, and perform some sort of optimisation to solve the emission problem, but the arrival of newer and more efficient units is a problem of a terminal nature. It means that the regional heat rate is such that the spark spread value for the older (higher-heat-rate) units is negative most of the time. That in itself does not necessarily mean that the unit is not economical, as it could command high prices for the limited number of hours that it would operate. The value of the spark spread varies considerably among the regions, as can be seen from Table 1. The prices of power and marginal fuel (natural gas) can be computed from (1), for example a 1,000 increase in the heat rate in NY Zone J diminishes profitability by US$8 (the fuel price) and therefore the power price can be calculated to be US$98.50.

A long-term load forecast depends on two unrelated components: seasonal temperature (Figure 7) and the growth rate of the

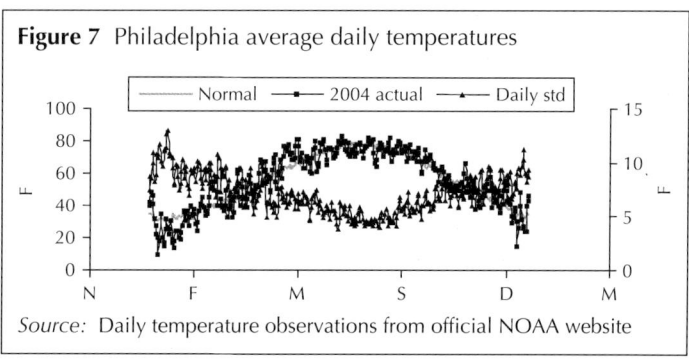

Figure 7 Philadelphia average daily temperatures

Source: Daily temperature observations from official NOAA website

customer base. The latter determines the type of load in the region and is based on overall economic and demographic changes. It can be estimated using appropriate econometric models and studies carried out by governmental and private agencies. To model load changes accurately, one has to model daily temperature changes beyond the 10-day forecast. It necessarily becomes a statistical model, which tends to favour normal (the smoothed daily averages of 1971–2000 period) weather, unless there are some global exceptions of the La Niña/El Niño type, when meteorological considerations have significant forecasting predictability.

In both cases, daily changes seem to be random. One can therefore model temperature as a periodic function with a random oscillation

$$TEMP(t) = N(t) + \eta(t, t-1, t-2) \tag{5}$$

where t corresponds to the day of the year, N is the normal temperature and $\eta(t, t-1, t-2)$ is an appropriate auto regressive moving average (ARMA)-type process. The persistence of weather affects the plant operational constraints, and results in a certain amount of memory carried by daily power prices. Combining (2) and (5) results in:

$$L(TEMP(t)) = f(\text{Max}(T_0 - N(t) - \eta(t, t-1, t-2), 0);$$
$$\text{Max}(N(t) + \eta(t, t-1, t-2) - T_0; 0))$$

The non-linearity of f makes impossible to find a closed formula for the average load, but a Monte Carlo simulation can be successfully implemented to provide a numerical distribution of the results.

The spread between the cost of a particular unit and the power price in the market is of primary importance in asset management (or in general in a tolling agreements). The term value of a unit is based on the expected average of daily spark spread options derived from (1)

$$E(S) = E(\text{Average }(\text{Max}(p_{\text{Daily}} - HR \times p_{\text{DailyFuel}} - OC - R, 0)) \tag{6.1}$$

where the average is taken over the delivery period. Alternatively, one can sell the unit forward, based on the value of

$$E(S) = E((\text{Max}(p_{\text{Forward}} - HR \times p_{\text{ForwardFuel}} - OC - R, 0)) \tag{6.2}$$

or a combination of valuations (6.1) and (6.2). The daily valuation (6.1) necessitates simulation of the daily forward prices for the period over which the average is taken, and as such depends also on the price modelling. The second valuation (6.2) is more straightforward.

Fundamentally, one should value the above option by simulating the impact of weather on the load, and the seasonal response in terms of the region's marginal fuel and the corresponding heat rate. Input data are either fuel- (prices and volatilities) or weather-related data in addition to the static information, like the grid topology, transmission constraints and the generation stack structure. The environmental components (SO_2, NO_x emission limits) introduce an additional cumulative dependence on the volumetric component or through the load function on the overall "cold" or overall "hot" season.

ENERGY MARKETS AND PRODUCTS

Energy markets have matured significantly since the mid-1990s. The most transparent sign of this is the fact that there are liquid products in the market that can be used to mitigate risk of a large percentage of transactions in the market. Hourly options are used in risk valuation of load following deals; daily and monthly options can be used for base load deals, and the futures and capacity markets provide information about long-term optionality. Certain regions, like the recently enlarged PJM, have introduced locational marginal prices (LMP) – an optimisation methodology, based on bidding structure and the topology of the grid. The ever-present problem of transmission constraints has received a new treatment with the introduction of financial transmission rights (FTR), which are tradable transmission congestion insurance.[2]

Given that even the simplest structured deal contains a number of the above-mentioned products, it is clear that valuation is usually a complex Monte Carlo procedure. We will describe the basic ingredients that need to be accounted for and a possible way of using them in modelling.

A customary and reasonable candidate for a monthly forward price is the average of daily marginal prices

$$p_{\text{Forward}} = E(\text{Average}(p_{\text{Daily}})) \tag{7}$$

where the expectation is taken under an appropriate measure and the average is taken over the contract delivery period (approximately 22 on-peak days). To analyse the relationship (7) from the point of view of asset management, we write

$$p_{\text{Daily}}(t) = p_{\text{Daily}}(p_{\text{MarginalFuel}}(t), L(T(t)), L(T(t)) - L^{\text{Cap}}(t)) \quad (8)$$

where, as before, $L(T(t))$ denotes the load, $T(t)$ denotes the daily temperature and t corresponds to delivery days. The first components, the forward fuel prices, are usually modelled as a lognormal process

$$\mathrm{d}p_{\text{Fuel}}/p_{\text{Fuel}} = r\,\mathrm{d}t + \sigma_{\text{Fuel}}\,\mathrm{d}z \quad \text{with } r = 0 \quad (9)$$

with σ_{Fuel} denoting the (lognormal) volatility.

The third argument in (8), $L^{\text{Cap}}(t) - L(T(t))$, is the most volatile and the most difficult to estimate. For the prompt and nearby contracts, its changes are driven by scheduled and unscheduled outages. For long-term contracts, that constitutes the back of the curve; its changes are driven by the difference between the growth rates of load and capacity in the market. The volatility in (8) has different sources (technological or related to investment capacity), but in all cases it influences profoundly the type of marginal units that are brought online. In the non-regulated environment, it creates a situation where the reason for having more capacity (namely price increases) actually diminishes as the capacity becomes available. The cycle characterised by delayed adjustments and periodic excess investments in response to changing load forecasts is known to have occurred in the regulated environment.

The response of price to load fluctuations becomes more pronounced as the reserve margins, $L^{\text{Cap}}(t) - L(T(t))$, are shrinking. A dramatic example of the response of power prices to the change in available capacity L^{Cap} (and to extreme temperatures and credit issues) happened during the week of June 26, 1998 in the Midwest. The next-day power prices had reached US$9000 per MWh, while the July 1998 forward price was at US$250 per MWh. However, in March 2005, Cinergy summer traded at US$60 with natural gas futures trading at levels 2.5 higher than in 1998 (US$7.30 versus US$2.70).

A simple and necessarily simplified model of the capacity constraints can be written as:

$$L^{Cap}(t) = L_0(1 - g(t)d\lambda(t)) \tag{10}$$

where $d\lambda$ is a jump process and $g(t)$ denotes the outage variable measured in percentage of the existing capacity. The multiplier L_0 corresponds to the available capacity at time t, while the parameters of the jump process should be obtained from a combination of the historical data and technical specifications, in addition to any pricing available in the immature outage option market.

POWER OPTIONS PRICING

The underlying components and the corresponding processes constituting (7) and (8) make it unlikely that the resulting distribution of daily power prices is lognormal. To the lesser extent, that objection holds for the forwards, but the data suggest that the Black–Scholes model is only an approximation to the dynamics of the market. Market prices and options quotes strongly suggest that there is a strong volatility smile as option pricing based on the variant of (9) results in strike-dependent volatility. It is commonly described as the "fat tail" property. The departure from the constant – independent of prices – volatility assumption, the cornerstone of the Black–Scholes model (9), is an indication that the market itself prices the probability of remote events higher than the corresponding lognormal distribution would suggest.

One of the issues of fundamental importance to asset management is daily and monthly options. These options provide hedges against both volumetric and price risk (through heat-rate instruments). Fundamentally, they are driven by weather expectation and weather volatility, and in the case of daily options by a forecast and its accuracy. They obviously differ in the exercise schedule: a monthly option expires before the term begins, while the daily option continues for every day until the month expires; the underlying for the monthly option is the forward price, as opposed to daily or day-ahead prices for the daily options. In addition, monthly and daily options traded in the market come with the regular options data: the underlying price, term, strike and volatility (or volatility smile). The market quotes option premiums for both types of option in terms of their volatilities, with the tacit understanding of premium computation via the Black–Scholes formula (see Table 2).

Any simulation based on the market data should be able to reconcile the volatilities of the term and daily options. In addition, the volatility smile (US$100 Jul/Aug. PJM option is traded in the market at US$6/MWh while the $B-S$ value is closer to US$5; historical 2004-based valuation is zero) should be automatically present. The

Table 2 Daily and monthly PJM and Cinergy options 17 March 2005

PJM term	PJM forward price (US$)	PJM daily bid vol (%)	PJM daily ask vol (%)	PJM monthly bid vol (%)	PJM Monthly ask vol (%)
Apr'05	58.00	55	63	33	40
May'05	59.75	57	65	33	37
June'05	62.00	57	65	32	36
Jul/Aug'05	76.25	60	65	33	35
Sep'05	61.00	45	55	28	31
Oct/Nov/Dec'05	58.00	40	50	26	28
Jan/Feb'06	70.00	45	49	29	31
Jul/Aug'06	73.50	37	50	25	29

Cinergy term	Cinergy forward price(US$)	Cinergy daily bid vol (%)	Cinergy daily ask vol	Cinergy monthly bid vol (%)	Cinergy monthly ask vol (%)
Apr'05	52.00	55	65	30	40
May'05	52.50	55	65	32	39
June'05	54.50	50	60	32	38
Jul/Aug'05	65.75	50	60	32	38
Sep'05	52.50	37	50	29	35
Oct/Nov/Dec'05	50.50	35	50	26	31

Source: ICAP Options Report.

Table 3 Cinergy summer options historical evaluation

Cinergy	On-peak forward (US$)	No transmission constraints ((US$))
Jul/Aug '98	120	50
Strike	200	
Average number of outage days 9	9	
Average historical outage price		375
Call option market premium		40

same requirements should be satisfied with the fundamental simulation: it should replicate the market volatilities based on the weather dynamics (and, of course, the fuel curves). The following example from the end of June of 1998 is instructive.

With the average unconstrained daily prices for the 45 days of summer 1998 (July/August) at about US$50/MWh (based on historical data, ie, previous summer), and with nine expected (historically based) outage/transmission constraints on hot days, the prices on those nine days should be such that the resulting average over the summer is equal to the forward market price (45 × US$120 = (45 − 9) × US$50) + 9 × (Hot Day price)). We conclude that the market expectation for the hot days is US$400. Given the extreme condition of those nine days, one can assume that the prices will be in excess of US$200. Therefore, the payout from the option will be (US$400 − US$200) × 9 = US$1,800 or US$1,800/45 = US$40 for summer daily power options, which coincides with the market valuation. Today's (March 2005) market values the US$200 Cinergy summer option close to zero.

EMERGING MARKETS AND ASSET RISK MANAGEMENT

A direct way of managing load fluctuation is provided by weather derivatives. The simplest and most common contract is based on cumulative heating (cooling) degree-days in the season. Given that the average loads correspond to the average seasonal temperatures, one can arrive at a US dollar price per HDD (or CDD), which makes the load-plus-weather-derivative combination delta-neutral. Specific end-user parameters influence both the US dollar price per HDD (CDD) as well as the load response function in (2). Increasing liquidity of the weather derivatives, as observed in the market, makes that hedging option more attractive, especially with the storable fuel suppliers. Recently started CME weather swaps provide a good platform for price discovery and eliminate the credit uncertainty. Most options are still OTC.

Table 4 represents a sample of weather contracts traded in the market. The geographical identifiers are followed by the term (and corresponding type of degree days) of the contract and the strike. The tenors of the weather contracts correspond roughly to summer and winter and therefore can be used as a load hedge.

FUNDAMENTAL ANALYSIS OF POWER PRICE MODELLING

Table 4 Summer weather options

Station	Option type	Strike	Tick (K)	Cap	Bid (K)	Ask (K)
Atlanta, GA	Put	1,550	5	1 m	90	120
Boston, BOS	Put	710	5	1 m	80	100
Baltimore, BWI	Put	1,200	2.5	750 K	200	260
Chicago, ORD	Straddle	825	2.5	500 K	245	330
Covington, KY	Call	1,100	5	1 m	140	170
Houston, IAH	Strangle	2,450/2,550	2.5	500 K	155	200
Kansas City, MCI	Strangle	1,375/1,225	2.5	500 K	120	220
New York, LGA	Strangle	1,170/1,270	2.5	750 k	160	185
Phoenix, AZ	Call	4,000	2.5	500 K	70	–
Philadelphia, PHL	Call	1,300	5	1 m	210	300
Raleigh, RDU	Put	1,425	5	1 m	95	130
Tucson, AZ	Strangle	2,825/2,950	5	1 m	300	345

Source: TFS brokers

There is a less liquid market of daily options. The typical structure is a string of daily options with daily strikes at a given temperature. The implied daily weather volatility of such instruments should be matched with the volatility implied from the daily power option. It is a very involved modelling task as the arbitrage between daily and monthly power options and the daily weather option is still only occasionally available.

We conclude with a few remarks on general asset management. The size and the efficiency of a plant determine its position in the stack, both at present and in the future. Given that the operating costs are much more stable than the corresponding power prices, the economic value of the asset can be estimated on the basis of the forward curves and their volatilities. One can view assets as options on power with strikes given by the product of the average generating cost and the average load. It is also clear that the allocation of generation into fixed-price, long-term power-purchase agreements or short-term contracts is important in extracting the optimal value of the power plant. Analogous valuation has to be carried out for the daily operations. Ultimately, considerations of that type, in conjunction with cross-region capacity/transmission availability, lead to decisions about building new generation facilities. The profitability of building a new plant can be thus viewed as a positive value of the average spark spread. It needs to be stressed, however, that the standard models for spread options, based on correlation of the underlying factors and their volatilities (in the sense used in the context of Black–Scholes modelling), need to be adjusted both for the granularity of the process and for the long-term capacity consideration. The interdependence of natural gas and power prices is highly non-linear and as such can be only approximated by the linear statistical correlation required by the Black–Scholes models. Power markets are inherently volatile, with a multitude of factors that need to be considered to adequately address and understand the risks involved. The modelling of energy has matured, as have the markets it models, but one can venture to say that a unified approach is yet to emerge.

1 PJM interconnection is a regional transmission organisation (RTO) comprising parts of Delaware, Illinois, Indiana, Kentucky, Maryland, Michigan, New Jersey, North Carolina, Ohio, Pennsylvania, Tennessee, Virginia, West Virginia, and the District of Columbia. Originally it was just Pennsylvania, New Jersey and Maryland – hence the name.
2 The transmission rights are discussed in the Chapter 7 by Martin Lin.

7

Management of Transmission in the Electricity Markets

Martin Lin

Electricity is an essential part of modern life, but most people's experience with electricity is limited to flipping a switch and paying a bill. Even for those in the energy industry, often electricity is still just another among several commodities, and many of the transactions may be purely financial. For this and other reasons, one can spend years in the industry but still simply accept many so-called "truths" as a matter of faith. Indeed, whether physical or financial, when it comes to power trading and marketing, the physical realities of the underlying system, especially the workings of transmission, are critical. A deeper understanding of these issues should replace oral tradition and unsupported but bold assertions.

As one might expect, what specialised practitioners spend years learning cannot be covered in a few short pages. However, a general appreciation of the fundamental concepts behind electricity, power systems and transmission can be. These fundamental concepts affect not only power system operations but also power delivery, pricing, and the value of financial instruments based upon electricity. This chapter addresses some major concepts in electricity, provides the motivation behind the adoption of alternating current (AC) as the system architecture for transmission of power, introduces some fundamentals of power transmission, relates these ideas to location-based pricing of power and discusses some market design and financial instruments related to power transmission.

ELECTRICITY AND THE AC SYSTEM
Four major concepts
Four of the most important quantities in electricity are voltage, current, impedance and power.

- *Voltage* is the electromotive force that acts upon an electric charge and tends to cause a charge to move. Voltage is also referred to as electric potential or potential difference. There are many ways to create voltage. One approach uses chemical reactions. A familiar example of this is the common battery. The chemicals inside a battery produce a potential difference between the positive and negative terminals of the battery. If nothing connects the two terminals to form a circuit, no charge will flow, but the voltage is still present. Voltage is sometimes likened to pressure when using water analogies for electricity.
- *Current* is simply the flow of charge. If the terminals of a battery are connected through a light bulb, as is done in a torch (or flashlight), the voltage of the battery would cause charge to flow through the bulb. This flow is a current. While conductors such as wires make current flow more easily, they are not always needed. If sufficient conditions exist, such as very high voltages, current can flow. Lightning is current without wires. Extremely high voltage "breaks down" the air, and the current flows as a bolt across the sky.
- *Impedance*: in the example of the battery and the light bulb, there is a clear interaction between voltage and current. Impedance characterises this relationship. That is, impedance is the relationship between current and the voltage applied. For an incandescent light, such as the one used in the flashlight example, the impedance is *resistance*. Resistance is often thought of as electrical "friction". With few exceptions, all substances have some electrical resistance. These exceptions are called *superconductors*, and even these superconductors lose their electrical resistance at only very low temperatures. Despite the fact that everything at normal temperatures has electrical resistance, sometimes that resistance is very low relative to other types of impedance, such as *capacitance* and *inductance*. Some effects of capacitance and inductance are discussed later, but it is important to remember that, while the literature may discuss normal electrical

components as having only capacitance or inductance, in reality all components have some characteristics of resistance, capacitance and inductance.
- *Power* is the rate at which work can be done or energy transferred. For electricity, in quantitative terms, power is the mathematical product of voltage and current. For a given amount of current, the greater the voltage, the greater the power. Likewise, at a particular voltage, more current equals more power. Though related, power and energy are not the same any more than speed and distance. Power, measured in megawatts (MW), and energy, measured in megawatt-hours (MWh), are often used interchangeably. This imprecision in language should not lead one to think they are actually equivalent.

Three important laws

Electricity is about interactions. The following three laws describe important interactions between the concepts already described as well as between electricity and magnetism.

Ohm's Law states that the resistance describes a linear relationship between voltage and current. That is, voltage (v) is the mathematical product of current (i) and resistance (r). Stated as an equation:

$$v = i\,r$$

Faraday's Law states that a changing magnetic field can induce a voltage. Faraday's Law is the principle upon which most generators are based. A typical generator will have a magnet turning inside coils of wire. Because the magnet is turning, the magnetic field surrounding the coils of wire changes. This changing magnetic field induces a voltage in the wires. If these coils are connected to a circuit, current can flow.

Ampere's Law states that a flowing current creates a magnetic field. A popular school project uses Ampere's Law in the construction of an electromagnet by winding a wire around a nail and then connecting the wire to a battery. Current flowing in the wire produces a magnetic field, allowing the nail to attract magnetic materials. The magnet that turns inside a power plant generator is an electromagnet based on the same Ampere's Law as the school project.

The motivation for AC power systems

The benefits of using AC for power transmission stem from the desire of power producers for low transmission losses, the need of users for safety and convenience, and the fundamental properties of electricity. Power is often produced in remote areas far from the demand centres. Hydroelectric dams at Niagara Falls produce considerably more power than the local users require, and the excess is consumed in other areas, including New York City. Even local power plants are frequently sited outside city limits for numerous reasons.

All transmission lines result in some energy lost during power transfer. Considering the losses related to electrical resistance, recall that power is the mathematical product of voltage and current, and combine that with Ohm's Law.

$$p = v\,i = (i\,r)\,i = i^2\,r$$

Note that power losses to electrical resistance are proportional to the square of current and proportional to resistance. Reducing current would reduce losses, as would reducing resistance. Since resistance is roughly proportional to distance, bringing the source closer to the destination would help; however, this is not always practical. Neither Niagara Falls and nor New York City is likely to move by much. Besides, reducing current would produce a much greater effect on losses.

Since power is proportional to both voltage and current, increases in voltage would result in similar reductions in current for a given power level. Higher voltage means lower current, which in turn means lower losses. This is the motivation for high-voltage transmission.

High-voltage transmission is typically at or above 69 kilovolts (kV) and can be at levels as high as 230 kV, 345 kV, 500 kV, 765 kV, or even higher. Living and working around such high voltages poses serious safety hazards. Even if we ignore these hazards, the insulation and size of components to operate at those voltages would be unacceptable. Therefore, household voltages in the US are typically 120 V, substantially lower than 69 KV.

Changing voltage levels, or transforming, can be done in both direct current (DC) and AC systems. However, DC transforming requires so-called active components that consume energy, adding to overall energy losses. Because of Faraday's and Ampere's Laws, however, transformers in AC can passively raise and lower voltages, allowing for the ability to transmit power at high voltages and consume power at low voltages.

In AC systems, the voltage reverses polarity and the current reverses direction regularly over time. The amplitudes of the voltage and current follow sine waves, and the nominal levels can vary widely. The frequency of the waves, however, is fairly constant. In the US, the frequency of this reversal cycle is approximately 60 times per second, or 60 hertz (Hz).

As current rises and falls, the induced magnetic field of Ampere's Law also rises and falls. Just as with the school project electromagnet, coiling the wire through which the current flows will concentrate the induced magnetic field. If this changing magnetic field is coupled with another coil of conductor, an electric potential will be induced in that second coil by Faraday's Law. Connecting this secondary coil to a circuit allows for current to flow. By varying the number of coils, or windings, on the primary and secondary sides, the relative voltages of the primary and secondary coils can differ. This difference is proportional to the number of turns in each winding, also called the *turns ratio*.

Passive transformation of voltages is why AC is the system of choice for power transmission. Many people mistakenly say that DC has greater losses over distance than AC, and even cite Thomas Edison's Manhattan DC power station as an example. Actually, DC and AC can have similar loss characteristics for comparable voltage and current levels. However, Edison already had problems with horses being shocked by his underground power lines when it rained, so higher voltages, and therefore lower losses, were not possible. Transforming voltages was the key requirement for a successful system architecture, and AC could do it passively.

Complications of an AC system

The AC story, however, is not all positive. There are some consequences that complicate the picture, including *synchronicity*, *stability* and *phase*. As previously stated, the US operates at a 60 Hz system frequency. This 60 Hz is only approximate, since the exact frequency will vary slightly above or below the nominal target of 60. When all parts of a network are operating at the same frequency, the system is considered *synchronous*. Any device that is to be attached to a system, such as a generating unit that is being brought online, must be synchronised first so that its frequency matches that of the rest of the system.

Synchronisation effectively couples all devices in a network. Consider some situations that affect the system frequency. Since electricity is very difficult to store in large quantities, for all practical purposes, generation must balance load in real time. The mechanical input power, such as steam from a boiler, must equal the electrical output load. When the load is less than the input power, the overall energy of the system increases, resulting in a higher frequency. When the load is greater than the input power, the overall energy of the system decreases, resulting in a lower frequency. The inability to perfectly match generation to load at all times causes the frequency of the system to vary.

Since an entire synchronised system must be at the same frequency, a generation/load imbalance in Florida will cause the generators and loads in New England to change their frequency. Typical systems have controls that automatically increase the power output of units if the frequency drops below 60 Hz while automatically decreasing the power output of units if the frequency rises above 60 Hz. In large electrical systems with widely dispersed generation and load, this can cause problems.

If there is insufficient generation at one end of a system, the frequency will drop. Units on the other end of the system may increase output in order to compensate. The units in the area where the imbalance originates, however, will also respond. The result can be overcompensation, resulting in too much generation. This will cause units in both areas to reduce output. They may again overcompensate. The result can be a transfer of power back and forth that results in the power repeatedly flowing between the two ends of the system. In subtle cases, the results can be physical damage to equipment. In severe cases, unstable operation can develop.

The situation described above does not happen in every system. Those systems with large generation and load centres separated by large distances of sparse generation and load are most susceptible. In those cases it may make more sense to divide the system into separate sections, perhaps with a DC tie between them. In DC systems, power can be very carefully controlled and, unlike with AC, there are no synchronisation or frequency issues. The added cost of active components for DC may outweigh the control issues related to frequency and stability.

This is the situation with the US. Due to geography and demographics, the nation has a great deal of fairly closely spaced generation and load centres along the West Coast and throughout the Midwest and eastward. There is a large section, particularly across the Rocky Mountains, with little load. If the US were one synchronous system, the problems discussed above would exist between the West Coast and the East. Consequently, the country is divided into different synchronous systems: the Western Electricity Coordinating Council (WECC, or simply the West), and the Eastern Interconnect.

Curiously enough, a subset of Texas is a separate synchronous system as well. The Electric Reliability Council of Texas (ERCOT) maintains DC ties to the East, but is not synchronously interconnected, though it could be without the same magnitude of problems the West would have. The reason ERCOT is separate is that the Federal Energy Regulatory Commission (FERC) derives its jurisdiction from the commerce clause of the US Constitution. Since ERCOT is entirely within the state borders, FERC has limited say in how ERCOT conducts itself.

The use of DC ties may help divide systems that have little justification for being fully interconnected in the first place. However, not all issues with AC can be addressed using that approach. Phase and reactive power, for example, exists only in AC systems and cannot be ignored when operating real power systems.

Because voltage and current are changing over time, they can become offset with respect to when each reaches the highest values (peaks) and lowest values (troughs) of the cycle. Moreover, when the voltage or current peaks or troughs can be offset by location. Therefore, voltage and current may peak/trough at different times at a given location, and voltage or current may peak/trough at different times across locations. These offsets are referred to as phase angles, phase differences and sometimes simply as phase.

Phase differences can have various causes. The impedances of system components can cause phase differences between voltage and current. Two major types of impedance that contribute to voltage/current phase differences are capacitance and inductance.

How generation and load are distributed throughout a synchronous system can influence voltage or current phase differences across locations in that system.

The existence of phase differences also creates another issue that is often misunderstood: reactive power. Capacitance relates to storage of energy in electric fields. Inductance relates to storage of energy in magnetic fields. As current flows through system components that have capacitance or inductance, some of the electrical energy is stored in fields. As the current changes direction, this energy is released back to the system. This storing and releasing can cause the voltage and the current to shift in time relative to each other, resulting in phase differences.

Power is the rate at which work is done or energy transferred. *Real power* relates to power applied to doing work such as lighting a light bulb or turning a motor. *Reactive power* relates to the storing and releasing of energy in capacitive or inductive fields. Though most people are concerned with doing real work, and are therefore focused on real power, the influence and interaction with real power or reactive power cannot be overlooked. Reactive power is sometimes called *imaginary power*, which only makes matters more confusing.

Though the name implies the power is fictional or unreal, imaginary power has very real consequences. Components must be sized to handle the total *apparent power*, which is a combination of real power and reactive power. As a result, more conductor or other materials must be used, and at a real cost. Furthermore, some components "consume" reactive power. What this means is that these components have impedances that create larger and larger phase differences. To prevent these phase differences from becoming too large and unstable, components with complementary impedances must be added to the system to counteract the phase difference and "supply" the reactive power. These components are very real and so are the associated costs.

LIMITS ON POWER TRANSMISSION
Primary limitations on power transmission
The physics of power systems, combined with the frequent desire and need to site generation away from load, has resulted in the high-voltage AC transmission system in place today. Long-distance

transmission of power is simply a necessity, and, while there are many benefits, there are limitations as well. The limits on power transmission fall into three primary categories: *thermal*, *voltage* and *stability*.

Thermal limits, as the name suggest, are related to heat limits. Electrical resistance converts electricity into heat, resulting in line losses, since this heat is not generally usable. As with many substances, the conductors used in power lines expand when heated. With towers or supports at a fixed separation, if a transmission line were to elongate, the result would be that the line sags. The losses to electrical resistance are proportional to the current flow, which is proportional to the total power flowing on the line. As more power flows, more current flows, and more heat is generated, increasing not only the temperature of the line but also the length. With sufficiently high power flow, the line could sag far enough to contact trees, buildings or other objects. This contact causes a short circuit that can disrupt power flow, if not damage the system.

Though thermal expansion/elongation of power lines can be calculated, and the distance to objects measured, precise and fixed thermal limits are difficult to establish. Trees grow, and, while transmission operators have programmes to maintain rights of way and trees, the exact amount of sag is hardly completely objective and unchanging. Additionally, the ability of the transmission line to dissipate heat is important. For a given power flow, the amount of heat generated may be known for a given line. However, the temperature is affected by how quickly that heat can be passed on to the atmosphere and towers or supports, and it is temperature that affects sag. Ambient temperature has a large effect, which is why there are often different thermal limits for different seasons. Anybody who has blown on a spoonful of hot soup should also know that wind speed affects cooling. In fact, logs relating to transmission operations have noted comments regarding raising or lowering transmission limits based on higher or lower than normal wind speeds.

The condition of the line is also a factor in heat build-up. Lines along coastal areas can become encrusted with salt and sand. This crust acts as a thermal insulator, decreasing the capacity of the line. If rain has been infrequent or insufficient, transmission operators sometimes actually wash the power lines to improve transfer capability.

Maintaining proper voltage levels is important to power systems and their users. If the nominal voltage drops, a condition known as a *brownout* can occur. The term brownout derives its name from the dimming of lights experienced under lower than normal voltage. Besides improper lighting, low voltages can cause motors to run too slowly, resulting in machinery failures among other problems. Electronic equipment is often sensitive to the input voltage and can fail if voltage variations exceed certain tolerances. The amount of power transferred between points in a system can affect the voltage levels throughout the system, with the receiving end of the transfer typically at a lower voltage than the sending end. To help control voltage, power transfers may be limited.

The term "power system stability" refers to the state where voltages, current flows, frequency and other major aspects of the network can remain within acceptable operating ranges, even when subject to disruptions and disturbances. Disruptions are common in large power systems. A generator may go offline, or *trip*, as a result of some failure. The sudden loss of its power should not cause a larger disruption system-wide. Power lines may also fail, perhaps because a storm toppled a tree into the lines. In a stable system, this event should also be contained locally with the rest of the system continuing to operate within limits. Due to automated operations in the system, however, not all conditions are stable.

Automatic generation control for frequency regulation has already been introduced. Other automatic systems are in place to maintain operations as well as protect system components. Circuit breakers, for example, can be triggered automatically when short circuits, also called *faults*, are detected in the system. By opening, circuit breakers can prevent an overload on a line or transformer from causing permanent damage. These automatic operations are beneficial to the system and desirable. Damage can occur after only a few cycles, and each cycle lasts less than 0.017 of a second. Clearly, human intervention is sometimes too slow to prevent damage, so automation is the only option. The downside, however, is that uncoordinated automatic actions can combine to cause massive outages.

The system must not be operated in such a configuration that a localised disruption such as a lightning strike or single outage can evolve into a widespread collapse. Power transfers can push system components and systems closer to the limits of stability. As a result, in some cases, stability transmission limits must be imposed.

Additional limits on power transmission
Thermal, voltage and stability limits constrain the amount of power a transmission line carries. A line limit may be stated as the most constraining of the three primary limits. Other operational practices may further reduce the permissible power flow.

It should be clear that many factors influence the amount of power that can flow on a power line without significant risk of damage or system collapse. Thermal limits may impose seemingly objective limits on lines, but, given the relative difficultly in defining purely objective measures for all of the other potential limits, transmission capacity itself can be rather ill-defined. Since *transmission limit* and *transmission capacity* tend to imply greater precision than is appropriate, the term *transfer capability* will be used instead.

Maintaining reliable and continuous operation of a power system is an extremely high priority for operators and regulators, not to mention suppliers and consumers. While enforcement of thermal, voltage and stability limits on transfer capability results in a certain level of reliability, additional limits are desired. One operational limit is the *first-contingency criterion*, also called the *single-contingency criterion* and sometimes $N-1$.

A *contingency* is any unplanned outage of a system element. This includes generators, transmission lines, transformers, large loads and other important elements. A system that meets the first contingency criterion will remain stable and without overload after any single contingency. This criterion can be more stringent than any of the primary limits, since a line that conforms to a primary limit may fail under a single contingency.

Consider a simple system with a generator in Area 1 connected to a load in Area 2 by two identical power lines that have a thermal rating of R MW. Under thermal limits, each line may flow R MW for a total transfer capability of 2R. However, an outage of one line will cause flows to redistribute such that the remaining line will carry the entire 2R MW, exceeding its thermal rating. Under normal operation, each line is not overloaded, but post-contingency, the remaining line is overloaded. Enforcement of the first contingency criterion would limit each line to at most 0.5 R MW of flow, even though the thermal rating is twice as high.

Implementing the single-contingency criterion offers a level of reliability greater than that achieved with primary limits alone. However, experiencing a simultaneous or near-simultaneous outage of two elements is very possible and could cause a system that is $N-1$-compliant to end up in a collapse. The ability to withstand such a *double contingency* is desirable. A triple contingency, though less likely than a double contingency, could also be considered. With each additional level of reliability, however, comes a cost. Moving from thermal limits to the first-contingency criterion cuts in half the transfer capability in the two-area example. That system could not operate under any double contingency, not to mention a triple contingency. In larger systems, enforcing a universal double- or triple-contingency criterion would restrict flows to a great extent and may provide only minimal incremental reliability.

To balance the desire for reliability against the very real cost of extreme limitations, many operators chose to satisfy the $N-1$ criterion along with selected so-called *credible* double and triple contingencies. Besides reducing the chance to under-utilise infrastructure, implementing the combination of $N-1$ with credible multiple contingencies reduces the computational burden operators and planners face. For a system configured with N components in service, there are N single contingencies. The total number of double contingencies is $N(N-1)$ or $N^2 - N$. The total number of triple contingencies is $N(N-1)(N-2)$ or $N^3 - 3N^2 + 2N$. A system with 1,000 components has 1,000 single contingencies, but 1,000,000 single and double contingencies. Including triple contingencies results in 998,002,000 total contingencies. The motivation for using credible multiple contingencies *versus* all multiple contingencies is clear. Credible contingencies are selected using historical experience as a guide as well as aspects of the physical system. The failure of a capacitor may be highly correlated with the subsequent tripping of a generator, for example. Two separate transmission lines that are carried on the same tower, even if only at a certain point, may fail together from the same lightning strike or car accident.

Available transfer capability

The influence of contingency selection makes calculating the total amount of power a line can transmit, or the *total transfer capability* (TTC), quite challenging. What is generally of interest to traders, however, is not so much the TTC but the *available transfer capability* (ATC). This is the amount of transfer capability remaining on a line and is sometimes referred to as the *incremental transfer capability* (ITC). Sometimes the ATC is equated with the first-contingency incremental transfer capability (FCITC). Strictly speaking, the FCITC is one way to estimate ATC; however, ATC is a broader concept. The ATC may include consideration to withstand certain selected multiple contingencies, while the FCITC satisfies only the $N-1$ criterion.

People speak of calculating ATC. More correctly, they establish or estimate it. This is because the exact ATC depends on the existence of some true upper limit, and, as discussed, there are many subjective considerations in determining any limit. A governing body can set forth rules, such as adherence to only the $N-1$ criterion, but any resulting limits are the consequence of how the organisation chooses to balance reliability, utilisation, efficiency, practicality and cost. Because any resulting ATC is based on how the rules are set, ATC is established. Since the actual configuration and dispatch of the system determines how much transfer capability is available on a given line, even calculating the ATC established by the adopted rules is difficult or impossible. System operators have a great deal of information regarding the state of the overall system, but not all voltages, power flows, phase angles and other values are known everywhere. As a consequence, at best, even the established ATC can only be estimated.

The influence of the state of the system on ATC points to another issue regarding transfer capability. In much of the existing literature, TTC and ATC are linked as follows. There exists some total transfer capability on a line, the TTC, and, after subtracting from that the existing flows and a reliability margin, one determines the remaining or available transfer capability, or ATC. While conceptually this appears to make sense, in reality this approach is flawed. The TTC of a line is not a fundamental quantity from which ATC is derived. The TTC of a line cannot be determined before the ATC. Since the voltage and stability limits of a line are dependent on the configuration of the system, such as which generators are online and at what levels those generators are dispatched, the TTC of a line is configuration-dependent.

With the configuration of the system known, or at least estimated, the ATC can be assessed using the appropriate criteria and rules. To this ATC, existing flows can be added to determine the TTC. Rather than having a known TTC from which an ATC is derived, the process actually first estimates a configuration-dependent ATC and then derives an implied TTC.

Even if an analyst were given the system configuration and a set of reliability guidelines, such as a list of contingencies to withstand, estimates of ATC can differ. This is because the typical approach to determining ATC is to increase transfers between points until some element of the system is overloaded or exceeds some other limits of operation – for instance, voltage levels or stability. In the simple two-area example, the overload of the lines between the areas clearly drives the transfer capability. In more complex systems, defining the sending and receiving points may be less obvious. In many markets, an area may comprise a set of system components, and a number of lines may connect such areas together as several geographically diverse points. When examining the transfer capability between areas of this type, how the power transfer is performed has a significant effect.

To simulate these effects and estimate potential states of the system, a software tool known as a *powerflow* or *loadflow* model is used. A powerflow model is loaded with information about the electrical system, including its electrical topology, which describes what components are present and how they are interconnected. The model also includes the electrical characteristics of lines such as impedance and thermal rating and the electrical placement of generating units and major loads. The entire set of input data for a powerflow is called a *powerflow case*.

To estimate ATC, an analyst must define an approach to simulating power transfers between areas in a powerflow model. One approach is to simulate an increase in generation in the sending area with the output of each unit rising by an amount proportional to its initial output, relative to the total output of all units in the sending area. To balance the increased power output, and to complete the other side of the transfer, all generation in the receiving area is decreased in proportion to initial generation output by the amount of the proposed transfer. The transfer is increased progressively until some system component overloads or fails some other constraint, for example voltage, stability or reliability. The maximum transfer is considered the ATC.

A shortcoming of this approach is that generation output changes are relative to the initial levels in the powerflow case. Large units at high output in the sending area contribute most of the generation in the transfer. In reality, transfers may be from specific units or from units of lowest marginal cost. These choices may have nothing to do with the assumed initial configuration of the powerflow case. Furthermore, additional generation in proportion to initial output levels may mean that certain units at or near their respective capacity will be asked to deliver more power than those units are actually capable of. The result would be an unrealistic simulation of a power transfer. To address this, a modified approach is to increase generation in the sending area in proportion to the remaining generation capability of the units. This avoids the generation capacity problem. Even more sophisticated approaches will increase generation in the sending area based on estimated marginal costs or production and unit capacities. Even with these potential improvements, an actual transfer may be assigned to units in a completely different manner, due perhaps to contractual or ownership arrangements not captured in the powerflow.

Besides so-called generation-to-generation methods, transfer simulations can be generation-to-load, load-to-generation or load-to-load. In these other approaches, instead of relying strictly on generation output shifts, load in the sending and/or receiving areas will adjust, typically in proportion to initial load level assumptions. Despite the aforementioned shortcomings, generation-to-generation estimations are more appropriate for most situations.

At the very least, ATC is influenced by the data of the underlying powerflow case, the assumptions regarding the configuration of the case, the adopted reliability standards and the methodology in simulating incremental transfers. In addition, these inputs are subject to interpretation and can vary with application and context. For example, many system operators develop powerflow cases for use in operation and planning. There will often be a set of cases for various timeframes, such as winter 2005–6 and summer 2006, and conditions, such as on-peak and off-peak. After ATC is established, operator judgement may influence the enforcement of any transfer limit. The overarching theme is that ATC is not a precisely defined number that is universally and objectively calculated.

TRANSMISSION LIMITS AND POWER MARKETS

What is of interest to many in the power industry is not the complexity associated with determining transmission limits but how these limits ultimately interact in the market and influence transactions and prices. The overall structure of the market determines how transmission affects outcomes, and these structures vary across North America. For much of its history, the power industry has operated as a natural monopoly.

Early evolution of the structure of power systems

High-capital costs, substantial barriers to entry, significant economies of scale and potentially destructive competition all underpinned the motivation in establishing a regulated monopoly structure. Under this framework, each utility was granted the exclusive right to provide electricity for a geographic territory, and any prudent investments on behalf of the ratepayer would be guaranteed a reasonable return. In exchange, each utility had an obligation to serve virtually any customer in the franchise territory, and any investments as well as rates and rate structure required approval from a regulatory body, usually a state public utility commission.

The emphasis under the regulated regime was on reliability and least-cost operation and planning. Transmission was required primarily to link power plants and load centres across utility territories. Enhanced reliability and lower costs, not trading opportunities, were the drivers behind interconnecting separate utilities via transmission.

If a unit were to fail in an isolated utility, the system frequency could drop dramatically as a small number of units responded to make up the lost output. If the units are insufficiently responsive to maintain or restore the frequency quickly enough, then additional units may trip, exacerbating the problem. To prevent cascading system failure due to such "underfrequency" issues, most systems commit, that is, start up and synchronise, more capacity than the minimum required to serve the expected peak load.

Whenever a unit is operating below its maximum output, the difference between its operating point and capacity is called *reserve*. One operating guideline requires the commitment of sufficient units so that the cumulative *spinning reserve*, excluding any contribution by the largest committed plant, exceeds the capacity of the largest committed plant. When this is the case, any single plant outage should not severely threaten the system. Additional guidelines may require a reserve amount exceeding a percentage of expected peak demand — for instance, 15 per cent.

Since large systems typically have more units and greater diversity of unit characteristics than small systems, they are usually able to more efficiently commit units to meet the reserve requirements than otherwise similar small systems. Transmission interconnection can make a set of previously isolated small systems act as if they were a single, large system. By allowing for *reserve sharing*, transmission can both enhance reliability and reduce costs.

Once large transmission interconnections were in place between synchronised utility systems, transfers of power between areas became possible. Traditionally, a utility operator would manage its control area to maintain proper frequency and ensure that local generation satisfied local load. Any instantaneous imbalance between supply and demand, measured as the *area control error* (ACE), would cause the system frequency to change and power to flow out of or into the control area. Standards were established to limit ACE and the associated *inadvertent interchange*.

Under certain situations, interchange between interconnected systems was desirable. Because generation costs vary by unit type, capacity, vintage and other characteristics, the cost profile of the portfolio of units in one system can differ from the portfolio of a neighbouring system. If one system has a lower marginal cost of generation than its neighbour, then transferring power from the lower-cost system to the higher-cost system reduces the combined cost of the two. Sale of *economy interchange* was an early form of power trading.

Differences in resource availability, and therefore differences in generation types, often led to opportunities for economy interchange. The Pacific Northwest is endowed with great water resources, and the hydroelectric potential far exceeds the regional loads. The large population centres in California would benefit from the abundant hydroelectric power to their north, and seasonal economy interchange was the norm.

During times of weather diversity, the relative loads between systems can differ substantially. For example, the Midwest may experience a heat wave at the same time the Northeast experiences moderate or cool weather. Loads in the Midwest may be unusually high at the same time loads in the Northeast are normal or even below normal. Since generation costs typically increase as more output is required from the system, when one utility is experiencing higher than normal loads, it is also experiencing higher than normal costs. In this example, the Midwest and Northeast can benefit from economy interchange as long as the weather situation persists.

Even when normal weather patterns exist, offsets in the diurnal cycle can lead to opportunities. Loads tend to shift at the same local time, even in different areas. This is because the workday ends at essentially the same local time everywhere, and the sun rises and sets at nearly the same local time for a given latitude. Consequently, systems to the East will peak before systems to the West. Economy interchange from West to East could reduce costs. Later, as the West is peaking, the East is already reducing demands as the peak has passed. Economy interchange, this time from East to West, could reduce costs.

Utilities continued to grow as electrification continued, demand increased and interconnections expanded. While legislation such as the Public Utility Holding Company Act of 1935 (PUHCA) and the Federal Power Act of 1935 (FPA) regulated certain aspects of utility company structure and operations, many other issues were left loosely coordinated, if at all. The Great Northeast Blackout of 1965 highlighted shortcomings of operating such an enormous, complex, integrated power system under that regime. The outage of a single line led to a series of overloads and trips that cascaded the Northeast into darkness in a matter of minutes.

Born of this experience was the North American Electric Reliability Council (NERC). NERC is divided into several *regional reliability councils*: East Central Area Reliability Coordination Agreement (ECAR), Electric Reliability Council of Texas (ERCOT), Florida Reliability Coordinating Council (FRCC), Mid-Atlantic Area Council (MAAC), Mid-America Interconnected

MANAGEMENT OF TRANSMISSION IN THE ELECTRICITY MARKETS

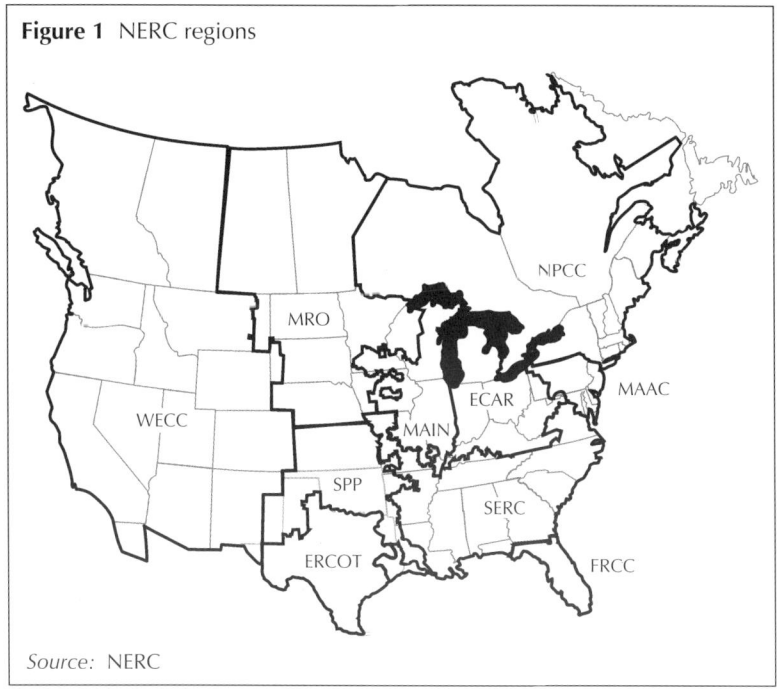

Figure 1 NERC regions

Source: NERC

Network (MAIN), Midwest Reliability Organization (MRO), Northeast Power Coordinating Council (NPCC), Southeastern Electric Reliability Council (SERC), Southwest Power Pool (SPP) and Western Electricity Coordinating Council (WECC). Along with its regions, NERC coordinates reliability standards, develops guidelines and monitors and analyses performance. Figure 1 shows the geographic boundaries of the regions.

Since its formation in 1968, NERC has been a voluntary organisation. Combined with federal, state and local authorities, NERC forms an integral part of the regulatory framework under which power systems operate. Unlike the governmental agencies, however, NERC guidelines are not mandatory. Following the 14 August 2003 Northeast Blackout, NERC has championed legislation to empower the Council with enforcement capabilities.[1] NERC also points to major changes in the overall power delivery structure as reasons for shifting from a voluntary organisation to a mandatory one. These changes began almost a decade after NERC was formed and continue to this day.

Restructuring begins

For many years, the transmission lines that interconnected utility systems were primarily used for enhancing reliability, sharing resources and economy interchange. Technological improvements in power plant design and systems controls allowed for the changes that began with the Public Utility Regulatory Policies Act of 1978 (PURPA). By permitting and encouraging qualifying facilities (QFs) comprising non-utility generators (NUGs) and co-generation units, the first steps towards major restructuring were taken by PURPA. With QFs, power plants could be owned by entities other than traditional utilities. While NUGs were small power producers, co-generation plants produced electricity along with other forms of energy, such as heat and steam.

Sometimes classified as combined-heat-and-power (CHP) units, co-generators were typically found at refineries and factories and driven by either waste heat or heat that was required for other uses. Costs for electricity from co-generation units were generally low since input heat costs were not fully allocated to the generation of electricity. Better still, PURPA required the local utility to purchase any excess generation from a co-generation facility at avoided cost.

With more and more capacity owned and operated outside the traditional utility framework, the stage was set for the next major regulatory change, the Energy Policy Act of 1992 (EPAct). EPAct created exempt wholesale generators (EWGs) and allowed for more diverse entities to operate in power markets while exempt from regulation under PUHCA. FERC issued Orders 888 and 889 granting non-discriminatory access to the bulk transmission system and requiring information systems to manage this access. Transmission operators could not offer services to their affiliates on terms that were different from those offered to any other qualified entity. This requirement prevented incumbent operators from erecting operational, cost or other barriers to competitors. Implementation of an open-access same-time information system (OASIS) by the transmission operator was required to provide near-real-time information regarding transmission availability and to ensure competitive access.

Implementing a power transfer for economy interchange was a relatively simple transaction between entities operating under a longstanding regulatory regime. As new players entered power markets, the complexity and number of requested power transfers increased. Power was being produced and sold by independent power producers (IPPs), and the buyers were not always traditional utility-type entities. Power marketers emerged to buy and sell power without necessarily being involved in its production or consumption. These power marketers would buy power in one area, sell it in another and arrange for its transfer via OASIS sites for the intervening transmission systems. Transactions for spot and forward power sales and purchases were executed, and market participants began to offer options and other derivatives.

Transmission asserts itself in the markets

As the markets evolved, the creative energies of entrepreneurial new entrants and the efforts of existing players to adapt were focused on designing, executing and delivering on the growing range of transactions. Though transmission operators worked to develop and launch electronic OASIS sites, insufficient attention was paid to how well the existing transmission systems, designed and built to support reliability and economy interchange in a regulated environment, would perform under the restructured demands. Technical realities and limitations of transmission were also ignored.

An example of how the industry designed transactions without full regard for these technical realities is the contract path. The transmission network in most regions is a mesh of interconnected lines that offer multiple paths for electricity to flow between any two points. Without calculations of a power flow, it is difficult to say how power will distribute among the lines. Incorporating power flows into transactions, however, can be cumbersome. To simplify, many transactions were based upon the notion that one could specify which lines were used to transfer power, the so-called contract path. With a designated contract path, the exact transmission providers, point-to-point paths and resulting tariff charges could be established. To complete the transactions, the transfer capability on the contract path could be requested, approved and reserved on the appropriate OASIS sites.

The ATC, or, worse yet, the TTC, along sub-segments of the contract path would be summed to determine whether sufficient transfer capability existed for a transaction. A problem arises, however, when such simplifications clash with the real system capability. Since power transfers do flow over lines not included in the contract path, one transfer may "interfere" with another along a different contract path. With several transactions, all with different *sources* and *sinks*, the actual sufficiency of the system to carry all of the transfers cannot be assured simply by checking the adequacy under contract paths.

In the initial years of restructuring, the mismatch between contract paths and actual flows did not pose significant problems. The volume of transactions had not yet outgrown the capacity. The summer of 1998 would refocus the industry on the issue of transmission. NERC had instituted a procedure known as transmission loading relief (TLR), under which security coordinators would monitor transmission loading and had the authority to hold or even reduce transactions if overloads threatened operation reliability. Monitoring all transmission system elements is not feasible, so a selected group of flowgates was determined as the subset to track.

When loading on a flowgate exceeded established limits, a security coordinator could issue a TLR. Each TLR event carries associated levels depending on the severity of the situation. Level 1 TLRs were only warnings, while Level 2 TLRs could hold existing transactions and prevent new transactions from starting. Level 3 and higher TLRs could require existing transactions to reduce in flow, ie, *curtail*. As the situation improved, the security coordinator could reduce the level of the TLR, finally issuing a Level 0, indicating the end of the TLR event.

High temperatures in the Midwest, combined with plant outages, created substantial differential supply–demand balance and therefore prices, between the Midwest and the Northeast in the summer of 1998. Many power transactions were executed to bring power from the Northeast to the Midwest. Combined with that were a large number of power price call options that had been written for delivery in the Midwest. The large transfers resulted in a TLR that disrupted the planned power transfers into the Midwest. The TLR isolated the Midwest from many potential supply regions, forcing demand in the area to find supply within the Midwest. Prices soared as the market scrambled to find local sources of power.

Writing call options had been a profitable way to collect option premiums without needing to deliver power. Suddenly, virtually every option was in-the-money and buyers were exercising. The original sellers of the options, however, had typically hedged themselves with call options of their own. As players went up the chain, eventually there were defaults. With one side of such back-to-back transactions gone, the other side was soon to default. Lack of clauses requiring liquidated damages in many contracts further encouraged defaults.

While such situations were not the primary cause of the price spikes, reports of defaults did fuel the confusion and emotional aspect of the price response. As weather returned to normal and even the most marginal of units was put into service, the event abated. In its wake was a stunned industry. Bankruptcies were filed and certain players left power trading and marketing altogether. Transmission adequacy had gained new found respect.

TRANSMISSION FLOWS AND LOCATIONAL PRICING

The interaction between transmission and power prices is not always as dramatic as it was during the summer price spikes in 1998. While TLRs are used for transmission congestion management in much of the Eastern Interconnect, several systems take different approaches. One approach is using *nodal* prices. To understand nodal prices, however, it is important to understand more about power flows.

As discussed earlier, power is proportional to current. Therefore, the way power distributes across the network of transmission lines is directly related to how current flows. The transmission system *topology* and impedances are the most important factors in determining how current flows in a system. Topology refers to the actual configuration of the transmission system elements, that is, what lines exist in the system and how they interconnect the other parts of the system. Impedance defines the electrical characteristics of the elements, including resistance, capacitance, and inductance. When discussing power system topologies, it is often convenient to refer to specific locations of the system, such as where a generator interconnects, where major transmission lines intersect or where large loads attach. Such locations are referred to as *nodes*.

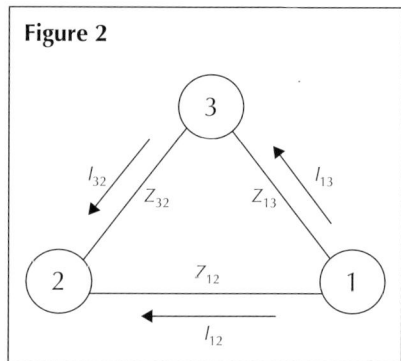

Figure 2

Consider the ubiquitous three-node system. This common example is simple enough to easily understand without losing sufficient complexity to illustrate important behaviours of power systems. As illustrated in Figure 2, the system comprises three nodes connected in a triangular shape by three lines of equal impedance. Two important principles of electrical engineering apply: Kirchoff's Voltage Law (KVL) and Kirchoff's Current Law (KCL).

KVL states that the sum of voltage drops around a closed loop is zero. This is related to the principle of conservation of energy. KCL states that the sum of the current into a node is zero. This is related to the principle of conservation of charge.

Before KVL can be applied properly, an algebraic sign convention must be adopted so that voltages can be summed. The standard sign convention is to define a voltage drop as positive in the direction of current flow. For Figure 2, using subscripts to refer to the from and to nodes, defining a positive current from Node 1 to Node 3, denoted i_{13}, means a positive voltage drop of v_{13} from Node 1 to Node 3.

Reversing the direction would reverse the sign. Therefore, from Node 3 to Node 1, v_{31} would be a negative voltage drop. Since it is often confusing to have variables be negative, often the direction of current flow, as indicated by the subscripts, is selected to make the voltage drop positive, or the negative sign is explicitly used and the subscripts are in the appropriate order for a positive variable. That is, if current were actually flowing from Node 3 to Node 1, rather than have v_{13} be a negative value, one could use v_{31} or $-v_{13}$. For current, just as with voltage, if flow in one direction is positive, then flow in the opposite direction is negative. Therefore, i_{31} equals $-i_{13}$.

Using the standard sign convention, and following the anticlockwise closed loop 1–2–3–1 for the 3-node system in Figure 2, KVL states:

$$v_{12} + v_{23} + v_{31} = 0$$

Note that, when summing along the closed loop from Node 3 to Node 1, the voltage drop in the equation is v_{31}, reflecting the move in the opposite direction to the current flow, i_{13}, illustrated in the diagram.

Recall from Ohm's Law that voltage is given by the product of current and impedance:

$$v = i\,z$$

Maintaining standard sign convention and combining the KVL result with Ohm's Law yields:

$$i_{12}\, z_{12} - i_{32}\, z_{32} - i_{13}\, z_{13} = 0$$

Suppose current I were injected into Node 1. Standard sign convention for current designates current flowing into a node as positive and current flowing out of a node as negative. At Node 1, then, current I is flowing into the node and currents i_{12} and i_{13} are flowing out of the node. KCL states:

$$I - i_{12} - i_{13} = 0$$

A power transfer from Node 1 to Node 2 would look like the current transfer in Figure 3. The current can be divided into two quantities: the current along the direct path between Nodes 1 and 2, i_d, and the current along the indirect path through Node 3, i_i. The KVL and KCL results from above can be applied to solve for these currents:

$$z\,i_d - z\,i_i - z\,i_i = 0$$
$$I - i_i - i_d = 0$$

Applying algebra to solve for the currents:

$$i_d = 2/3\ I$$
$$i_i = 1/3\ I$$

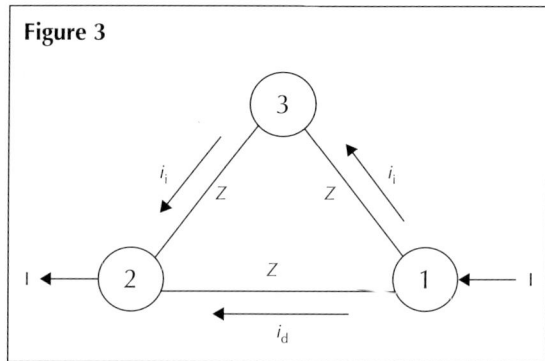

Figure 3

Note that Node 3 carries one-third of the transfer whether it is formally involved in the transaction. A contract part may be specified from Node 1 to Node 2, but fully one-third will still flow along the indirect path. It is a result of the topology and the impedance that only two-thirds of the power flows along the direct path. Kirchoff's Laws, not contract law, govern the power flow.

Though the effect of Kirchoff's Laws on power flows in a realistic system may be difficult to calculate without sophisticated software, the impact of those laws are even more complex on nodal pricing. Nodal pricing, as the name suggests, assigns a price for power at the nodal level. Because this makes prices location-specific, variations of nodal pricing are sometimes called locational marginal prices (LMP) or location-based marginal prices (LBMP). Since LMP and LBMP can sometimes imply particular characteristics of systems that have adopted them, nodal pricing will refer to the general principle without respect to an individual implementation or jurisdiction.

Consider again the three-node system. Suppose Generator S1 with capacity 110 MW and offer of US$50/MWh is located at Node 1, and Generator S3 with capacity of 150 MW and offer of US$60/MWh is located at Node 3. Further suppose Load D2 of 220 MW is located at Node 2. This system is illustrated in Figure 4. The line from Node 1 to Node 2 is the direct path for power injected by S1 to supply D2, and the lines from Node 1 to Node 3 and from Node 3 to Node 2 form the indirect path. For power from S3 to D2, the line from Node 3 to Node 2 is the direct path, while the lines from Node 3 to Node 1 and Node 1 to Node 2 form the indirect path.

MANAGEMENT OF TRANSMISSION IN THE ELECTRICITY MARKETS

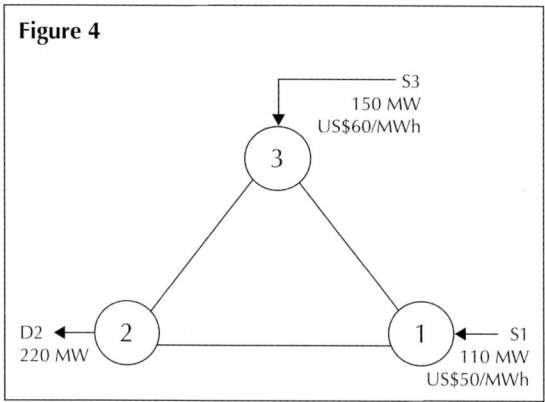

Figure 4

To serve D2 at lowest cost, power offered from S1 is taken first, but the 110 MW is insufficient to supply D2. S3 supplies the remaining power for D2. The resulting power flows are shown in Figure 5. These flows conform to the results found in the power flow discussion regarding the split of current along direct and indirect paths. The nodal prices are determined by finding the lowest cost to supply an incremental demand at each node. Since S3 is the only supply left and the system is unconstrained, the price at each node is US$60/MWh.

Suppose that the capacity of the line from Node 1 to Node 2 is limited to 100 MW. The solution just found is no longer feasible. The only option is to reconfigure generation by shifting output from S1 to S3. For each megawatt of output reduced at S1, two-thirds of a megawatt is unloaded from the congested line, since it is the direct path. To balance supply and demand, that megawatt reduction at S1 must be replaced by a megawatt increase from S3. This will load the congested line by one-third of a megawatt, since the line is on the indirect path for S3. Therefore, for each megawatt shifted from S1 to S3, a net one-third of a megawatt will be reduced. Since the overload is 10 MW, a total of 30 MW must be shifted to satisfy the constraint on the line.

What are the resulting nodal prices? In this case, unlike in the unconstrained case, S1 has additional generation capacity. Since serving incremental demand at Node 1 will not load any of the lines, S1 can supply that increment without violating the constraint

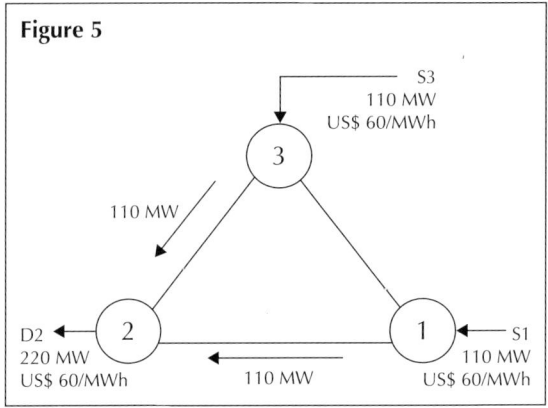

Figure 5

on the line from Node 1 to Node 2. The nodal price at Node 1 is US$50/MWh. Node 3 cannot be served by S1, even though it is the lowest offer supplier, because doing so would increase loading on the constrained line in the indirect path from Node 1 to Node 2. S3 is the only supply option, so the price at Node 3 is US$60/MWh.

Node 2 is more complicated. When supplying Node 2, the constrained line is on the direct path for S1 and the indirect path for S3. By decreasing the output of S1 by one increment and increasing the output of S3 by two increments, the loading on the constrained line will remain unchanged while an additional increment is supplied to Node 2. One more megawatt at Node 2 costs two megawatts from S3 and avoids the cost of one megawatt from S1. The net cost then is US$70/MWh. The constrained results are illustrated in Figure 6.

In the constrained case, the line between Nodes 1 and 2 is considered congested since flows would have been higher if no constraint were imposed or if the constraint were sufficiently higher. Directionally, the congestion on the line is said to go from Node 1 to Node 2, since additional flows in that direction would overload the line. The congestion cost is typically calculated as the difference between the nodal prices at the endpoints. In this example, the congestion cost is US$20/MWh. Note that the receiving node should have a higher nodal price than the sending node.

This simple example reveals some important results regarding nodal prices. First, when the system is unconstrained, nodal prices

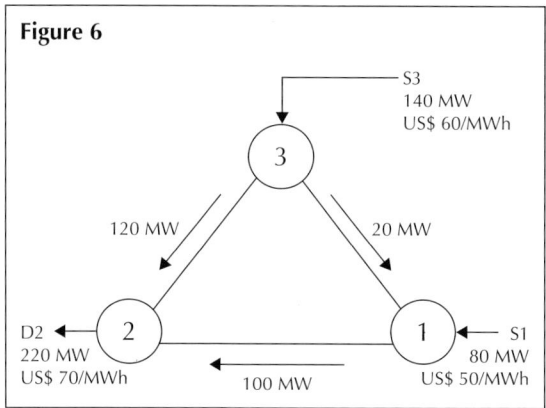

Figure 6

should be equal at all locations, ignoring losses. Second, nodal prices are not necessarily capped by the highest offer. Notice that the Node 2 price is greater than either of the offers in the system. Despite assertions otherwise, an offer cap is not the same as a nodal price cap!

Real power systems are more complex than the three-node example. One other assumption in the example system is that the transmission lines are *lossless*. How actual systems treat losses varies. The New York Independent System Operator (ISO) includes a component for marginal losses in its LBMP. Other ISOs such as the PJM Interconnection (PJM) are contemplating implementation of marginal losses in their LMP. One implication of losses is that the price may not be equal at all locations under conditions with no transmission congestion. This is because prices can differ by the losses from location to location.

DEALING WITH THE RISK OF TRANSMISSION CONGESTION

Since transmission congestion can have a significant impact on pricing, there is great interest in managing the risks associated with congestion. In systems where TLRs are used, the tools available to directly address congestion risk are limited. By assessing potential problems and carefully monitoring potential TLRs, parties transacting among TLR systems can estimate the likelihood of disruption. Powerflow software can help in these determinations, as can careful experience operating in those areas. The size of an energy

PANEL 1 FOCUS ON PJM

Among the various US independent system operators (ISOs) utilising transmission congestion instruments, PJM is one of the oldest and most developed. In 1997, PJM was the first ISO approved by the Federal Energy Regulatory Commission (FERC). In 2001, PJM became the first regional transmission organisation (RTO). Since that time, PJM has grown to integrate the transmission systems of Allegheny Public Service, Commonwealth Edison, American Electric Power, Dayton Power & Light and Duquesne Light. Though not yet completed at the time of writing, expansion of PJM to include the Dominion system is planned. Because of the prominence of PJM, this discussion will focus on this single, albeit evolving, system. While each ISO offers detailed training courses and materials explaining its own intricacies, what is presented here is only an overview of selected issues in PJM.

Congestion charges and FTRs

PJM operates several markets for products including energy, ancillary services and transmission, and any entity that meets certain qualifications, such as credit, may participate in these markets. The energy markets are those for which hourly nodal prices, called *locational marginal prices* (LMPs) in PJM, are computed and posted for both real-time and day-ahead. As described in the section on nodal prices (see "Transmission flows and locational pricing"), transmission congestion manifests via price differences. For an entity transferring power from one location to another, the per-megawatt congestion charge, if applicable, is the difference between the LMP at the destination, or *sink*, and the LMP at the origin, or *source*. Financial transmission rights (FTRs) can be used to hedge day-ahead market congestion charge exposure.

Because transmission and congestion are directional, congestion charges are directional as well. When congestion exists from source to sink, congestion charges are positive. That is, market participants holding positions flowing power from source to sink when congestion exists in the same direction face a positive congestion charge that must be paid to PJM. If congestion realised in the market is from sink to source, however, the congestion charge is negative – paid from PJM to the market participant. Regardless of which locations are designated the source and the sink, positive congestion is always in the direction from the lower price location to the higher price location.

Another way to think about congestion and direction is to consider injection and withdrawal of power. If an entity, such as a generator, were to inject power into the grid at some location, the entity will receive from PJM the LMP of that location. On the other hand, if the entity were to withdraw power from the grid at some location, the entity would pay to PJM the LMP of that location, or is some cases, the average of the LMPs at several locations around the withdrawal location. Generation and load must be balanced across PJM, but any given market participant can be unbalanced with respect to generation and load. Therefore, a generator without matching load could be dispatched into PJM and paid the LMP at the injection location. If that generator were matched to a withdrawal, however, the combined positions of injection and withdrawal would represent the transfer of power from the injection location of the generator to the withdrawal location. In this case, the entity would receive the LMP at the injection location, the source, and pay the LMP at the withdrawal location, the sink. If the sink LMP were higher than the source LMP, the net revenue will be payment to PJM. This payment represents the congestion charge for transferring power in the direction of the congestion.

In cases where the LMP at the source is higher than the LMP at the sink, the congestion charge is negative. Those who flow power in this direction would receive the congestion charge, rather than pay it. Sometimes this flow is also called counterflow, since it is flow in a direction opposite the congestion.

When PJM first introduced FTRs, holders of FTRs received payments equal to the calculated congestion charge realised in the real-time energy market. Later, payments to FTR holders were based on congestion charges in the day-ahead energy market. Regardless of whether FTRs were settled against the real-time or day-ahead market, the holder was always paid the congestion charge. Because of the directional nature of congestion and FTRs, holders would sometimes face a liability. Should the congestion be in the opposite direction of what the market participant holds in FTRs, the congestion charge would be negative, and the resulting payment to the FTR holder would also be negative, meaning that the holder would owe money to PJM. In effect, these FTRs were like a financial LMP swap where the holder pays the LMP at the source and receives the LMP at the sink.

When used to hedge power flows, these FTRs were still useful, even though the FTRs could require payments to PJM. This is because the underlying flow would be subject to a negative congestion charge whenever the FTR required payment to PJM. The FTR payment to PJM would offset the congestion charge received from the flow. With only a few exceptions, the holder of a portfolio of flows and FTRs would not be exposed to congestion risk as long as the flows and FTRs were in the same market (real-time in the earlier years and day-ahead more recently), have the same sources and sinks, and were for comparable quantities.

Despite the application of FTRs to hedging congestion risk, a large segment of the market participates in FTRs on a speculative basis. Positions in selected FTRs are taken with the expectation, or at least the hope, that congestion will be in the direction of the FTRs and payments can be received from PJM. If congestion were in the opposite direction, the speculator would not only lose the money paid to acquire the FTRs but also owe PJM payments. PJM has added another class of FTR that helps avoid this outcome.

The two primary classes of FTRs now available are *obligations* and *options*. Obligation FTRs operate like the original instruments, where holders can receive or owe payments. Option FTRs entitle holders to congestion payments when such payments are positive, but do not obligate the holder to pay the charges when congestion is in the opposite direction.

From the discussion thus far, FTRs are clearly point-to-point instruments that are defined in part by a source–sink pair. The validity period of an FTR is also an important specification. PJM currently offers annual and monthly FTRs. Annual FTRs start on 1 June and run though 31 May of the following year. Monthly FTRs apply only during a single calendar month.

Acquiring FTRs
FTRs are distributed primarily through auctions, though some market participants can be allocated FTRs. PJM conducts a series of auctions for different FTRs, starting with annual FTRs and then moving to monthly ones. Annual FTRs are auctioned via a multi-round process that starts in mid-April. For each of the four weekly rounds, participants submit bids and offers for various source–sink pairs and receive notice of the results within days of the completion of the round. Holders of FTRs acquired prior to a given auction round, via allocation or the previous annual FTR auction for example, may offer those FTRs in that round. Each round requires the application of a simultaneous feasibility test (SFT). The SFT combines all bids and offers with other data related to the expected configuration of the system to determine which bids and offers are feasible. The SFT is similar in many respects to powerflows, though the approach is modified to specifically test the system for FTRs. The SFT determines which FTRs are awarded, how much of each FTR is awarded and the price of each FTR. The auction rounds conclude in May and the awarded FTRs start shortly thereafter on 1 June.

Monthly FTR auctions are single-round and begin in the middle of the month prior to the FTR start. For example, June FTRs are auctioned in mid-May. As with the annual FTR auctions, monthly FTR auctions process bids and offers through an SFT. Also like the annual FTR auction rounds, monthly auctions allow participants to offer FTRs previously acquired, such as those from an annual auction. Besides through auctions, FTRs may be acquired in the bilateral FTR secondary market. Only existing FTRs may be transferred between participants in the secondary market.

Congestion revenues and revenue adequacy

As described above regarding injections and withdrawals, PJM pays the LMP at injections and receives the LMP at withdrawals. When congestion exists in the system, differences in LMP result in more total receipts across withdrawal locations than total payouts at injection locations. This so-called "overcollection" is considered a congestion charge. The collection of these congestion charges by PJM from users of the day-ahead energy market is only part of the story. PJM must pay FTR holders an amount equivalent to the congestion charge for the source–sink pair. These payments are made from the pool of congestion charges collected.

Under normal conditions, the total of all congestion revenues collected in a period throughout PJM are sufficient to cover all FTR payment owed to holders in that period. This state is referred to as revenue adequacy. Under certain situations, however, congestion revenues will be insufficient to meet the required FTR payments. PJM can be revenue-inadequate when an FTR path is oversubscribed. The SFT typically prevents any FTR path from being oversubscribed. However, the physical status of the system can change between the execution of the SFT used during an FTR auction and the time when day-ahead prices and congestion are realised.

Recall that June FTRs are auctioned in the middle of the preceding May. In late May or during June, a component failure of the system can reduce the transfer capability of the transmission network. The SFT for the June FTR auction normally limits the quantity of FTRs awarded for a source–sink pair to the feasible transfer capability between that pair, but this limit is based on an expected state of the system that is now incorrect due to that component failure. While the per-megawatt congestion charge and FTR payments are equal, the quantities against which the charges and payments are applied are not. This results in a revenue imbalance where collections of congestion charges fall short of payments owed to FTR holders.

Suppose 100 MW of FTR are auctioned for a path. Because of a component failure, say a transformer outage, the transfer capability for the path is reduced, or *derated*, to only 50 MW. If the congestion charge is US$10/MWh, PJM will collect (50 MW) × (US$10/MWh) or US$500/h. The payment to FTR holders, however, is (100 MW) × (US$10/MWh) or US$1,000/h. The payments exceed the charges because there are more FTRs awarded than transfer capability for day-ahead flows.

When PJM is revenue-inadequate, FTRs are derated and the payments are reduced on a pro rata basis to cover the shortfall. One implication of this potential derating is that FTRs may not fully hedge a flow position in the day-ahead energy market. Considering the example above, suppose a participant holds 10 MW of FTR to hedge a 10 MW flow in the day-ahead market. The total congestion charge for the 10 MW would be US$100/h. The FTR should pay the holder US$100/h keeping the participant whole. However, because of the derating, the FTR payment is reduced to US$5/MWh for a total payment of only US$50/h. Despite acquiring FTRs to hedge the congestion risk, the participant is still exposed to derating risk. Furthermore, since derating is applied across all FTRs, regardless of which path is the specific source of the derating, a participant faces derating risk even if the component that failed is not on the source–sink path of any FTRs that the participant holds.

In this example, it may appear that PJM is collecting plenty of money. After all, the participant above is paying a net US$50/h into PJM. Of course, this ignores what positions other market participants may have. Some of the FTR holders may be speculators with no flow positions. In this case, PJM will collect no congestion charges but still be required to pay on the outstanding FTRs. Also, another participant may hold 10 MW of FTR and have planned to flow 10 MW, but, because of the reduced transfer capability, the participant may flow only 2 MW. Here, the congestion charges collected are only US$20/h, but the un-derated FTR payment would still be US$100/h.

Many market participants dislike the potential for FTR derating. However, overall revenue balance must be maintained. Within a month, excess congestion charges from some periods are used to offset situations when the system is revenue-inadequate in other periods. If, in the example above, the transformer outages that are the root of the derating were to last only for a few days, then revenue inadequacy would last for only a few days. During the rest of the month, PJM could potentially collect more in congestion charges than it is required to pay FTR holders, and this collection would make up, at least in part, for the previous shortfall. Reduction of FTR payments happens only if the entire month is net-revenue-inadequate.

Since derating is undesirable, PJM not only applies excess congestion charges from some periods to offset shortfalls in other periods within the same month, but it also applies revenues from the FTR auction in certain circumstances. If the available auction revenues are still insufficient, the FTRs are derated for that month.

Auction revenues
PJM has created auction revenue rights (ARRs) as a way of allocating the net revenues from FTR auctions. As discussed above, FTR auction revenues can be used to support FTR revenue adequacy. Only after all ARR holders have been fully funded will any excess auction revenues be made available for FTR holders to address revenue inadequacy.

Holders of firm transmission and users of network transmission service are entitled to ARRs. Load-serving entities (LSEs) use network service to bring power from sources to their respective loads. Initial ARR allocations are based, in part, on the load an LSE is expected to serve and the historical utilisation of network service. An entity that is entitled to an ARR allocation may elect to receive a corresponding FTR instead. As an alternative, the entity may choose to offer the ARR in the ARR auction, or, of course, keep the ARR. Those not entitled to an ARR allocation may bid for ARRs in the auction. As with the annual FTR and monthly FTR auctions, an SFT is applied to the ARR auction. Holders of ARRs receive the net FTR auction revenues.

Numbers and FTRs

The amount of data associated with the ARR/FTR markets is significant. For 15 April 2005, PJM published LMP for 6,639 separate locations in the day-ahead market. These locations include aggregations, such as zones and hubs, as well as individual nodes with voltages as low as 2 kilovolts (kV). These LMP must be published for every hour and every listed location, though PJM does change the list of locations from time to time. If an FTR were available from every location to every other location for which an LMP is published, more than 36 million FTR source–sink pairs would be needed. Even treating source–sink pairs equivalent to their sink–source counterpart leaves a staggering number of potential FTRs. This still ignores the existence of obligations as well as options. To keep the auction tractable, PJM publishes a list of valid paths and nodes. Most recently, there are 18,050 valid option paths and 6,587 valid obligation nodes.

The number of valid option paths is far smaller than that of total potential paths, while the number of valid obligation nodes is almost nearly the total of all reported nodes. Given the complexity of the SFT with respect to option FTRs, the relatively limited list of valid option paths is not surprising. In contrast to option FTRs, obligation FTRs are computationally simpler in the SFT. Furthermore, obligation FTRs are essentially the same as the original FTRs with which PJM has acquired many years of experience. These facts make the relatively larger number of valid obligation FTRs reasonable. Just as the locations for which LMP are published can change, so can the list of valid FTR paths and nodes.

Despite the complexity of ARR/FTR markets, as of 23 March 2005, there were 219 participants registered with PJM for the FTR auctions. Affiliated firms, however, may each be included, so the total number of completely independent and unique participants is somewhat smaller. Relatively new among the participants are financial firms that were not traditionally or previously in the energy industry. These firms are typically active as speculators in the FTR markets.

> PJM has a tradition of making data available to the marketplace. By combining the day-ahead energy market price data with the FTR auction results, the performance of any participant in the FTR market can be calculated. This, however, is a substantial amount of data. Furthermore, the use for FTRs by the participants is not necessarily clear. While the FTR positions may lose money, other positions in different markets, such as energy, may make the overall portfolio and strategy a winner. Nonetheless, the ability to know the positions and performance of FTR portfolios of every competitor is interesting, if not a little unnerving.

trade could be decreased to reduce exposure to potential congestion. TLR events have different levels, and the transactions that are potentially affected by a TLR have a priority associated with them. Transactions supported by firm transmission have higher priority than transactions supported by non-firm transmission. Among transactions with the same firmness of transmission, priority is higher for longer-term rights than shorter-term ones. Therefore, a transfer flowing under one-year firm transmission will have greater priority than a transfer flowing under monthly firm transmission. A Level 3 TLR should not curtail transfers backed by firm transmission. If congestion is anticipated, firm transmission and longer durations may be used to reduce the chance of being curtailed. Anecdotal evidence suggests that some of the annual firm transmission services purchased in the past were acquired strictly to increase the priority of transfers during the summer months with no intention of being used for transfers during the rest of the year.

Other systems offer financial transmission instruments that effectively pay the holder the equivalent congestion cost experienced by users of the system. Nodal price systems have such instruments. Zonal systems typically do as well. With zonal pricing, many locations are grouped into a zone and a single price applies across all locations of the zone. Debates continue to rage over whether nodal or zonal is better or more appropriate. Regardless of the answer, the simple fact is that both systems exist. Furthermore, implementation of each class of systems varies within the class. Zonal systems include the California ISO and

MANAGEMENT OF TRANSMISSION IN THE ELECTRICITY MARKETS

ERCOT, while examples of nodal systems are PJM, New York ISO and ISO New England.

Transmission congestion instruments vary by name and nature across the systems that use them. Variations include financial transmission rights (FTRs) in PJM and ISO New England, firm transmission rights (FTRs) in California ISO, transmission congestion contracts (TCCs) in New York ISO and transmission congestion rights (TCRs) in ERCOT. Note that FTR can mean different things in different systems. Each ISO that offers these transmission congestion instruments conducts training for interested parties. Given the intricacies of each respective market and constantly evolving rules, this chapter will not attempt to describe each instrument in detail. Some important general features will be discussed, however.

Transmission congestion instruments are typically only financial in nature and do not grant the holder physical rights to the system. These instruments pay the holder in relation to the amount of congestion calculated for the system. They may be held between specific locations, like PJM FTRs, or between zones, like California ISO FTRs, or along a set of lines, like TCRs in ERCOT. They may pay out based on congestion in day-ahead markets, like FTRs in ISO New England, or in real-time markets, like TCRs in ERCOT. They may be an option, paying the holder only under certain conditions but never requiring payment to the ISO, such as TCRs in ERCOT, or they may be obligations that pay the holder or require payment from the holder, depending on the direction of congestion. Some systems offer both options and obligations.

Transmission congestion instruments are typically available via an auction conducted by the ISO. Some rights are acquired along with, but tradable separately from, physical transmission rights. Rights obtained directly from the ISO are normally settled with the ISO, though secondary transfer of rights can often be registered with the ISO. While secondary transfers are facilitated sometimes, typically, only those transfers registered with the ISO will be settled with the ISO. Otherwise, the holder of record will receive payments or demands for payment.

When offered at auctions, congestion instruments usually have different durations such as one year or one month. As a result, auctions are held regularly. In ERCOT, an annual auction is conducted in December of the prior year, while monthly TCRs are auctioned for each month during the prior month. Despite these calendar limitations, secondary transfers can usually be done for shorter time periods, such as single days or even single hours. In practice, while allowed, such subdivided transfers rarely occur, if ever. In fact, secondary transfer of congestion instruments for any duration is uncommon. Nearly all primary purchasers hold the instruments until expiry.

Transmission congestion instruments can be used to hedge congestion exposure. If XYZ Corp had sold 10 MW of power forward at Node 2 for US$60/MWh and purchased power from S1 at Node 1 for US$50/MWh, a gross margin of US$10/MWh could be realised on the transactions in the unconstrained case. For a 16-hour on-peak block, this totals to US$1,600.

Suppose, however, that the line between Nodes 1 and 2 were constrained as in the nodal price example. With a congestion charge of US$20/MWh to move power from Node 1 to Node 2, XYZ Corp will lose US$10/MWh, for a total of US$1,600 during the on-peak block. To hedge the power position, XYZ Corp could acquire 10 MW of transmission congestion instruments. The instrument would pay the hold, XYZ Corp, the congestion charge. Regardless of what the congestion charge turns out to be, the instruments would pay XYZ Corp exactly the right amount.

Holding 10 MW of instruments does not necessarily hedge all transmission risk for XYZ Corp. The instrument will pay for congestion charges, but events that can create congestion, such as line outages, can also affect marginal losses. In systems where marginal losses are charged, XYZ Corp may still face some risks. Also, XYZ Corp must be careful to acquire the correct instruments. In the three-node example, the correct instruments and amounts are trivial to determine. In ERCOT, however, more care is needed. Because power flows distribute among transmission lines and because TCRs apply to specific transmission paths, a portfolio of TCRs is actually required to properly hedge congestion risk.

If XYZ Corp were to flow 100 MW of power from the South Zone to the North Zone, simply purchasing 100 MW of TCRs along Sandow–Temple, also called the South-to-North commercially significant constraint (CSC), would not be a proper hedge for interzonal congestion. Due to distribution of power flows, a 100 MW transfer from South Zone to North Zone will not flow exclusively on the Sandow–Temple CSC. The amount is closer to 40%, though the exact amount will vary. Holding 100 MW of TCRs is too much for that CSC, but holding only 40 would be too few. This is because some of the 100 MW transfer will flow on other CSCs. TCRs for those CSCs must be held as well. ERCOT publishes before each auction what the so-called *shift factors* are that determine how interzonal transfers will distribute among the CSC pathways. In the past, TCRs for each CSC were auctioned sequentially, complicating the task of obtaining the proper portfolio to hedge a transaction. ERCOT has implemented a combinational auction, sometimes called a combinatorial auction, that allows bidders to specify a portfolio of TCRs desired.

While nobody should obtain and use transmission congestion instruments strictly on the information contained in this chapter, it should be clear that market particulars are critical to the proper use of these instruments. Lessons learned and approaches used in one system cannot necessarily be directly applied to another. Superficial similarities can be misleading.

CONCLUSIONS

For the more than four decades between PUHCA and PURPA, regulation of the power industry had been relatively stable. In the decade since the EPAct, the industry has experienced tremendous change. Restructuring has not been without its growing pains. While some call for a return to the past, the real focus should be on moving forward with our new found, though expensive, wisdom. Debates will continue just as the demand for power does. As we forge ahead, a realisation and respect for how physical constraints and financial markets interact are fundamental to meaningful restructuring.

---

1   "Final report on the August 14, 2003 blackout in the US and Canada: causes and recommendations," US–Canada Power System Outage Task Force, 5 April 2004, available at http://www.nerc.com/~filez/blackout.html.

Section 2

Modelling and Market Realities

Introduction

Vincent Kaminski

The second part of the book illustrates how the general concepts introduced in the first section are applied in the practice of the merchant energy business, and how the industry is handling unique challenges resulting from the growing complexity of the business and its dependence on exogenous factors such as weather and changing market design.

Giulio Federico and **Adam Whitmore** point out that stochastic processes developed for highly efficient financial markets cannot be used for the energy markets without making adjustments for the complexities of market design, evolution of the regulatory framework and the potential for strategic behaviour of different agents operating in these markets. The long-term tenors of many energy transactions expose the parties to the risk of dramatic changes in the market rules, as examples such as the UK after transition to NETA and California after introduction of the semi-deregulated market designed around the Power Exchange have demonstrated.

Several of the chapters included in the first section emphasised the importance of weather as the fundamental driver of the electricity prices, primarily through the impact on load and, to a much lesser extent, on the efficiency of the generation units and transmission lines. It is obvious that the traders and risk managers in the merchant energy industry have to understand the weather dynamics as well as master the art of translating the weather information into statements about market developments. Most energy trading operations employ several meteorologists to study weather patterns, collect and disseminate weather forecasts across the organisation, and track hurricanes in the energy-producing regions. This is a very difficult task and one has to be very efficient to gain competitive advantage, given that most market players have access to the same information and use the same models. A positive contribution to the organisation can be made only by accelerating the process of weather-forecast formulation and interpretation, disseminating the available forecasts in a more timely way, especially when the different models produce conflicting information. The traders and risk managers can use the weather forecasts effectively only if they develop an understanding of meteorology, the data sources and different weather patterns, such as ENSO, PDO and NOA, understand that forecast precision varies in different seasons and learn how to translate weather forecast into market-moving information. In a nutshell, one has to understand both the importance of weather information and the current limitations of this science. **Daniel Guertin** provides a review of the current state of weather-forecasting tools and their usefulness for the merchant energy business. The chapter contains an extensive review of weather information sources, very important for smaller organisations that cannot afford their own weather staff.

One of the greatest challenges of the energy business is that, in addition to price risk, both the producers and end-users of energy face volumetric risk. This risk manifests itself in fluctuations of the market demand, often on very short time scales, that cannot be managed effectively – or at all in the case of electricity – from the existing inventories. The energy markets have developed a number of instruments, such as swing options that allow managing both risks in the context of the same contractual arrangements. One of the most complicated contracts in this area is the so-called full-requirement transaction, under which a power marketer makes a commitment to satisfy full energy needs of his or her customer. Typically, such contracts apply to a slice of the full system load of a local utility or a specialised service, such as POLR (provider of last resort) load. The contract design can be very complex, exposing the power marketer to price risk as well as volumetric risk, resulting from the weather fluctuations and the vagaries of customer behaviour. In many cases, a power marketer faces exposure to customer migration, both in (more end-users signing up for their service), as well as out (end-users leaving the service and choosing other competing suppliers). The chapter on the full-requirements transaction was contributed by **Yan Gao, Harald Ullrich** and **Krzysztof Wolyniec**. The authors start with the review of different contractual structures used for these transactions and review the regional differences between the ways such contracts are originated (through auctions or requests for proposals (RFPs)). The discussion of these transactions is followed by the review of different valuation and hedging methods. The authors review the indirect fundamental method that is based on explicit modelling of the factors, such as load (or the weather as its main driver), fuel prices, generation and transmission system, and unit outages. The indirect parametric method is based on capturing the relationships between the variables listed above through a statistical method. The alternative to this approach is a direct approach based on an estimation of the parameters of the equation representing the cost of serving a shaped load, that is to say the load that is characterised by a non-rectangular shape, reflecting intraday fluctuations of electricity consumption. The challenge of hedging these contracts results from the fact that a power marketer has to address both load and price uncertainty and absorb additional risks resulting from the evolution of the regulatory framework and the changes in the competitive landscape.

The next chapter, written by **Boris Chibisov, Alexander Eydeland**, and **Krzysztof Wolyniec** is devoted to the description and valuation of highly specialised options used in the electricity markets, so-called heat-rate options. Heat rate represents the thermal efficiency of a power plant, the amount of energy measured in BTUs necessary to produce one kilowatt hour of electricity. Alternatively, the implied heat rate is defined as the ratio of power prices and fuel prices, in most cases natural gas, which is the marginal fuel in many US power pools. It is difficult to underestimate the importance of the heat-rate construct for the power markets. Heat-rate options are used to lock in power plants' economic results (this option can be used as a hedge instrument by producers) or can be used as building blocks to structure a virtual power plant, a contractual equivalent of owning a generation unit without all the operational risks associated with ownership of a physical asset. The importance of this chapter goes beyond the discussion of this specific instrument. The authors offer many important insights into valuation of the heat rate options that apply to valuation of different types of spread options that are very popular in the commodity markets. One of the most important inputs to pricing such options is correlation – a very simple concept but sometimes very difficult to estimate numerically from the historical data and apply in practice.

The high level of price volatility and high volumetric risk make the energy business a natural field for the application of different contracts and financial instruments that are designed to transfer risks to the institutions with the sufficient financial resources and skills required to operate in these very difficult markets. In spite of high potential demand for energy-related derivatives, the trading volumes in the US and European markets shrank significantly after 2001, following the bankruptcy of Enron and other energy merchants. One of the underlying factors behind the crisis of the merchant energy business can be found in archaic, highly inefficient and expensive approaches to credit risk management that the industry adopted in the 1990s. The chapter contributed by myself and **Vasant Shanbhogue** reviews the current credit risk practices in the US energy business and discusses recent initiatives, related to multilateral clearing and netting, which are quite correctly seen as a way to overcome current difficulties. The industry has been, however, rather slow in embracing this approach for a number of reasons discussed in the chapter. The second part of the chapter looks at the tools that can be used to assess the credit risk embedded in a portfolio of energy contracts. Implementation of a system designed to assess numerical levels of such exposure is a critical requirement for any energy-trading operation.

The next chapter, by **Laura L. Brooks**, expands the narrower definition of credit risk to discuss the concept of economic capital required to transact in the power markets. The concept of capital adequacy is not new but the merchant energy industry has ignored it as the events of the last few years have demonstrated. Economic capital is required to sustain profitability and absorb operational and trading losses, to grow earnings, and to meet daily liquidity obligations. Given the characteristics of the energy markets, the capital requirements can be quite significant and have to be estimated through a rigorous and fairly sophisticated system.

The chapter reviews the concept of regulatory capital defined for the banking industry in the Basel Agreement and extensions and modifications of this concept formulated by the Committee of Chief Risk Officers (CCRO). The CCRO addressed in its white papers both economic (regulatory) capital and capital required for financial liquidity; reviewed methods for aggregating market, credit and operative risks; and the methods for estimating each of these component exposures.

The final chapter summarises the main conclusions of the empirical study of bidding strategies used by the generators operating under the SMD, or *standard market design*, paradigm, in the New York Day Ahead Market, based on the data from the summer of 2001. The study, undertaken by ESAI (Energy Security Analysis, Inc), shows the importance of empirical analysis of the bidding strategies of generators that are influenced by the specific types of plants (base load, mid-merit, or peaking) and by the characteristics of the demand curve for each of those types of plants. The chapter contributed by **Paul Flemming** offers important insights into the behaviour of the managers of power plants that in many cases may be suboptimal and may reflect the rules of thumb and intuition of the plant managers. The review of the results of the study is supported by the summary of the power pool design based on the concept of locational marginal prices.

8

The Importance of Market Structure and Incentives in Determining Energy Price Risk

Giulio Federico; Adam Whitmore

CRA International (UK) Ltd; Deloitte*

Managing energy price risk requires assessment both of the price outlook to which a company is taking exposure and of the risks this creates. If either of the assessment of price or the associated risks is not robust, investments and contract terms may be misevaluated, opportunities may be missed, controllable risks may remain unmanaged and poor decisions are likely to be made.

Risk managers in the electricity and gas industries need to be especially sensitive to the possibility of incorrectly estimating price behaviour, because these markets usually embody characteristics that make them difficult to treat using standard statistically based models. Models based on stochastic processes have been developed mainly to represent fully competitive markets. In practice, gas and electricity markets are often not fully competitive. Price behaviour may be quite different from that which would prevail in a perfectly competitive market, creating different types of risks. Consequently, models designed to deal with less competitive markets are needed. Furthermore, any one model may be insufficient to capture pricing behaviour over the relevant time horizon. Assets in the energy sector are long-lived and the commercial and regulatory structures that shape prices in the market may undergo marked changes during the asset's life.

---

*The views expressed in this chapter are the authors' and do not necessarily reflect those of the institutions they represent.

Consequently, an approach to modelling that takes account of the possibility of structural change is needed.

This chapter outlines an approach to analysing price that takes into account the need to understand the commercial and regulatory factors that shape the market structures, the imperfectly competitive nature of the structure of many gas and power markets and the possibility that these structures may change. If such issues are ignored or treated inadequately, then answers may be misleading. Consequently, price-risk modelling needs to become more broadly based than the use of traditional analysis of stochastic processes alone would allow.[1]

This chapter is organised in two parts. The first provides an overview of why the use of numerical rather than analytical modelling approaches to risk should be preferred in the energy sector, creating a need for reliable methods of modelling input price behaviour. It argues that the most appropriate models of price behaviour in gas and power markets will be those that explicitly take account of market structure and the incentives on players. It also outlines the aspects of energy market structure that shape price behaviour and describes the importance of potential changes to these structures. The second part reviews the techniques from industrial economics that can be used to model energy price risk, showing that resulting price behaviour can be very different from that found in fully competitive markets. Applications of these techniques to issues of horizontal concentration, vertical integration and regulation are illustrated in this part of the chapter.

THE NATURE OF ENERGY PRICE RISK AND IMPLICATIONS FOR MODELLING

The preference for using numerical modelling rather than closed-form algebraic solutions for risk evaluation in energy markets

Powerful analytical techniques have been developed over several decades for modelling competitive markets. Net present value (NPV) analysis and standard portfolio theory are now widely applied. They can be adapted to value a single deal, such as a gas sales contract, and to examine the effect of a single deal on the distribution of value for a portfolio of assets, such as a producer's portfolio of gas sales contracts. The original Black–Scholes framework and the variants subsequently developed have been applied to option pricing in traded markets. Similar techniques can, in certain circumstances, be adapted to value the real options embedded in some contracts.

The application of such techniques usually involves modelling price as a stochastic process. Some of these processes yield elegant analytical solutions that provide insights into the drivers of value. For example the Black–Scholes formula clearly highlights the importance of volatility in determining option value and also indicates the importance of the risk-free discount rate.

However, it is increasingly difficult to derive analytical models as price processes become more complex. Energy markets are often characterised by mean reversion, skewed distributions and instances of extreme values. Closed-form algebraic solutions become difficult, and in some cases impossible, to derive, or require assumptions that are inappropriate to the problem in hand.

Difficulties in deriving analytical solutions also arise from the complexity of options sometimes implicit in contract terms. These may be, in effect, multiple or compound options. For example, swing in a natural gas contract (the inclusion in the contract of elements that provide some flexibility in the timing and quantity of delivery) offers multiple American options, the profitability of which is not independent. Similarly, decisions to expand production are often compound options because the exercise of one option creates or closes off others. In some cases options may be exercisable only within certain periods.

Most fundamentally, the asset being valued may not be subject to an exogenous price and so the assumption of any kind of independent stochastic process becomes inappropriate. For example, a decision to expand capacity may itself affect the price in the market. Consequently, models in which price is endogenous to the decision-making process must be used.

To deal with these problems, more general numerical techniques involving Monte Carlo and tree-based approaches have been developed. Their use is described elsewhere in this volume and in standard texts. They allow complex price processes and non-standard distributions to be modelled. They also make more tractable the analysis of complex terms such as compound options. They even allow for endogenous pricing to be incorporated. These strengths make them the tool of choice for much practical risk management in the energy sector.[2]

The generality of numerical techniques is a great strength but creates a corresponding problem. Because they are able to model almost any form of price risk it is critical that the appropriate choice of price behaviour be made. It is this choice, rather than the derivation of risk measures from underlying price behaviour, that presents the greatest potential difficulties. In some cases standard stochastic processes can be assumed. However, other models of price behaviour more accurately characterise many energy markets and if standard stochastic processes are used in such markets then materially misleading answers will result. It is to this we now turn.

The need for pricing models that reflect market structure and the incentives on market participants

The most robust method of modelling price behaviour is to take an approach based on economic principles – constructing a model based on an analysis of how players are likely to behave taking account of the structure of the market and the incentives they face. Models based on assumptions of competitive market structures form a subset of models within the wider set of incentive-based economic models. An approach to modelling based on economic principles will indicate when models that do not rely on assumptions of full competition are required. This will assist in identifying risks, such as the possibility of increased price volatility or a risk that quoted prices may not be realisable in practice due to a fall in market liquidity.

In the case of markets with little or no historical data, such a model may be the only possible means of defining the price series. Where extensive historical data are available, an economic model will indicate the range of stochastic processes that should be surveyed and increase confidence in a choice of model made on statistical grounds. It may also aid estimates of parameters derived from statistical analysis. An economic approach to modelling will help understanding of when a proposed price model is likely to break down as the result of changes in market structure. This may be crucial to understanding risks. Simple extrapolation of present behaviour may lead to misleading expectations of future price behaviour.

In contrast, if analysis takes account of the possibility of underlying structural change it may help identify risks that would otherwise be missed. Such change may be endogenous to the market, for example resulting from new investments, or exogenous, for example being imposed by regulatory bodies. The importance of structural change in assessing price behaviour is illustrated in Panel 1 with reference to the UK wholesale gas market.

PANEL 1 THE STRUCTURE OF UK GAS PRICES

The structure of the forces shaping the wholesale gas price in the liberalised UK market has undergone at least one major change in recent years, following the opening of the Bacton–Zeebrugge interconnector. A further structural change now appears to be under way, as the UK becomes a net importer of gas. A third structural change may happen within the next ten years as increasing imports of LNG have the potential to decouple gas and oil prices. Thus, within the life of a typical gas field, pipeline or power plant the forces shaping price are likely to have undergone two or three major structural changes. Even if there is stable and well-defined price behaviour within each market structure, price behaviour will change significantly over time because of these structural changes.

A competitive gas market emerged following the liberalisation and regulatory reform in the UK gas market in the late 1980s and early 1990s. There were many producers and many buyers, leading to effective rivalry in the market, which was largely isolated from markets in Continental Europe. Consequently, price was formed by the competitive interaction of supply and demand with the UK.

This changed in October 1998 with the opening of the interconnector. This allowed for trade between buyers and sellers in the UK and in Continental Europe and therefore to a convergence between UK and European prices. Gas prices in Europe tend to follow oil prices because of the indexation terms in long-term contracts. Consequently, after the opening of the interconnector the annual average UK price tended also to follow oil price, although with significant seasonal variation and a discount of typically some 2p/th to Continental European prices on an annual average basis. Figure A shows the way that UK gas prices have tracked oil prices since the opening of the interconnector.

The discount to Continental European prices has been consistent with a continuing surplus of indigenous UK gas. Broadly speaking, the UK price has in effect been netted back from the Continental European price by transport costs from UK NBP to Continental European delivery points. This is likely to change as the UK becomes a net importer of

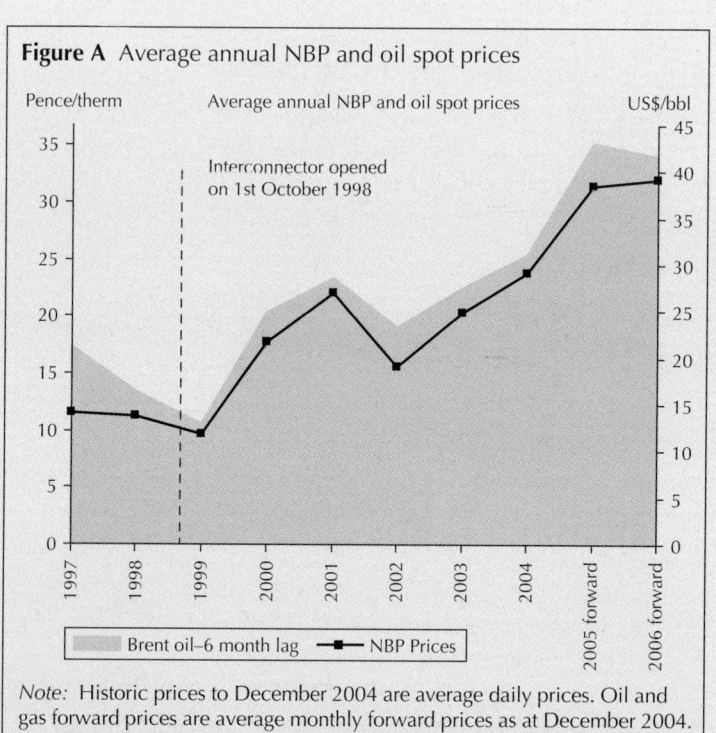

Figure A Average annual NBP and oil spot prices

Note: Historic prices to December 2004 are average daily prices. Oil and gas forward prices are average monthly forward prices as at December 2004.

gas. The costs of delivery to the UK do not differ greatly from the costs of delivery to elsewhere in northern Europe, and prices would therefore be expected to be similar provided that the market remains effectively competitive.

Although the relationship between UK and Continental European prices seems likely to persist, the way in which European gas prices themselves are formed might change. As the costs of delivering LNG to Europe fall relative to new pipeline gas, large amounts of additional LNG are expected to be available to the European market. Sources of LNG are more diverse than those of pipeline gas, so there is greater potential for gas-to-gas competition to set the price. It is possible that any gas-to-gas competition may lead to an influence from prices in the US market as routing of LNG cargoes to secure the most favourable prices leads to some tendency for prices on each side of the Atlantic to become correlated. However, there are factors that may prevent the decoupling of gas and oil prices in Europe, including the continued importance of inter-fuel competition in final markets, especially in industry.

The use of an economic approach to modelling to lend support to the choice of a standard stochastic process can be illustrated by consideration of standard geometric Brownian motion as a model of equity price behaviour. Investors will, on average, expect a return on capital, so there will be a drift term in the process to give a positive return on average over time. There are many independent influences on the price of an equity with no single factor predominating, so the shape of the distribution of returns is normal, with large price shifts rare compared with small ones. There is no systematic dependence on previous events that is not reflected in present prices, so no correlation between successive price movements and the assumption of a Markov process (where price movements are not dependent on previous price history) is consequently appropriate. The large number of buyers and sellers also implies that trading will be liquid and that the market price will be realisable in most circumstances. Consideration of the underlying economic model also indicates when there may be a structural break in price behaviour, for example in the case of an acquisition of the company for which the equity price is being modelled.

Similarly, consideration of an economic model indicates why mean reversion may be a more appropriate model of energy prices than geometric Brownian motion. Random influences such as short-term weather variations will affect prices in the short term but over the longer term prices will be drawn back to levels determined by supply-and-demand fundamentals. These levels may vary seasonally. In these circumstances a model is needed in which there are normally distributed shocks from a wide variety of independent events but, instead of affecting the price permanently, as in a random walk, their effect decays through time and price returns to longer term levels. Support from economic modelling for the choice of a mean reverting process is especially important as statistical tests tend to find it difficult to distinguish unambiguously between a random walk and mean reversion except when mean reversion is strong or there is a great deal of historic data.

Most importantly, application of economic modelling to examine the effects of market structure and the incentives on players also indicates that there are many circumstances in which the use of stochastic processes is not adequate. First, markets may be horizontally concentrated, leading to players' actions having an influence on the price and so invalidating the fundamental assumption that there is an exogenous price path. Horizontal concentration is the norm rather than the exception in gas and electricity markets in at least some parts of the chain. At the transmission and distribution level there are typically natural monopolies or oligopolies in networks. In wholesale power markets high transport costs lead to national markets being isolated from the type of international trade that characterises most commodity markets. Economies of scale in production and the legacy of historic patterns of ownership often lead to few players being present in each national market. Even in comparatively competitive markets, such as the UK wholesale gas market, the provisions of some types of services may be concentrated, with a single facility accounting for a significant proportion of the total capacity. Retail supply to small consumers is concentrated in many European markets, affecting also the concentration of buyers in the wholesale markets for electricity, especially towards the peak and for flexible load.

Such horizontal concentration can greatly affect price behaviour. First, it will tend to lead to higher prices. Second, changes in the capacity or cost structure of a single player or entry by a new player will affect price in ways that would not be the case in a competitive market. Third, volatility of prices will also be affected, leading to very different values for options.

Second, vertical integration or existing contracts may influence wholesale markets. Generators may have contract cover or retail assets that insulate part of their revenue from the effect of price fluctuations in wholesale markets. In some instances, electricity companies may be fully vertically integrated, implying that the wholesale price will be largely a transfer price between two parts of the same organisation. Likewise, gas producers may have retail sales that make a wholesale price less relevant. For example, gas producers' marketing activities in Europe may weaken their dependence on a border price for gas. Changes in vertical relationships, such as expiry of contracts, can lead to large price shifts in wholesale markets. Such changes may be sudden and in either direction, depending on the nature of the change in incentives. The size and probability of these jumps will have a large effect on value, and may lead to subsequent knock-on effects on market structure.

Finally, regulation may also continue to affect prices more than in most commodity markets. Regulation is likely to have a continuing influence on price in much of Europe for the foreseeable future. Network access terms, retail price regulation, stranded cost recovery mechanisms and environmental policy will all affect the outlook for prices. Even in the liberalised UK market, regulation continues to be an important influence, as shown for example by the radical change in market design introduced by the regulator in 2001 (*New Electricity Trading Arrangements*), and the forced divestments in generation capacity by the incumbent producers that preceded it (see Panel 2 for an analysis of these reforms).

Regulatory policy may lead to changes in market structure and trend breaks that are significant drivers of risk. These can take a

---

PANEL 2 PRICES IN THE ENGLAND AND WALES POWER MARKET

The behaviour of wholesale electricity prices in England and Wales since liberalisation illustrates the effect of several of the price drivers in concentrated markets discussed in the main text. Changes in concentration, market design and other regulatory intervention since liberalisation in 1990 have closely affected price determination in the market.

Starting from a situation where the two main incumbent generators, National Power and PowerGen, controlled almost all the price-setting capacity in the market in the early 1990s, regulatory measures were taken to increase competition. The two incumbents were forced to divest plant in 1996 (6GW, under a lease arrangement with Eastern) and sold further plants in 1999/2000 (an additional 10GW). These divestments significantly lowered the market concentration as measured by HHI.[i] As the price trend shows (see Figure A), the second round divestments (and the associated fall in the HHI) appears to have led to a large fall in prices, despite broadly stable fuel prices and plant margins. In contrast, neither the first set of divestments (which increased the cost of the divested plant under an earn-out system) nor the significant base load CCGT entry that occurred throughout the 1990s (more than 20GW) had led to significant price reductions. This is because the main price-setting generators (including Eastern after 1996) continued to face an inelastic residual demand curve especially in high-demand periods, given the absence of effective competition at the margin. In spite of the high levels of independent entry and the first round of divestments, the HHI of price-setting marginal plant was still above 2,000 by 1999.

The prevailing market design in the 1990s arguably also facilitated the exercise of market power by generators. Between 1990 and 2001 the market cleared in a compulsory, day-ahead, uniform-price pool,

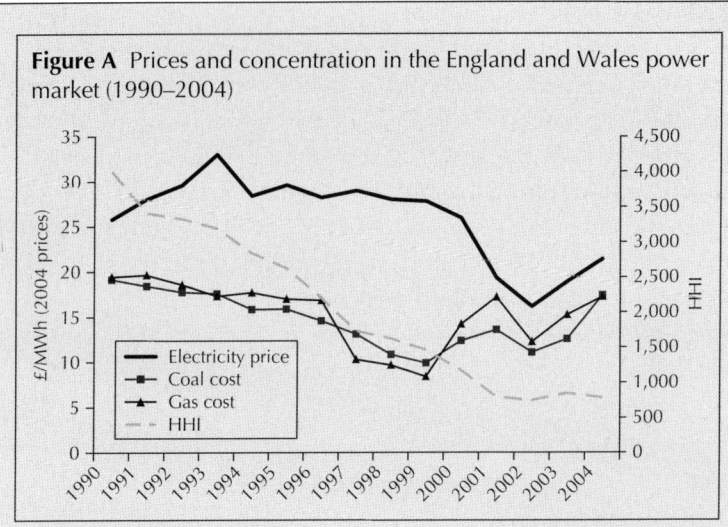

Figure A Prices and concentration in the England and Wales power market (1990–2004)

where all plants called to produce earned the same price. This market design was relatively transparent (increasing the scope for tacit coordination through repeated interaction), and it also allowed generators to compete in "quantity-withdrawal" strategies (see Panel 4 on Cournot competition). In particular, under the Pool, producers could set relatively high marginal prices with their marginal plants, and protect their infra-marginal volumes with lower bids (see Federico and Rahman (2003) and Klemperer (2003) for a discussion of this effect). The abolition of the Pool in 2001 and its replacement with a decentralised system of contract trading (NETA) can therefore be expected to have had a further downward effect on market prices. However, the main price fall preceded the introduction of NETA and it is difficult to find firm evidence NETA actually had a downward effect on prices (see, for example, Evans and Green (2003)). In the medium term there are also ambiguous effects on competition and prices due to NETA, for example due to the lack of liquidity in spot and short term markets and its possible effects on entry.

[i] HHI is the Herfindahl–Hirschmann index, which is the sum of squares of market shares.

number of forms. First, changes to existing regulations, such as the level of network access charges or the severity of environmental regulation, such as the Large Combustion Plant Directive, can affect the position of competitors in the market. Secondly, the introduction of new regulations and the details of their design can create new forces in the market. Thirdly, regulation can change the structure of the commercial interaction by intervention to change ownership or constrain behaviour. An example of the impact of regulatory forces on a market is shown in Panel 3, which describes how some of the details of the EU Emissions Trading Scheme (EU ETS) are likely to influence wholesale power markets.

PANEL 3 THE IMPORTANCE OF THE STRUCTURE OF POLICY INSTRUMENTS: THE EU EMISSIONS TRADING SCHEME

The EU ETS came into force on 1 January 2005. The scheme covers large-combustion plants (those with rated thermal input exceeding 20 MW), which includes power plants and other large emitters such as oil refineries and bulk chemical plants. The scheme covers over 40% of CO_2 emissions in the EU. The scheme has a number of detailed rules, some of which vary among member states. In many cases these affect the operation of the scheme and consequently price formation. We describe the effect of some of these detailed rules here to illustrate how details of policy and regulation design can influence traded markets. Correspondingly, when regulations change, or new ones are introduced, the outlook for prices and risk can change.

Banking and borrowing
Scheme rules enable emitters to use allowances from one calendar year to meet obligations in subsequent calendar years. This is usually referred to as *banking of allowances*. For example, an allowance for 2005 can be used to meet the obligation for 2006. This essentially limits the price rise between years to the time value of money. If allowance prices were to drop below the forward price for the following year (adjusted by the appropriate interest rate) there would be a simple arbitrage opportunity from selling forward and retaining the allowance.

A further constraint is imposed by the rules allowing allowances for one year to be used to meet the previous year's obligation, which in effect gives the ability to borrow allowances from the subsequent year. This places a corresponding limit on the extent to which the price can rise above that of the following year.

The rules of the scheme thus tend to limit the differences in price between years within a phase of the scheme. However there are no similar arrangements that allow banking and borrowing between Phase 1 of the scheme (which runs from 2005 to 2007) and Phase 2 of the scheme (which runs from 2008 to 2012). This means that there can be a substantial discontinuity in the price of allowances between Phase 1 and Phase 2 of the scheme.

Loss of allowances on closure
If a generating plant remains on the system, it retains the right to collect valuable allowances free of charge, which it would lose on closure. The value of these allowances may be a substantial proportion of its annual avoidable operating costs (excluding fuel and other short-run variable operating costs), depending on the price of allowances. Profitability for a plant that remains open may thus be significantly improved. Consequently, capacity will be incentivised to remain on the system longer than it otherwise would in order to capture the value of allowances.

These incentives to retain capacity on the system could affect wholesale electricity prices. Capacity that would otherwise close – because avoidable costs are high relative to revenue – may stay on the system, increasing the capacity margin. This could lead to prices being lower than they otherwise would be. It is even possible that such a weakening of prices could forestall efficient new entry.

Indicative calculations lead to an estimate of the magnitude of this effect of approximately £0.5/MWh on annual base load prices at an allowance price of €9/tCO_2.

Allocation rules for Phase 2
If allocations in Phase 2 depend on emissions in Phase 1, then incentives for abatement could be reduced. Each additional MWh generated in Phase 1 would still lead to an increase in costs due to the requirement to buy allowances. However, there would be an offsetting future benefit in the form of an increased future allocation of allowances in Phase 2. As a result the net cost of increasing emissions may be reduced (or, in extreme cases, eliminated) by the increase in the future allocation of free allowances.

Incentives to abate emissions could be retained to a greater extent with an updated baseline if the future allocations of allowances are on the basis of output (MWh) rather than emissions (tCO_2). An allocation of future allowances based on output creates the same benefit (in the form of the value of future allowances) for a gas plant as it does for a coal plant, because a MWh generated from each type of plant increases the future allocation of allowances by the same amount irrespective of the fuel source. However, generators must continue to purchase allowances to match their emissions. Consequently, the costs of a coal plant rise by more than those of a gas plant and the costs of less efficient plant rise by more than those of more efficient plant. The differential in net costs between coal and gas plant is thus retained when future allocations are output based, and with it the incentives for abatement of emissions.

The detailed rules for the allocation of allowances will thus affect which plants will generate and how electricity prices are set.

These features of gas and electricity markets give a fundamentally different outlook on price from that derived from models assuming more fully competitive market conditions. The form of price behaviour is different from that represented by conventional models and can change in ways that such models cannot represent effectively. Consequently, these features of the market have significant implications for assessment of risk profiles and valuation of investments and contract terms. In the next part of this chapter we describe how these market characteristics can be modelled using economic techniques and their effect on prices determined.

ILLUSTRATION OF APPLICATIONS OF ECONOMIC TECHNIQUES
The need for models from industrial economics and game theory

There is no single model or type of model that captures all the possible features of energy markets. Instead, industrial economics provides an approach to modelling that allows the issues raised by the presence of horizontal concentration, vertical integration and regulation to be addressed explicitly.

The approach employed in industrial economics typically first considers the objectives market participants pursue. Usually profit maximisation is assumed, but this will not always be appropriate (for example, in the case of state-owned entities). Moreover, fully profit-maximising behaviour may be subject to regulatory constraints, either directly or because of the threat of regulatory intervention if outcomes are unacceptable to the regulator.

The different possible strategies that market participants might adopt are then defined and an evaluation made of the extent to which different strategy choices affect the market participants' realisation of their objectives, such as profitability. The strategies examined will usually include ranges of offers of price and quantity to the market. They may also take account of other possible actions such as acquisition and consolidation.

Analysis of these strategies must also take account of the external constraints and incentives that players face. This will often be in the form of contractual structures, either implicit or explicit. These contracts may be with other commercial entities in the market, or imposed by regulatory policy.

Most importantly, the choice of which strategy to pursue for any given player will be made taking into account expectations about what other rival firms will do. Analysing the effect of the interaction between players' strategies is at the heart of the modelling of horizontally concentrated markets (typically relying on game theoretical tools).

The formal analysis of the strategies available to market players, and the key determinants of possible strategies, allow identification of an outcome or range of outcomes where all players pursue their objectives as best they can in the environment they face. A market outcome for which all players are realising their objectives given what their competitors are doing is referred to as an *equilibrium*. This equilibrium need not be static: it may be a path of outcomes through time.

Game theory provides the formal tools to examine such strategic interaction and predict market outcomes. Game-theoretical tools can significantly enhance the modeller's understanding of effects on prices and risk of market features such as concentration, frequency of player interaction, vertical links and regulation.

More formally, game theory identifies the possible equilibrium strategies that can emerge as strategic players interact with each other trying to maximise their payoffs. Formally, if S_i represents the set of possible strategies available to player i, $s = (s_1, s_2, \ldots s_i, \ldots s_I)$ denotes a given strategy profile, which combines the strategies adopted by each player, and $U_i(s)$ represents the payoffs to player i from a given strategy profile, then the optimisation problem faced by each player i is the following:

$$\max U_i(s) \text{ with respect to } s_i \in S_i \qquad (1)$$

A market equilibrium will result when all players are optimising their payoff given what other players are doing. There may be several such equilibriums in a particular strategy space. This type of equilibrium (a Nash equilibrium) is self-enforcing as no player has an incentive to deviate from it. More formally, a strategy profile s^* is a Nash equilibrium if for all players i

$$U_i(s_i^*, s_{-i}^*) \geq U_i(s_i, s_{-i}^*) \quad \text{for all } s_i \in S_i \qquad (2)$$

where s_{-i} is the strategy profile that includes all players' strategies except for player i.

THE IMPORTANCE OF MARKET STRUCTURE AND INCENTIVES IN DETERMINING ENERGY PRICE RISK

The remainder of this chapter illustrates the broad approach from industrial economics described above with specific examples. These examples are not intended to be comprehensive: rather they are intended to show how this type of approach can give a materially different assessment of risk from that which would arise from the adoption of analysis derived from the assumption of fully competitive markets.

Modelling prices in horizontally concentrated electricity and gas markets

In a market with few large players the interdependence of behaviour needs to be modelled explicitly because it affects both price levels and price volatility. In these markets, players will typically face unilateral incentives to optimise their market profits by raising prices above competitive levels (eg, as given by short-run marginal costs). The profitability of any price increase for a given firm will depend on a number of factors, and most crucially on (i) the volume base profiting from any increase in the market price (which is directly affected by the firm's market share); and (ii) the volume loss resulting from a price increase, which will depend on the price elasticity of the residual demand faced by the player (ie, aggregate demand net of the supply offers of all other players).

Price raising can be achieved by a player increasing its price offers relative to its variable costs. Alternatively, market agents can withdraw outputs from the market to drive prices up. These two strategy choices (bid mark-ups and capacity withdrawals) can be used to achieve the same impact on the market price, but also have implications on the reactions to the price increase by rivals firms. This is because the shape of the supply function offered by a player will affect the slope of the residual demand faced by its rivals, and therefore will also affect their incentives to price above costs. In particular, the steep bid functions implied by a quantity-withdrawal strategy will tend to increase the incentives by rival firms to also price above costs.

A stylised example of formal models of strategic behaviour is shown in Panel 4, which illustrates the standard Cournot model of quantity competition (which implements the quantity-withdrawal strategies discussed above). The results of Cournot modelling indicate that with strategic interaction prices will be above those

PANEL 4 QUANTITY COMPETITION IN AN ELECTRICITY WHOLESALE MARKET

As an illustration of the kind of modelling that can be applied to concentrated markets we consider a wholesale electricity market dominated by two main players, i and j. Assume that each player maximises its payoffs (or profits) by changing the capacity it offers into the market at cost.[i] The optimisation problem faced by each producer k can be expressed as follows:

$$\max_{q_k} \pi_k = q_k p(Q) - c(q_k) \quad \text{for } k = i, j \quad (4.1)$$

where q_k is player k output
$Q = q_i + q_j$ = industry output
$c(q_k)$ = player k cost of production
$p(Q)$ = inverse demand function faced by the two players, giving the market price as a function of their total output.

Setting the first order conditions to zero defines the *reaction functions* for each player, which determine the optimal level of output of each player as a function of the output of the other player:

$$q_i^*(q_j) = -\frac{1}{p'}\left[p(Q) - c'(q_i)\right]$$

$$q_j^*(q_i) = -\frac{1}{p'}\left[p(Q) - c'(q_j)\right] \quad (4.2)$$

The intersection(s) of these two functions represent(s) the Nash equilibriums of the "game" played by the two generators, defining the optimal quantity produced by each generator and the resulting market prices. These can be obtained explicitly by making assumptions about the demand and cost functions. For instance, assuming linear functions of the form $p(Q) = a - bQ$ and $c(q_k) = cq$, Equations 4.2 yield the two reaction functions:

$$q_i^*(q_j) = -\frac{a - bq_j - c}{2b}$$

$$q_j^*(q_i) = -\frac{a - bq_i - c}{2b} \quad (4.3)$$

which in turn yield the following market equilibrium:

$$q_i^* = q_j^* = \frac{a - c}{3b} \quad (4.4)$$

$$p(q_i^*, q_j^*) = \frac{a + 2c}{3} \quad (4.5)$$

> The model therefore predicts that, for $a > c$, prices lie above their competitive level c, and is able to quantity this difference as a function of the demand level.[ii]
>
> Generalising this result for n symmetrical players, one obtains:
>
> $$q_n^* = \frac{a-c}{b(n+1)} \qquad (4.6)$$
>
> $$p^* = c + \frac{a-c}{n+1} \qquad (4.7)$$
>
> which shows that the difference between market prices and costs fall with the number of players and eventually vanishes for a large n[iii] (see Figure 1).
>
> ---
>
> [i] This sort of interaction is referred to as a Cournot game. Cournot models have been directly applied to electricity markets (see, for example, Ocaña and Romero (1997) and Borenstein and Bushnell (1999)). A variant of the Cournot model, supply function equilibriums, was used to model the England and Wales Pool (Green and Newbery, 1992).
> [ii] The condition $a > c$ is not restrictive as it is necessary for the market to be viable, since it implies that the maximum willingness to pay in the market is higher than the marginal cost of production.
> [iii] $\lim_{n \to +\infty} p^* = c$.

experienced under competitive conditions. The gap between competitive and strategic prices depends directly on the number of firms in that market. As the number of players increases the difference between market prices and costs falls, eventually vanishing for a large number of players n (see Figure 1).[3]

This model is a highly stylised representation of a real electricity or gas market. For example, it does not reflect cost differences between players. In practice, in most power markets more concentration of high-cost price-setting capacity will tend to lead to higher prices than concentration of base load infra-marginal capacity, by affecting the slope of the residual demand faced by players with market power (see discussion in Panel 2). In gas markets elasticity may change along the demand curve due to the influence of the price of competing fuels. There may also be the potential for price

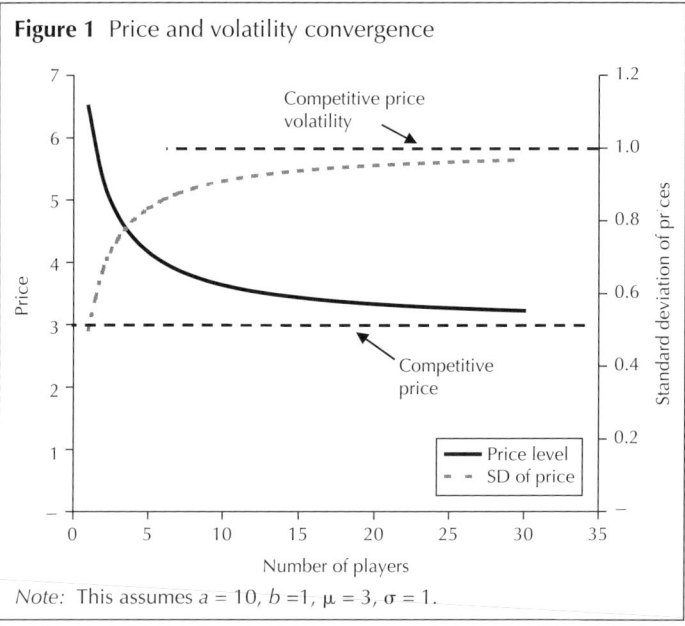

Figure 1 Price and volatility convergence

Note: This assumes $a = 10$, $b = 1$, $\mu = 3$, $\sigma = 1$.

discrimination by producers, for example by volume. The potential for entry, for example by new suppliers of LNG and the terms of existing contracts, will also play an important role.

Despite the stylised character of the simple Cournot model presented in Panel 4, the results capture important features of such markets and show how they may differ from more fully competitive markets. For example, the results show that price may be more sensitive to changes in market concentration than to changes in marginal costs. This will affect the outlook for price and the assessment of the probabilities of magnitudes of price changes. Both of these can have a major effect on contract terms. For example, the value of a contract clause that limits downside risk may be greatly increased by the perception of a significant probability of a large price fall due to regulatory action, designed to decrease concentration in production.[4] Consequently, risk managers' decisions are likely to be significantly affected by the use of this type of model.

Price volatility in concentrated markets

A game-theoretical approach to price determination in a concentrated energy market also provides insights into price volatility. For a given distribution of cost shocks, the price volatility obtained in a market with few players will often tend to be lower than that prevailing in a market with numerous players.

Reduced price volatility in concentrated markets results from players interacting strategically and finding it optimal to partially absorb a cost shock. If the cost shock is positive they cannot afford to pass all of it through to consumers, as this would cause an excessive reduction in the residual demand they face given the existing price-cost mark-up, and would therefore reduce profits. Alternatively, if the cost shock is negative, players can afford not to pass on all of the cost reduction to consumers.

This is clear from Equation (4.7) in Panel 4. If the marginal cost parameter c is distributed normally, with mean μ and standard deviation σ, prices in a competitive environment are also distributed with standard deviation σ, while in a Cournot interaction with n players prices will have a lower volatility, $(n/(n + 1))\sigma$, which converges to σ only for n large (see Figure 1).

The nature of price volatility in energy markets also depends on the nature of the game market agents play. This would create a degree of "strategic" price volatility as players randomise their strategies when interacting with each other.

The resulting differences in volatility will affect risk-management choices. For example, if volatility is decreased the option fee implicit in a one-way contract-for-difference (CfD) price may be correspondingly reduced. Combined with consideration of changes in price levels due to market concentration, this may give a very different risk profile from that implied by conventional estimates. Reduced volatility may reduce short-term risk but medium-term risk may be increased by the possibility of regulatory intervention. This may lead to very different valuations for similar contracts covering different timescales. It would also give valuations different from those that would be given by conventional methods, which would suggest smooth increases in option value with time horizon.

Repeated interaction between players

Market agents in energy markets interact with each other frequently, and over long time periods. The models of static optimisation described above may not fully capture the kind of price equilibriums that can emerge as a result of dynamic interaction.

Game theory can also be a useful tool in predicting the kind of equilibriums that are likely to emerge in a repeated game between strategic players. In this case, the payoffs from deviating from a given dynamic equilibrium will depend on both the short-term gain of doing so (such as increasing one's output when other players price high), and the long-term implications of such behaviour (such as undermining a high-price equilibrium). The notion of Nash equilibrium in a repeated game therefore needs to account for the effects that a strategy played by a market agent *today* has on possible strategies that can emerge *tomorrow*. In this context a Nash equilibrium is one where players optimise their behaviour over the horizon of the game given what other players are doing and given what they would be doing in case of defection by players from the Nash equilibrium.[5]

In general, more equilibriums are attainable in a repeated game between players. A market outcome that is not viable in the short run, because players have short-term incentives to deviate from it, can be sustainable in a longer-time horizon as players realise that they benefit from not undermining implicit cooperation over their pricing or production decisions. Key drivers of equilibrium in this context are the players' discount rate (which determines how much weight they attach to future payoffs), the frequency of the strategic interaction and the visibility of other players' moves. The payoff structure typically corresponds to that of the standard prisoners' dilemma model in game theory (see Figure 2).

Players in energy markets often interact with each other very frequently, for example at least on a daily basis in electricity pools, and the relevant discount factors that apply to their strategic decisions will correspondingly discount future payoffs relatively little,

Figure 2 A "pricing" game

		Player A	
		Price at cost	Price at p^*
Player B	Price at cost	0 , 0	π_H , π_L
	Price at p^*	π_L , π_H	π_C , π_C

where $\pi_H \geq \pi_C \geq 0 \geq \pi_L$
1st term in each box is Player B's payoff, 2nd term is Player A's payoff

THE IMPORTANCE OF MARKET STRUCTURE AND INCENTIVES IN DETERMINING ENERGY PRICE RISK

tending to make a repeated equilibrium more sustainable. Other features of energy markets that also make coordination easier to sustain include the fact that the product is usually homogeneous with little potential for differentiating on quality, that cost structures of firms may be similar and that demand is mature and tends to be inelastic.

However, other features of markets may make coordination more difficult to sustain. If their strategies are not visible to other players, cooperation may be harder to sustain because the incentive to "cheat" increases as punishment by other players is delayed. For example, high-price equilibriums may be more difficult to sustain in a bilateral contracting market than in a transparent daily wholesale pool. Regulatory changes to the way firms interact (eg, the abolition of a compulsory organised market, and the introduction of decentralised bilateral trading) can have significant implications for the feasibility and likelihood of a high-price repeated game outcome, and a corresponding impact on price levels and volatility.

Other significant constraints on incumbent producers' price behaviour, even in a repeated interaction, include threats of regulatory intervention and the threat of entry by other players. This last factor is discussed below.

Other market dynamics
Entry constraints. Realistic analysis of strategic interaction in energy markets must take account of the potential for entry by independent players in a market. Even if in single or repeated interaction incumbents are able to sustain high mark-ups over their costs, it may not be optimal to sustain them if they attract excessive entry. This will be the case if incumbents think they can affect entry decisions by lowering current prices below the level that could be sustained in a repeated interaction. If this is the case wholesale prices will lie relatively close to the average cost of entry, after accounting for the degree of risk present in the market.

On the other hand, the potential for entry will not significantly affect market prices under two separate conditions.

❑ Under conditions of perfect information on players' payoffs and cost structures. This is because entry will be driven by relative

costs only, and the current price will not be able to act as a credible signal to entrants about the conditions that are likely to prevail in the market once they enter it.
- ❑ Even under conditions of asymmetric information, incumbents may prefer to trade off future payoffs for current payoffs, accepting a loss in their market share for a higher short-run price, which induces entry. If this is the case the incumbents effectively renounce the entry-deterrence properties of the market price and concentrate on maximising short-run payoffs.

Demand. Changes in costs and demand through time may affect the market's strategic equilibrium. For example, as demand grows it may become optimal for some players to add new capacity. However, if capacity increments are large, as in the case of long-distance gas transmission pipelines, only one player may have the opportunity to add capacity at any one time. Once this investment is made, it may create new options, such as the ability to add additional compression to the pipeline. This leads to the need to analyse the consequences of future decisions in considering present strategies. Consequently a model is required in which a series of moves and countermoves are specified to determine the longer-term outlook for price (usually referred to as an "extensive form" model). This will normally require the application of tree models to solve for the equilibriums by backward induction.

Contracts and vertical integration

As noted above, energy markets are commonly characterised by vertical links between players, which can take the form of either contracts (such as contracts for differences between electricity generators and retailers) or outright vertical integration (integration between production and transportation and marketing in the gas industry, for example).

These contractual links have an effect on pricing behaviour, especially in intermediate markets. For instance, pricing behaviour in an electricity wholesale market is affected by the presence of CfDs that limit players' exposure to the wholesale price and therefore change their pricing incentives. Similarly, for a vertically integrated player a pool or short-term contract market price may be not more than a "transfer" price between arms of its business.

The effects of these contractual links on pricing behaviour can be modelled using the game-theoretical approach outlined above. As an illustration, consider the hypothetical duopoly in the generation market modelled in Panel 2, and assume that each generator k holds a contract for differences on their output with a contract level of α_k and a strike price of f_k. The optimisation problem they face when competing with each other by changing their output levels becomes

$$\underset{q_k}{\text{Max}}\ \pi_k = q_k\left[p(Q)-c(q_k)\right] - \alpha_k\left(f_k - p(Q)\right) \quad \text{for } k = i, j \quad (3)$$

where the first term in the profit equation captures the standard definition of profits and the second term represents the difference payment, which depends on the relative levels of the strike price in the CfD and the pool price.

Equation 3 leads to revised first-order conditions. In particular, solving using the linear functions used in Panel 4 and imposing symmetry between players (such that $f_i = f_j = f$ and $\alpha_i = \alpha_j = \alpha$), one obtains the following market equilibrium:

$$q_i^* = q_j^* = (a-c)/3b + \alpha/3 \quad (4)$$

$$p(q_i^*, q_j^*) = (a + 2c - 2b\alpha)/3 \quad (5)$$

Equations 7 and 8 show that, for positive contract cover, output in the industry increases and prices consequently fall. This is because the generators' payoffs from quantity withdrawal decrease with contract cover since the price they obtain for output under contract is independent of the market price. As long as contract cover is below outturn output, prices will lie above the competitive level. If contract cover is set at the competitive output level for each generator, prices will be equal to variable costs c (see Figure 3).[6] However, price volatility does not change as a result of contract cover, as there is still more cost absorption at the margin in an oligopoly model relative to a competitive benchmark, as discussed above.

ENERGY MODELLING

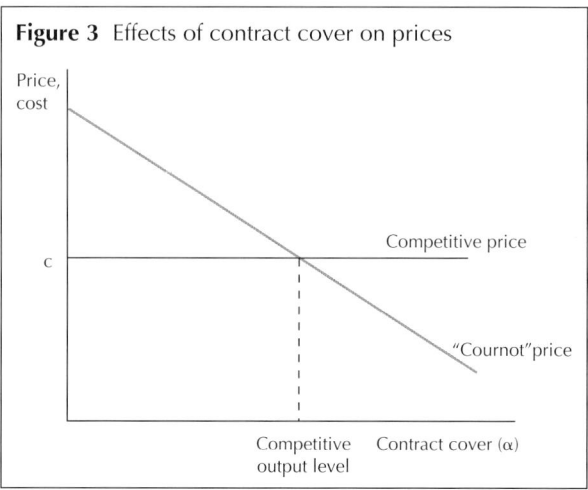

Figure 3 Effects of contract cover on prices

The results described above may change if players think that the price they set in the wholesale market does affect the strike price of contracts they hold or the retail price their downstream business receives. If this is the case, some of the incentives towards high prices will be restored and effects of high levels of contract cover α mitigated.[7]

Modelling contracts in the way discussed here generalises to other forms of vertical links, including outright vertical integration. If a player is integrated downstream the variable α can be interpreted as the amount of output that is directly sold to final consumers or other retailers. As with a contract, the price received on this output may be independent of the level of the wholesale price (at least for some period of time, or over a given share of retail demand), thereby affecting generators' bidding incentives.

Such vertical links may change through time. For example, contracts may expire or regulations limiting vertical integration may be relaxed. These can lead to potentially large price shocks with corresponding risks to those players exposed to wholesale market prices. They may also create different risk profiles for different players, which may in turn affect their incentives, for example for bidding in a pool. The effect of all contracts held by all players will need to be considered. If contracts have been imposed by a regulator or agreed as part of a package of liberalisation measures, their terms may be widely known; in other cases they will remain confidential and estimates of players' contractual positions will need to be made.

Regulation

Government regulation is a prominent feature of energy markets and it can have significant implications for the nature of the strategic interactions at work in these markets. Regulatory risk can therefore be a significant component of the overall market risk, and its effects can be understood in relation to the kind of strategic modelling outlined above.

In particular, regulation can affect a number of the structural features of energy markets, which, in turn, determine the nature of price formation in these markets. For example, regulatory intervention can reduce the degree of horizontal concentration via divestment measures. As described above, this may lead to a downward price shift and an increased pass-through of cost shocks. It may also affect the degree of vertical integration present in the industry, and thereby influence players' pricing incentives.

Regulation may directly change the payoff structure of players, for example by introducing payment flows that operate in parallel with market payments such as compensation for costs incurred before liberalisation. There may also be direct price controls in some markets (such as retail markets), which can affect the price level and price dynamics in other related markets such as wholesale markets. In some cases wholesale prices may themselves be regulated. This may affect not only the annual average price level but also the shape of prices within the year, depending on the form of the price control.[8] In other cases regulation may seek to promote market contestability and entry, thereby mitigating the effects of market power on outturn prices.

The strategic modelling approach described so far can provide the modeller with the tools necessary to assess the impact of regulatory constraints of this kind and of possible regulatory change. An illustration of this kind of modelling is provided by the example of the mechanism for the payment of competition transition charges that have been at work in the liberalised Spanish wholesale electricity market since 1998 (see Panel 5).

PANEL 5 CONTRACTS AND BIDDING IN THE SPANISH ELECTRICITY POOL

The Spanish electricity pool started operation in January 1998 on the basis of a similar set of trading rules to the ones applied in the England and Wales pool. The two largest players in the market, Endesa and Iberdrola, had shares of production of around 45–50% and 30–35% each at the time, and to date still account for 65% of total generation (excluding "special regime" renewable output).

Partly in an attempt to mitigate market power, stranded-cost recovery contracts were imposed on market players at the time of liberalisation, linking the payment of stranded cost compensation to the level of prices in the pool. These contracts, referred to as competition transition charges (CTCs) reallocate regulated industry retail margins between players in fixed shares, subject to the total amount of funds dispersed being lower than a given amount. These fixed shares were set to leave Endesa "short of generation" (with a higher CTC share than market share) and Iberdrola "long in generation", introducing conflicting incentives on the main players with respect to the overall wholesale price level (see Figure A).

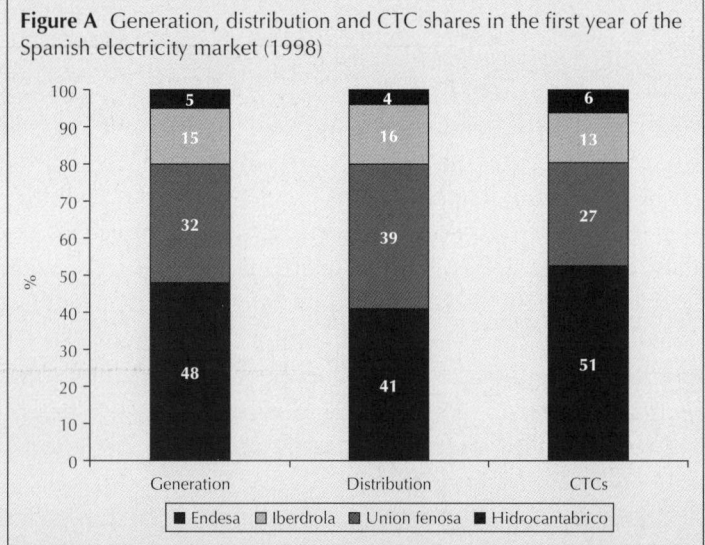

Figure A Generation, distribution and CTC shares in the first year of the Spanish electricity market (1998)

The CTC mechanism also includes a price cap, whereby if any generator's average revenue is above the cap (which is set at 6 PTA/kWh, or €3.6c/kWh) that generator's total CTC funds (to be recovered over a

12-year period) are reduced by an amount equal to revenue in excess of the level of revenue associated with the price cap.

CTCs have similar effects on players' bidding incentives to CfDs (being in effect a difference contract on the pool price) and can be modelled as follows, using the techniques described in the main text. The interaction between Endesa and Iberdrola can be analysed by examining optimal production by each player, given what the other is doing. At a stylised level each player k can be modelled as facing the following optimisation problem:

$$\max \pi_k = p(Q)\beta_k Q + \alpha_k (T - p(Q))Q - c_k(\beta_k Q) \quad (5.1)$$

where β_k is player k's market share
α_k is player k's share of CTC funds
T is the netback regulated retail tariff (net of transmission and distribution charges)
$(T - p(Q))$ represents total industry regulated retail margins.

Maximising profits for both Endesa and Iberdrola delivers the following equilibrium condition:

$$(\beta_E - \alpha_E) - (\beta_I - \alpha_I) = \frac{-1}{p'Q}\left[(\alpha_E - \alpha_I)(T - p) - (mc_E - mc_I)\right] \quad (5.2)$$

where subscript E denotes Endesa
subscript I denotes Iberdrola
mc_k represents the marginal cost of player k given its output level

The equilibrium condition reveals that players' CTC shares tend to drive their market shares, subject to cost differences. It shows that the net generating position of the two players (described by the left-hand side of Equation 5.2) is determined by their relative CTC shares (the term $(\alpha_E - \alpha_I)(T - p)$) and their relative cost position $(mc_E - mc_I)$. The player with a higher CTC share (or more "contract cover") will tend to have a higher market share, unless its marginal costs are substantially higher than its rival's costs.

This approach reveals that Endesa, with a CTC share of 51%, will face incentives to increase its market share to its CTC share, thus taking market share from Iberdrola. This tendency may be mitigated by variations in cost conditions (which exist in particular as a result of the significant amount of hydroelectric energy controlled by Iberdrola) and by other regulatory incentives. Evidence from the Spanish pool in the first year of operation (1998) confirmed this theoretical prediction and displayed a striking convergence of the two players' market shares towards their CTC shares (see CNE, 2000).

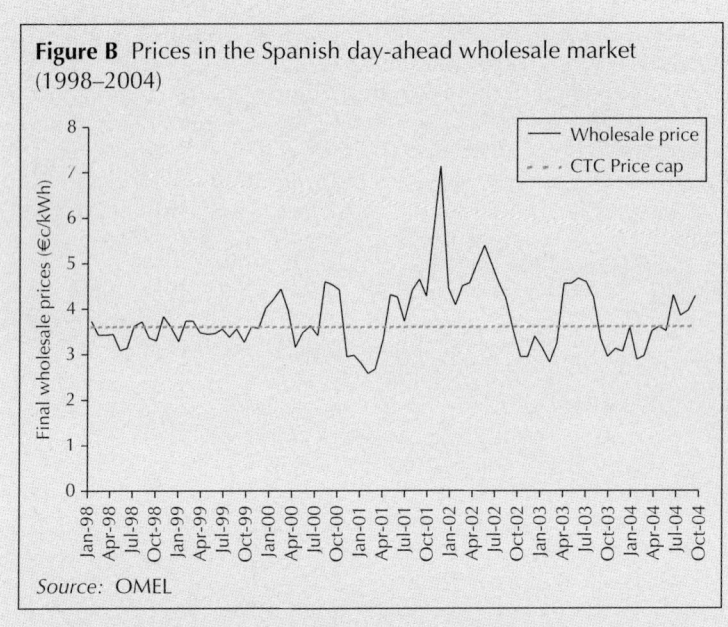

Figure B Prices in the Spanish day-ahead wholesale market (1998–2004)

Source: OMEL

Moreover, during the early years of the market, wholesale prices were set very close to the level implied by the CTC price "cap", even though players may have been capable of sustaining a higher price equilibrium (at least in the short run, see Figure B). More volatility around this level has been experienced since 1999, partially as a result of the entry of players without CTC entitlements, the reduction in the share of total demand still under the regulated tariff, and variations in demand and hydroelectric conditions (see, for example, Crampes and Fabra (2005)).

The example of the Spanish electricity wholesale market (especially in its early years of operation) clearly shows how regulatory intervention can have a significant impact on market outcomes and indirectly on the price risk faced by players. This effect can be understood and quantified using models of strategic interaction, which can also help in identifying the timing and nature of possible regime shifts driven by regulatory change.

CONCLUSION

In this chapter we have argued that standard models of pricing in competitive markets are inadequate to deal with some common features of energy markets. Concentration of production implies that it is frequently inappropriate to regard players as price takers. The vertical relationships and regulatory influences that tend to characterise energy markets also affect price behaviour. These features can change, requiring an approach to understanding price behaviour that can take account of the possibility of such change.

There is no single model or class of models that addresses these problems. Rather, what is required is an approach to modelling that applies principles of economic analysis to understand the implications of market structures for prices. This approach will lead to traditional stochastic process models for price behaviour under certain conditions. However, these conditions are rare in energy markets and in most cases price behaviour will tend to be very different from that which would prevail in a fully competitive market, in terms of price levels, price volatilities and, potentially, market liquidity and the realisability of quoted prices.

An understanding of the mechanics of non-competitive markets is particularly important in the context of risk management for understanding the likelihood and nature of possible price regime shifts. If a market's structural characteristics (such as the degree of horizontal and vertical integration) are changing, an assessment of risk and risk management measures based on historical data will be misleading, and models of strategic behaviour will provide a better model of price formation.

When a robust price outlook has been derived for a given market using the type of approach described in this chapter, the powerful numerical risk management techniques that are available can be used to derive risk measures. These results can be used to inform negotiations on the terms of any specific deal that is under consideration. Such an approach to risk management will yield materially better results than the application of standard tools to circumstances for which they were not designed. This should lead to significantly improved decision making.

1 We take as given that players seek to manage their risks. We do not discuss the validity of the various motivations for risk management.
2 Both Monte Carlo and tree-based models may become computationally cumbersome, although the power of modern desktop computers makes this much less of a problem than it once was. They also yield insights less readily than an analytical result. However, in most cases these drawbacks will be outweighed by their flexibility and generality.
3 $\lim_{n \to \infty} p^* = c$.
4 Decreases in concentration may not reduce price if this is already limited by other factors discussed below.

5 A variant of Nash equilibrium (subgame perfect equilibrium), considers credible "punishment strategies" only if there is defection from an equilibrium.
6 This can be derived by setting $p = c$ in equation (5) and solving for α.
7 For instance, if players conjecture that $f = \beta + \gamma\, p(Q)$, then the effects of α on prices and outputs need to be scaled down by $(1 - \gamma)$, increasing prices and lowering output, other things being equal.
8 For example, the price undertaking introduced in the England and Wales Pool between 1994 and 1996 limited time-weighted and demand-weighted price separately. This created incentives to raise peak prices above base-load prices.

REFERENCES

Borenstein, S. and J. Bushnell, 1999, "An empirical investigation of the potential for market power in California's electricity industry", *Journal of Industrial Economics*, **47(3)**, pp. 285–323.

Comisión Nacional del Sistema Eléctrico (CNSE), 2000, "El funcionamento del mercado eléctrico en el año 1998", Madrid.

Crampes, C. and N. Fabra, 2005, "The Spanish electricity industry: *Plus ça change…*", *Energy Journal*, forthcoming.

Evans, J. and R. J. Green, 2003, "Why did British electricity prices fall after 1998?", Cambridge Department of Economics Working Paper, n. 326.

Green, R. J. and D. M. Newbery, 1992, "Competition in the British electricity spot market", *Journal of Political Economy*, **100(5)**, pp. 929–53.

Federico, G. and D. M. Rahman, 2003, "Bidding in an electricity pay-as-bid auction", *Journal of Regulatory Economics*, **24(3)**, pp. 175–211.

Klemperer, P. D. and M. Meyer, 1989, "Supply function equilibriums in oligopoly under uncertainty", *Econometrica*, **57(6)**, pp. 1243–77.

Klemperer, P. D., 2003, "Why every economist should learn some auction theory", in M. Dewatripont, L. Hansen and S. Turnovsky (eds), *Advances in Economics and Econometrics*, Cambridge University Press.

Ocaña, C. and A. Romero, 1998, *A Simulation of the Spanish Electricity Pool*, Comisión Nacional del Sistema Eléctrico.

9

Impacts of the Weather on Energy Demand and Supplies

Daniel Guertin

Sempra Energy Trading

Energy usage across North America and around the world is highly dependent on both the global economy and global weather patterns. In the US, for example, total natural gas demand in 2003 was estimated at nearly 22 trillion cubic feet (Tcf), based on data obtained from the United States Department of Energy's Energy Information Administration (EIA). Of the nearly 22.0 Tcf of natural gas demand in the US in 2003, about 8.2 Tcf of gas (37%) was consumed in the residential and commercial sectors. Of the four main gas-consuming sectors in the US – residential, commercial, industrial and power generation – demand in the residential and commercial sectors typically exhibits the strongest correlation to seasonal temperatures and temperature departures from normal. This relationship is especially strong during the heating season, when natural gas is the fuel of choice of many end-users for space-heating purposes. In 2003, it is estimated that natural gas demand during the heating season (November–April) accounted for nearly 80% of the yearly demand in the residential sector and 72% of the yearly demand in the commercial sector. The other six months of the year accounted for less than 30% of the total demand in the residential and commercial sectors, owing mainly to base-load demand.

In the past decade, there has been a steady increase in the number of gas-fired power generation units across the US. As a result, gas demand for power generation has steadily increased from nearly 3.2 Tcf in 1990 to more than 5.0 Tcf in 2000 (EIA data). The growth rate of natural gas demand in this sector has slowed in recent years as a result of more efficient gas-fired units being constructed, but the overall demand in this sector is still significantly higher than it was just 10 years ago. There is a relationship between temperatures and gas demand for power generation during the winter months, since many residential and commercial end-users rely on electricity to heat their homes and businesses, but winter electricity demand is eclipsed by summertime gas demand for power generation. This mainly stems from widespread air-conditioning load during the warm summer months, not only in the residential and commercial sectors, but in the industrial sector as well. In 2003, for example, nearly 4.9 Tcf of gas demand was recorded in the US, with about 50% of this occurring during the five-month May–September period.

In addition to the weather, energy demand is also sensitive to local and national economies. A strong economic correlation to gas demand is typically observed in the industrial sector, which includes factories, automobile plants and the fertiliser and petrochemical sectors. The same also holds true for power-generation demand, where year-to-year energy demand closely follows the strength of the economy. However, despite year-to-year economic-based demand fluctuations, the strongest and most significant yearly demand fluctuations are mostly weather-based. For example, the difference in national gas demand between a very mild winter and a colder-than-normal winter can be astounding, on the order of 800–1,000 billion cubic feet (Bcf). During the first quarter of 2003, gas demand in the residential and commercial sectors totalled nearly 3.88 Tcf, which was significantly higher than the same period in 2002, when gas demand in these two sectors only totalled 3.39 Tcf. The difference in gas demand of nearly 500 Bcf can be explained almost entirely by the weather patterns across the US during the respective periods. In 2002, the first quarter was exceptionally warm across the major population centres of the central and eastern US, with temperatures averaging 3–6°F above normal (1971–2000 30-year normal) from Chicago to New York. A year later, temperatures across the same geographic region averaged almost 3–6° below normal.

Such year-to-year fluctuations in temperatures are quite common across the US, especially in the winter months. As a result, strong year-on-year changes in natural gas demand are usually observed. Weather-driven demand changes are also observed in the power-generation sector, although not nearly so extreme as the

Figure 1 Temperature anomalies for Q1 2002 and Q1 2003. Of particular note is the year-on-year temperature difference across the major metropolitan areas of the central and eastern US

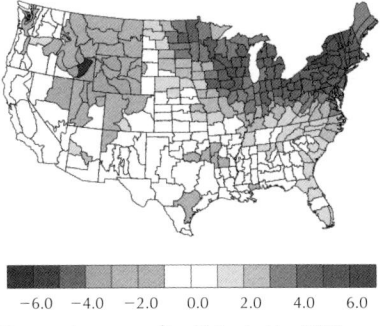
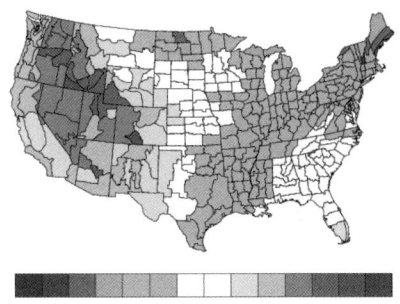

Temperature anomalies (F) Jan to Mar 2002 versus 1971–2000 longterm average.

Temperature anomalies (F) Jan to Mar 2003 versus 1971–2000 longterm average.

Note: The darker shading in the eastern US in the left panel represents above normal temperatures, and the lighter shading in the eastern US in the right panel represents below normal temperatures.

Source: NOAA-CIRES Climate Diagnostics Center, Boulder, Colorado, from their Web site at http://www.cdc.noaa.gov/

residential and commercial sectors during the heating season. In 2002, the May–September period averaged above normal across the vast majority of the US. During the same period, gas demand for power generation totalled nearly 3.0 Tcf. In 2004, however, temperatures were well below average across the central US, and temperatures in the east were only slightly above average. Cooler weather in the May–September 2004 period (relative to 2002) was one of the factors that led to weaker gas demand for power generation in 2004. Preliminary EIA data show gas demand for power generation in May–September 2004 totalling only 2.64 Tcf.

Similar demand cycles are also observed in Europe, although not to the extreme that they occur in North America since yearly natural gas demand is stronger in the US than it is in Europe and year-to-year changes in the weather patterns tend to be more extreme in North America.

Real-time and same-day power demand and prices are strongly impacted by temperature, and, to a lesser extent, wind, cloud cover, humidity and precipitation. These variables are strongly correlated to electricity demand, with humidity showing a stronger

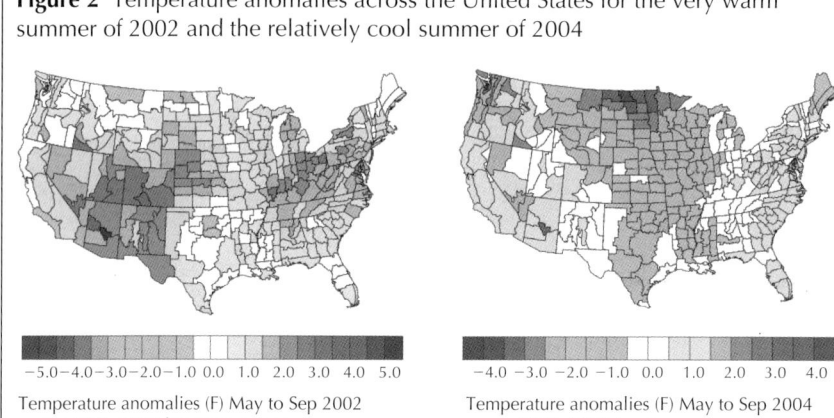

Figure 2 Temperature anomalies across the United States for the very warm summer of 2002 and the relatively cool summer of 2004

Temperature anomalies (F) May to Sep 2002 versus 1971–2000 longterm average.

Temperature anomalies (F) May to Sep 2004 versus 1971–2000 longterm average.

Note: Most of the shading over the US in the left panel represents above normal temperatures, and the shading in the central US in the right panel shown below normal temperatures.

Source: NOAA-CIRES Climate Diagnostics Center, Boulder, Colorado, from their Web site at http://www.cdc.noaa.gov/

correlation to demand in the summer and wind showing a higher correlation in the winter. High humidity levels in the summer raise the heat index, which makes it feel warmer than the actual air temperature, while wind in the winter affects the wind chill, which makes it feel colder than the actual air temperature. Of all the energy commodities that are traded in the futures markets in the US, electricity prices usually show the strongest correlation to the weather, since supply must be generated on a real-time basis to meet the demand.

Heating oil is another commodity that demonstrates a strong correlation between demand and weather. This is especially true in the northeastern US, where heating oil is the fuel of choice for space-heating purposes for many homes and businesses. Heating oil demand tends to spike when exceptionally cold weather develops in the northeast, which leads to strong increases in heating oil prices. Additionally, severe weather in the northeast can curtail heating oil supplies by making oil tanker travel (on the oceans and rivers) and tanker truck travel (on the highways) difficult, which may lead to delays in the arrival of heating oil shipments to the key population centres of the northeast.

WEATHER EVENTS

The discussion thus far has mainly focused on the impacts of weather on energy demand. Compared with the impacts on demand, the average yearly weather-related impact on energy supply tends to be rather small. However, there are, on average, a few weather-related events per year that impact natural gas, oil and electricity supply. Hurricanes and tropical storms in the Gulf of Mexico can severely damage natural gas and oil platforms in the offshore regions of Texas, Louisiana, Mississippi and Alabama if the tropical system is powerful enough (generally packing sustained winds of 110 miles per hour). At the very least, these systems can temporarily curtail operations in the Gulf of Mexico as platform owners evacuate their platforms of personnel as a precautionary measure. It is rare that a storm develops and moves through the Gulf of Mexico that is powerful enough to impact natural gas and oil operations for a significant length of time (weeks or months), mainly since most rigs are engineered to withstand hurricane-force winds and swells.

However, there have been years when hurricanes did develop or move through the Gulf that were powerful enough to cause long-term damage to the natural gas supply infrastructure. Recent examples are Hurricane Andrew (1992) and Hurricane Ivan (2004). Five months after Hurricane Ivan, nearly 0.5 Bcf/d of natural gas was still offline in the Gulf of Mexico as a result of Hurricane Ivan, and the storm itself was responsible for the loss of 150–200 Bcf of natural gas from the US supply system.

Other weather-related impacts on natural gas supply include the occasional temperature-related pipeline and rig problems in western Canada and the Gulf of Mexico, which occur when the temperatures drop so low or so quickly that they compromise the integrity of the supply system.

Electricity can be generated using coal, petroleum products, natural gas, water, wind and a limited amount of renewable resources. The supplies of most generation fuels are not impacted by the weather, with the exception of water and occasionally natural gas. Hydroelectric generation in the US is mainly confined to the mountainous regions of the west, and hydroelectric generation in Europe is mainly confined to Scandinavia and the certain regions of the Alps. Annual hydroelectric generation is highly dependent on precipitation, both in the form of liquid and solid (snow). Frozen precipitation

Figure 3 National Hurricane Center track for Hurricane Ivan. Hurricane Ivan passed through the eastern part of the U.S. Gulf of Mexico's natural gas producing region in September 2004, causing heavy damage to many natural gas and oil platforms and curtailing natural gas production for several months

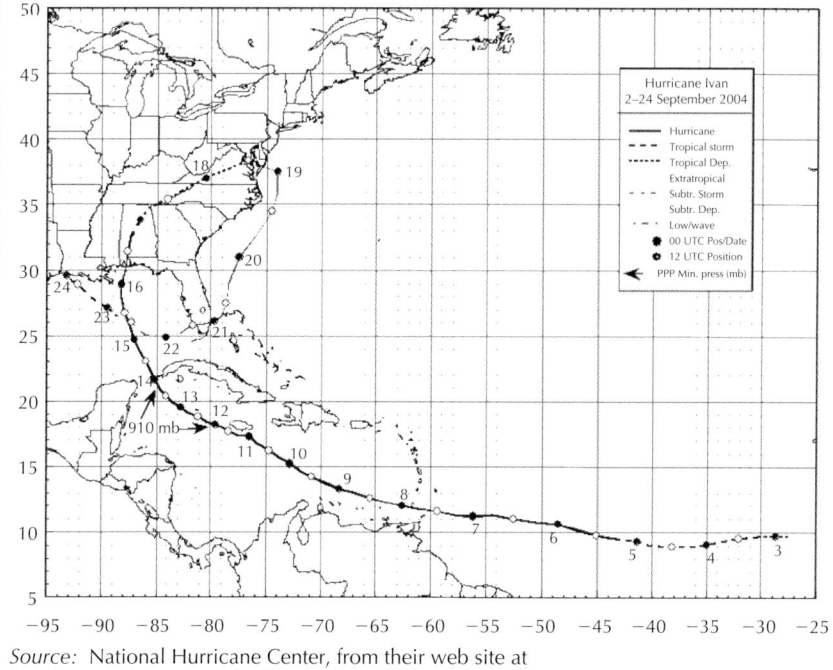

Source: National Hurricane Center, from their web site at http://www.nhc.noaa.gov/index.shtml

such as snow is vitally important, since a deep snow pack usually ensures adequate water supply for the spring and summer months, while liquid precipitation is important for both short-term hydroelectric generation and longer-term hydroelectric generation to the extent that the water can be stored in reservoirs.

Wind is becoming an increasingly greater supply source for electricity in both the US and Europe. Wind power is widespread in Europe, especially in Germany, since atmospheric conditions in northern and central Europe are very suitable for the use of wind as a source for power generation. In the US, the states of California and Texas currently have the most installed wind-energy capacity, but many other states across the west, Rocky Mountains, Plains and northeast also have some wind-energy capacity.

In extreme cases, when temperatures have been exceptionally warm for an extended period of time, power-generation facilities have been adversely affected. In the past, this has occurred when the water intake sources have warmed to the point that they could no longer adequately cool the power-generation units. At times when this occurs, during periods of extreme heat, power-generation operators have to curtail their operations, which reduces electricity supply for the given location. Additionally, if the cooling water intake source is a river or a lake, then prolonged drought conditions can lower the water levels to a point that the cooling source is rendered useless. These problems are primarily associated with power-generation units that use either coal or nuclear energy as a primary source of fuel.

The impact of global weather patterns on energy demand is well documented in both the preceding discussion in this chapter and in various industry publications. However, the main drivers of global and regional weather and temperature patterns are not easily understood, and it is the goal of this chapter to identify the main drivers of weather patterns across the globe and their impacts on energy supply and demand. Furthermore, it is the aim of this chapter to discuss the predictability of global weather patterns, in both the short-term and longer-term time periods.

GLOBAL CIRCULATION, CYCLONES, AND AIR MASSES

Due to the earth's shape, axis and rotation, the sun does not heat all areas of the planet equally. The tropics are heated more than the middle and higher latitudes, and the surface heating across the northern hemisphere is strongest during the spring and summer months. The earth is dependent on both ocean and air currents to balance the energy between the tropics and the higher latitudes. In the absence of these currents, the tropics would become progressively warmer, while the higher latitudes would become progressively colder.

One such current is the Gulf Stream, which is an ocean current that transports very warm water from the Gulf of Mexico and the southern North Atlantic Ocean into the northern parts of the North Atlantic Ocean. Another example is the jet streams, which are strong currents of air that circle the globe at an altitude of about 30,000 feet. There are two main jet streams in the northern hemisphere, the subtropical jet stream and the polar jet stream. The subtropical jet stream separates the tropical regions, which generally extend from the equator to about 30 degrees latitude, from the mid-latitudes, which extend from about 30 degrees latitude to about 60 degrees latitude.

The vast majority of the population in the northern hemisphere lives in the middle latitudes. The Polar jet stream separates the mid-latitudes from the polar regions. The polar jet stream is driven mainly by the north-to-south temperature contrasts in each hemisphere. Thus, this particular jet stream tends to be much stronger during the winter months when the hemispheric temperature contrasts are the strongest. The polar branch of the jet stream is highly variable and can undergo major fluctuations in both time and space. It is along this jet stream that low-pressure systems develop, which aids the northward and southward transport of energy in both hemispheres.

Cyclones are a very important part of the earth's energy transport. Cyclones, which are more commonly known as low-pressure systems, are usually accompanied by fronts and act to send warmer air from the tropical regions poleward and from the polar regions into the middle and lower latitudes. Between these areas of low pressure, there tend to be large and extensive regions of relatively high pressure. The weather associated with these areas of high pressure is highly dependent upon where the high-pressure areas originate, especially during the winter months.

Temperature and precipitation patterns such as these across the globe are vitally important for energy demand and occasionally energy supply. During the winter months, temperatures are largely determined by where air masses originate. Simply put, an air mass is defined as a large body of air in which temperature and moisture variables are relatively constant, and air masses are usually accompanied by high pressure at the surface. Air masses may extend over several hundred miles. In the winter, air masses originate over the oceans or over land, and may originate in the tropics or mid-latitudes, or over the Arctic regions. In general, maritime air masses in the winter tend to bring relatively mild temperatures to regions in the middle and higher latitudes, while continental polar and Arctic air masses tend to bring cold and dry weather patterns to the middle latitudes. Air masses in both the northern and southern hemispheres are constantly moving and are usually separated by fronts (areas of relatively low pressure). When a colder air mass is moving into a region with relatively warm temperatures, the front is called a cold front, and when warmer air is overtaking and eroding a colder air mass, the boundary between the two air masses is referred to as a warm front. In the tropics there are relatively few fronts since temperatures tend to be rather constant. However, other areas of low pressure (such as tropical storms and hurricanes) usually develop and are often much more powerful than mid-latitude cyclones. Source regions of air masses and where these air masses move are both highly dependent on the global weather patterns, and many times weather patterns in the mid-latitudes are a function of what is occurring in the tropics and in the Arctic regions.

GLOBAL AND REGIONAL WEATHER PATTERNS

As noted previously in this chapter, year-to-year fluctuations in energy demand tend to be much stronger in the winter than the summer. This is mainly due to the high degree of weather variability that is observed from year to year across North America, Europe and Asia. Comparatively speaking, the weather fluctuations in the summer months tend to be much lower, mainly since temperatures in the mid-latitudes for the June–August period tend to stay within a rather tight range. The weather and temperature patterns that impact land masses in the northern hemisphere are strongly dependent on where the air masses originate, and the origin of air masses is largely dependent on atmospheric and oceanic pressure and temperature anomalies. The US Climate Prediction Center (CPC) defines an atmospheric anomaly as the arithmetic mean between the value of a variable at a given place and the long-term average of that variable at that place and time of year. The long-term average can range in time from a 10-year average to a 50-year average, although 30-year averages tend to be the most common. Examples of anomaly-based climate indexes that impact the short-term and seasonal patterns across North America include the Pacific Decadal Oscillation (PDO), El-Niño-Southern Oscillation (ENSO), Arctic Oscillation (AO), and the North Atlantic Oscillation (NAO).

In the middle latitudes, weather patterns tend to move from west to east. In other words, there is a prevailing westerly wind direction across much of the US, Europe and Asia. In the tropics, there is a prevailing easterly wind direction, which is why many tropical cyclones that develop in the lower latitudes initially move towards the west or west-northwest, then usually curve northeastward upon reaching the mid-latitudes.

Since most of the population in the northern hemisphere lives in the middle latitudes, there has been a lot of research over the past several decades on climate variability and its impacts on temperatures across North America, Europe and Asia. Of particular note are the oceanic temperature oscillations in the Pacific Ocean and their impacts on North America. Since weather patterns across North America tend to move from west to east, there is a strong relationship between Pacific water temperature anomalies and atmospheric pressure anomalies and temperature anomalies over North America. The two main Pacific-based anomalies that impact the US are the PDO and the ENSO.

The PDO describes a water temperature anomaly pattern in the Pacific Ocean that tends to occur in 20–30-year cycles. This area of research is relatively new in the field of atmospheric science, the term Pacific Decadal Oscillation was first coined by scientists at the University of Washington in the 1990s. The two phases of the PDO are the positive (warm) phase and the negative (cold) phase. During the positive PDO phase, above-average water temperatures tend to persist in the Gulf of Alaska, along the North American West Coast, and across the tropical Pacific Ocean, while below-average water temperatures tend to develop over the north-central and northwest Pacific Ocean. During the negative PDO phase, the opposite anomalies are typically observed. There has been an observed relationship between these PDO phases and the seasonal weather patterns over the northern hemisphere, especially North America since this land mass is in the closest proximity "downwind" of the Pacific Ocean.

Since the late 1940s, the PDO has been in two distinct phases. The PDO was mainly in its positive phase from the late 1940s to the late 1970s, and has been in its positive phase since the late 1970s. Within both of these 20–30-year phases, there were short-term reversals in the phase of the PDO, which are not uncommon. The primary temperature and precipitation impacts associated with both PDO phases are primarily observed in the middle and higher latitudes of North America, and the relationship tends to be stronger in the winter months than in the summer. During the northern hemisphere winter (December–February), there is a fairly strong relationship between the prevailing PDO phase and temperatures in eastern

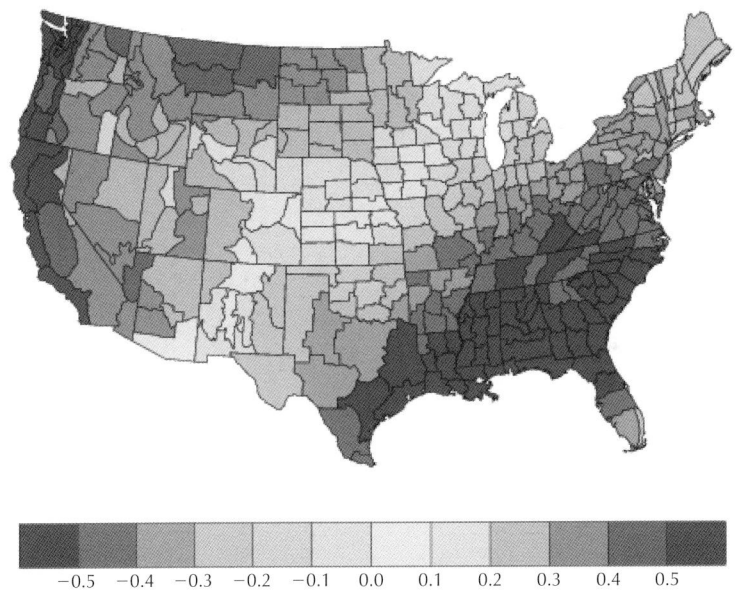

Figure 4 Correlation between the mean December–February PDO phase and mean December–February temperature anomalies across the United States. Dark grey shading shows where wintertime temperatures tend to be colder than average during the positive PDO phase. The strongest relationship between the positive PDO phase and cool anomalies is in the Southeast. The shading along the West Coast shows where temperatures tend to be above average during the positive PDO phase

Note: Correlation temperature Dec to Feb with Dec to Feb PDO 1949–04.
Source: NOAA-CIRES Climate Diagnostics Center, Boulder, Colorado, from their Web site at http://www.cdc.noaa.gov/

North America, with the positive phase often leading to below-normal temperatures in the east and the negative phase often leading to above-normal temperatures. There is also a moderately strong relationship between the phase of the PDO and precipitation patterns across the US, especially in the Pacific Northwest and along the East Coast. The positive phase of the PDO often leads to below-average precipitation across the northwest. This affects energy supply, if precipitation in the northwest is sufficiently weak during the winter months, then hydroelectric generation for the winter and the following spring/summer will be curtailed from British Columbia southward to northern California.

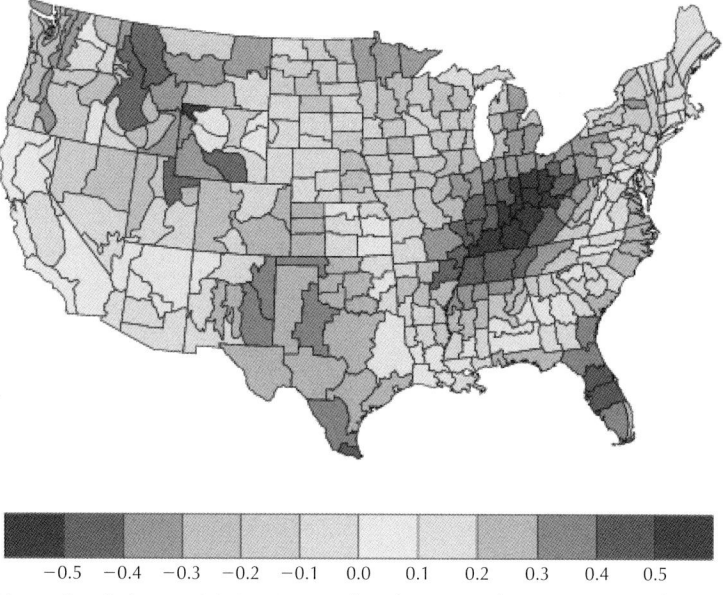

Figure 5 Correlation between the mean December–February PDO phase and mean December–February precipitation anomalies across the United States. Dark grey shading shows where wintertime precipitation tends to be drier than average during the positive PDO phase, while lighter grey shading shows where wintertime precipitation tends to be above average

Note: Correlation precipitaion Dec to Feb with Dec to Feb PDO 1949–04. The two main regions where precipitation tends to be below average during the positive PDO phase are the midwest and the northwestern US. The medium shading in Texas and Florida indicates a correlation to above average precipitate.

Source: NOAA-CIRES Climate Diagnostics Center, Boulder, Colorado, from their Web site at http://www.cdc.noaa.gov/

The PDO is a phenomenon that impacts the entire Pacific Ocean basin, and in turn impacts seasonal temperatures over much of North America. Closely related to the PDO, although on a smaller temporal and spatial scale, is the ENSO cycle, which develops in the equatorial Pacific Ocean and impacts the US, Mexico and parts of South America, primarily during the December–April period.

The term "El Niño" is used to describe above-average water temperatures across the central and eastern part of the tropical Pacific Ocean, while the term "La Niña" describes below-average water temperatures over the same geographic regions. El Niño and La Niña

Figure 6 Pacific Ocean water temperature anomalies for January–March 1998 and January–March 1989. The water temperature anomalies for the January–March 1998 period show the exceptionally strong El Niño cycle that was occurring at that time, while the anomalies for January–March 1989 show the strong La Niña cycle that was occurring. El Niño and La Niña cycles usually reach their peak intensity during the Northern Hemisphere winter

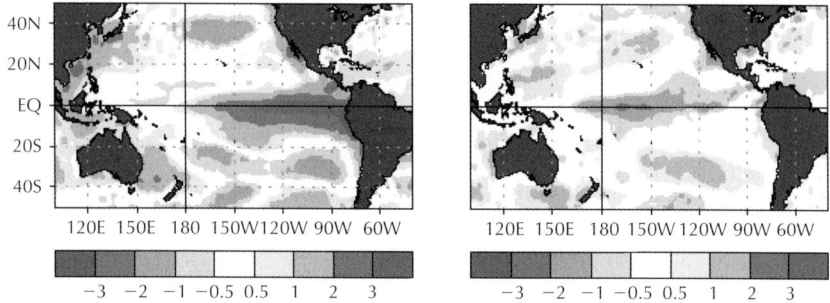

Note: Ocean temperature departures (°C). The shading along the equator in the left panel shows where water temperatures were above average, while the shading in the right-hand panel shows where water temperatures were below average.

Source: United States Climate Prediction Center, from their web site at http://www.cpc.ncep.noaa.gov/products/analysis_monitoring/ensocycle/ensocycle.html

cycles tend to last for 6–18 months. These cycles are opposite extremes of the ENSO cycle. A typical El Niño or La Niña cycle begins in the Northern Hemisphere spring, reaches its peak during the following winter, then weakens about three to six months later. However, there have been occurrences of El Niño and La Niña that have lasted for several years. In general, El Niño cycles tend to occur at irregular intervals of two to seven years. In the past 50–60 years, however, El Niño cycles have occurred more frequently during the positive phases of the PDO (beginning in the late 1970s and continuing into the early 2000s), while La Niña cycles have occurred more frequently during the negative phases of the PDO (1950s, 1960s and 1970s).

The two main variables that are used to measure the phase and strength of the ENSO cycle are ocean water temperature anomalies along the equator and air pressure anomalies between the western tropical Pacific Ocean and the eastern part of the basin. The fluctuation in air pressure anomalies between the western part of the basin and the eastern part is called the Southern Oscillation, and the Southern Oscillation Index can be used to gauge the strength of a developing El Niño or La Niña. During El Niño cycles, the Southern

Oscillation is in its negative phase, which means that surface air pressures are abnormally high near Indonesia and abnormally low near the coast of South America. During the positive Southern Oscillation phases, the pressure anomalies are reversed and La Niña conditions usually occur. The Southern Oscillation Index (SOI) measures the large-scale fluctuations in the air pressure over the tropical Pacific Ocean. The phase of the SOI can vary from week to week, but prolonged periods of the negative SOI phase usually correspond to El Niño conditions, while prolonged periods of the positive phase correspond to La Niña cycles.

During an El Niño cycle, the normal circulation patterns over the eastern tropical Pacific Ocean change, which leads to changes in the subtropical branch of the jet stream over the US. Above-average water temperatures and lower-than-normal surface air pressures lead to enhanced rainfall over the eastern part of the tropical Pacific basin, while dry conditions often develop over the western part of the basin (sometimes leading to droughts in Australia and Indonesia). The increase in thunderstorm activity over the tropical Pacific Ocean associated with El Niño conditions invigorates the southern branch of the jet stream over the eastern part of the basin and across the southern US, especially during the winter and spring months. The stronger southern branch of the jet stream often leads to much stronger winter storms over areas of the US that are not used to strong storms in the winter, including California and the southwest.

Along the Gulf Coast and Texas, El Niño cycles usually result in above-average winter precipitation, which often leads to flooding problems. Severe and unusual weather during moderate to strong El Niño cycles is not limited to North America. Impacts of El Niño are often observed in South America and Southeast Asia, mainly in the form of comparatively warm and dry conditions. Warm and dry conditions in Brazil can impact Brazil's hydro-based electricity sector, and dry conditions in Southeast Asia primarily impact agriculture interests. Global impacts of El Niño generally increase as the cycle itself strengthens, and impacts tend to be greater during the northern hemisphere winter season.

El Niño impacts the weather across North America during the winter months in two ways. First, as discussed above, El Niño cycles tend to strengthen the southern branch of the jet stream, which leads to stronger storms and above-average precipitation across the south

Figure 7 Global December–February weather patterns that are usually observed during ENSO warm episodes (ie El Niño cycles)

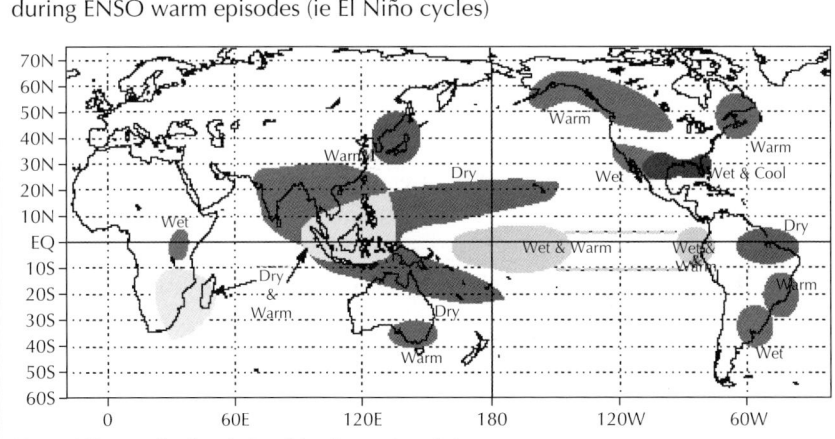

Note: Warm episode relationships December–February.
Source: United States Climate Prediction Center, from their web site at. http://www.cpc.ncep.noaa.gov/products/analysis_monitoring/ensocycle/ensocycle.html

and along the East Coast. Second, El Niño cycles often lead to a stronger west-to-east flow across North America during the winter months. When this occurs, the North American continent tends to be flooded with relatively mild Pacific maritime air, which leads to above-average temperatures across the northern tier of the US and southern Canada. During La Niña cycles, the impacts on the US during the winter months tend to be less severe. La Niña cycles, if sufficiently strong, often lead to mild and dry conditions across the south, above-average precipitation in the northwest, and below-average temperatures over western Canada. In general, El Niño cycles have proved to cause much more severe weather episodes in the US, especially across California, the southwest, and the Deep South.

The relationship between the Pacific Ocean and the seasonal weather patterns across North America has been well documented in the scientific community. In addition to the PDO and the ENSO, there are two other climate indexes that are worth discussing since they both have profound impacts on seasonal temperatures across North America and Europe. These indexes are the Arctic Oscillation (AO) and the North Atlantic Oscillation (NAO). The AO tends to impact temperatures across the northern hemisphere, especially in the winter, while the NAO mainly impacts the eastern US and Western Europe.

The AO is defined by the United States Climate Prediction Center as a measure of the difference between the mean sea level pressure anomaly over the Polar basin and the corresponding average anomaly in a ring surrounding the Polar basin at middle latitudes. The phase of the AO is considered to be positive when pressures are below average across the Arctic Circle and above average over the surrounding mid-latitude region. The AO is considered to be negative when the opposite pressure anomalies occur. Similar to the PDO and the ENSO cycles, the impacts of the AO are the strongest during the winter season and relatively weak during the summer months.

The negative phase of the AO usually indicates the presence of high-latitude blocking patterns over the northern hemisphere. High-latitude blocking patterns often lead to below-average temperatures in the mid-latitudes for two reasons. First, high pressure, both aloft (in the upper atmosphere) and at the surface, promotes the development of strong Arctic air masses in the far northern latitudes. Second, high-latitude blocking patterns result in a southward displacement of the Polar branch of the jet stream, which allows strong Arctic air masses to move southward from the Arctic Circle into the major population centres of North America, Europe and Asia. During the positive phase of the AO, there is usually a belt of high pressure in the mid-latitudes and a deep vortex in the vicinity of the Arctic Circle. This pattern usually keeps all of the cold arctic air locked up near the North Pole and prevents it from breaking southward.

The AO affects temperatures across the Northern Hemisphere. Closely related to the AO, but not exactly the same, is the NAO, which mainly impacts eastern North America and Western Europe. The phase of the NAO is considered positive when surface pressures are below average in the vicinity of Greenland and above normal throughout the central North Atlantic Ocean. The NAO is considered negative when high-pressure anomalies exist in the vicinity of Greenland and low-pressure anomalies extend from the East Coast of the US eastward into Western Europe. This type of pattern is often referred to as a "Greenland block", indicating a blocking high-pressure system over the far North Atlantic Ocean. During the negative phase of the NAO, temperatures are normally below normal in the eastern US and Western Europe, since the Polar jet stream pattern is forced to buckle southward across these two regions. Additionally, slow-moving storm systems over the

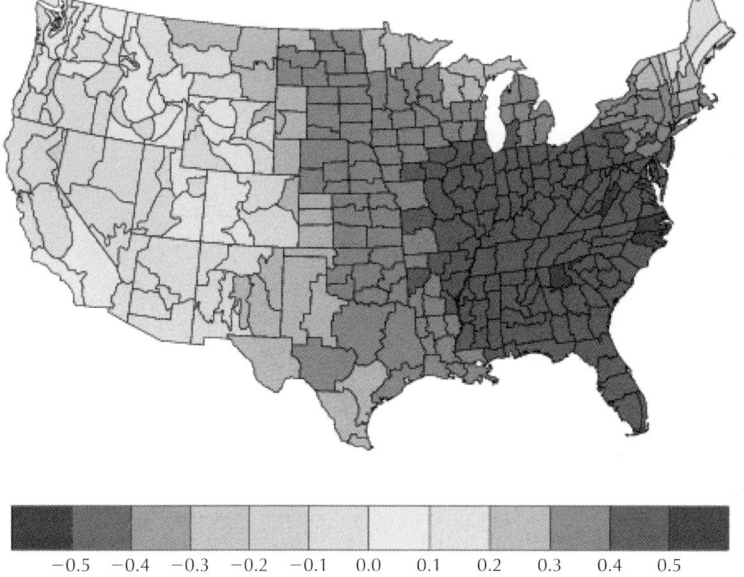

Figure 8 Correlation between the mean November–March Arctic Oscillation (AO) phase and mean November–March temperatures across the United States. A mean positive AO phase is strongly correlated with above average temperatures in the East, and vice versa

Note: Correlation temperature Nov to Mar with Nov to Mar AO 1958–01.
Source: NOAA-CIRES Climate Diagnostics Center, Boulder, Colorado, from their Web site at http://www.cdc.noaa.gov/

eastern US often lead to above-average precipitation along the East Coast. In contrast, the Atlantic storm track over Scandinavia is effectively blocked, which leads to below-average precipitation across the key hydroelectric-producing regions of Norway and Sweden. The duration of AO/NAO cycles varies. If the magnitude of each specific cycle is sufficiently strong (on the order of two or more standard deviations from average), then the cycle can last several weeks. On the other hand, weak cycles (less than one standard deviation from average) usually last for only a matter of days before switching back to neutral or the opposite phase.

Understanding the PDO, ENSO, AO and NAO is particularly important as they have profound impacts on temperatures across the northern hemisphere, and thus energy demand. Strong warm phases of the ENSO cycle usually bring very mild wintertime

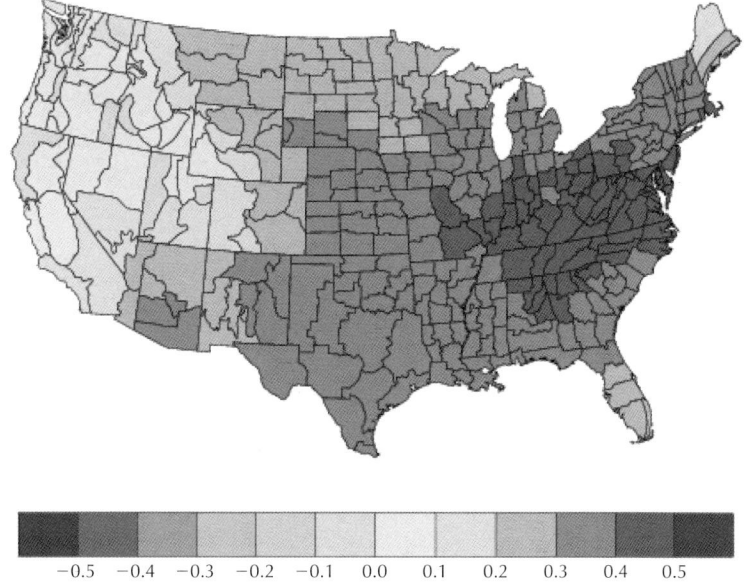

Figure 9 Correlation between the mean December–February North Atlantic Oscillation (NAO) phase and mean December–February temperatures across the United States. A mean positive NAO phase is strongly correlated with above average temperatures in the East, and vice versa. The correlation between the NAO and eastern US temperatures is the strongest during the core winter months of December–February

Note: Correlation temperature Dec to Feb with Dec to Feb NAO 1949–04.
Source: NOAA-CIRES Climate Diagnostics Center, Boulder, Colorado, from their Web site at http://www.cdc.noaa.gov/

temperatures to North America, leading to lower heating demand in the major metropolitan areas of the Midwest and east. During the exceptionally strong 1997–8 El Niño, for example, temperatures across the US for the November–March period averaged 7% above normal. It is estimated that this deviation from normal led to a reduction in residential/commercial natural gas demand of nearly 300 Bcf compared with a normal (30-year average) winter. In winters when there is neither a strong El Niño nor strong La Niña, temperatures across North America are largely influenced by the phase and the strength of the Arctic Oscillation. For example, neither a strong La Niña nor El Niño developed in the tropical Pacific Ocean during

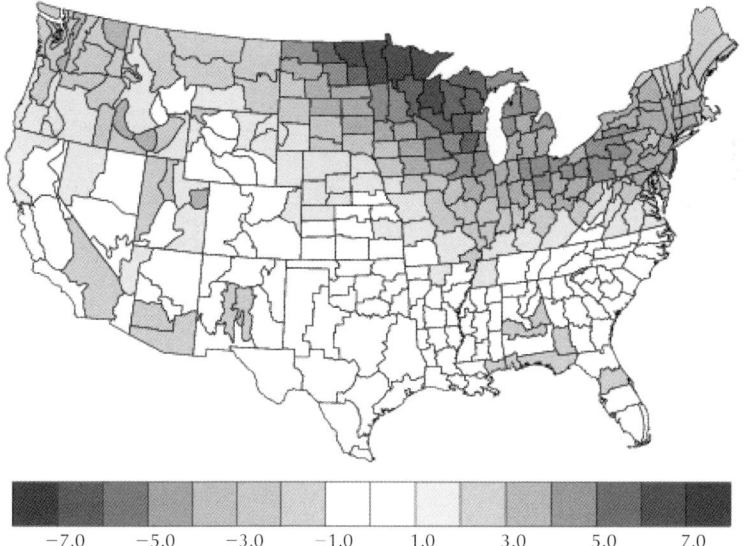

Figure 10 Temperature departures from average during the 1997–1998 winter season. Strong El Niño conditions were occurring in the tropical Pacific Ocean during the 1997–1998 winter, and temperatures were well above normal across the northern United States. This temperature pattern is a class El Niño pattern for the United States

Note: Temperature anomalies (F) Nov to Mar 1997–98 *versus* 1971–2000 longterm average.

Source: NOAA-CIRES Climate Diagnostics Center, Boulder, Colorado, from their Web site at http://www.cdc.noaa.gov/

the 2000–1 winter season, but the Arctic Oscillation was strongly negative for much of the November–March period. The 2000–1 winter season was about 8% colder than normal for the US, which likely led to an increase in residential/commercial gas demand of 350–400 Bcf compared with a normal winter. In contrast, the following winter season was a strongly positive AO winter, and temperatures for the November–March period averaged about 12% warmer than average. It is estimated that residential/commercial gas demand for the US dropped by nearly 1,000 Bcf in the 2001–2 winter compared with the same period in 2000–1. This radical change in the weather was mainly due to the severe shift in the phase of the AO from the 2000–1 winter season (strongly negative) to the 2001–2 season (strongly positive).

ENERGY MODELLING

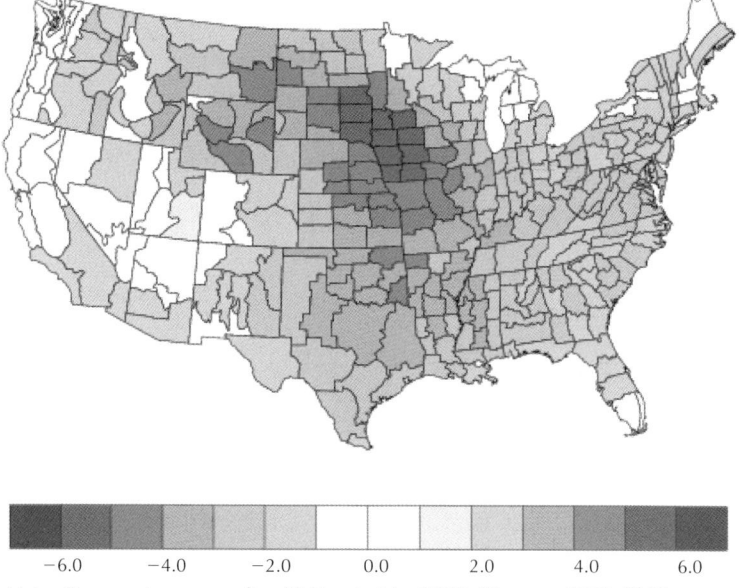

Figure 11 Temperature departures from average during the 2000–01 winter season. Pacific influences (e.g. El Niño and La Niña) were relatively weak during the 2000–01 winter, but a strongly negative AO contributed to widespread cold anomalies for the November–March 2000–01 period

−6.0   −4.0   −2.0   0.0   2.0   4.0   6.0

Note: Temperature anomalies (F) Nov to Mar 2000–01 *versus* 1971–2000 longterm average.

Source: NOAA-CIRES Climate Diagnostics Center, Boulder, Colorado, from their Web site at http://www.cdc.noaa.gov/

Long-term predictability (weeks to months) of the key climate variables ranges from somewhat predictable for the PDO and ENSO cycles to very difficult for the AO and the NAO. Anomaly patterns associated with the PDO and ENSO usually persist for several months and, therefore, are more stable than the mainly atmospheric-based variables such as the AO and the NAO. ENSO patterns tend to be somewhat predictable, since they follow a rather stable pattern of strengthening in the summer and autumn, reaching their peak intensity in the winter, and usually weakening thereafter. Thus, during the autumn, researchers and forecasters have a general idea of what to expect in the tropical Pacific Ocean during the winter season. The ENSO usually does not have much

Figure 12 Temperature departures from average during the 2001–02 winter season. Pacific influences (e.g. El Niño and La Niña) were relatively weak during the 2000–01 winter, but a strongly positive AO contributed to widespread warm anomalies for the November–March 2000-01 period, especially in the Midwest and East

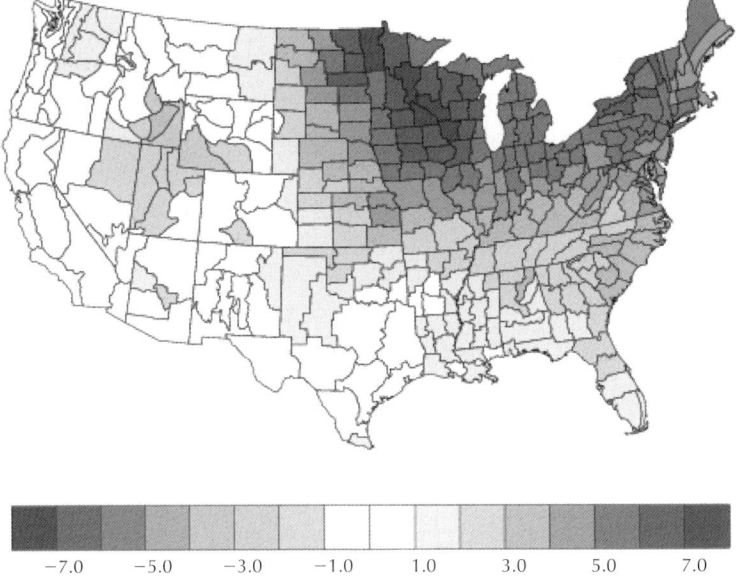

Note: Temperature anomalies (F) Nov to Mar 2001–02 *versus* 1971–2000 longterm average.

Source: NOAA-CIRES Climate Diagnostics Center, Boulder, Colorado, from their Web site at http://www.cdc.noaa.gov/

of an impact on summertime temperatures since the tropical Pacific Ocean is usually in a state of flux during the spring and summer months. The PDO has been shown to be fairly stable over the course of several years, although short-term fluctuations in the mean phase and strength can sometimes occur. Even with these fluctuations, the PDO does not change radically from week to week, so prediction of the PDO phase is often possible going forward several weeks or months.

The atmospheric-based variables such as the AO and NAO tend to be much more chaotic over the course of a season. Therefore, prediction of these variables tends to be very difficult going forward more than a week to ten days, and forecasts are based more on forecast models than anything else. Thus, there are many limitations in predicting the AO and NAO, and accurately forecasting these variables on a seasonal basis is difficult at best. It is hoped that, as research into the field of meteorology advances in the coming years and decades, predictability of the AO and NAO will also improve. As predictability of the AO and NAO both improve, seasonal forecasting (forecasting for several weeks or months into the future) will also dramatically improve. The Pacific-based variables (PDO and ENSO) are somewhat stable and the impacts of ENSO and PDO phases are relatively well understood, so improvement of the predictability of the AO and NAO will improve the overall skill level of seasonal forecasts. Until that time, however, seasonal forecasting will continue to exhibit relatively low skill and should be used mainly as a probabilistic tool.

SHORT-TERM PREDICTABILITY OF GLOBAL WEATHER PATTERNS

Seasonal and long-term forecasting both have their limitations since most of these forecasts rely on accurately predicting climate indexes and understanding their impacts on weather patterns in the middle latitudes. At the very best, seasonal forecasts assign a probability that a particular weather pattern will verify in the weeks and months to follow. Seasonal forecasts can rarely quantify temperature and precipitation departures from average for specific locations for long periods of time. Short-term forecasting, on the other hand, tends to be much more accurate since there are a number of short-range and medium-range forecasting models available to forecasters and the general public alike. Weather-forecast models have become more sophisticated and more accurate with the advancement of computer technology in the past 10–20 years. Weather-forecast models are compiled using a vast array of data points and physical equations to generate forecasts across virtually the entire earth, so computer processing speed has become a key component in generating timely forecasts using numerical weather models.

Predicting short-term weather patterns (weather that is occurring within three days) has become much more accurate since the late 1970s and 1980s. Improvement in the accuracy of short-term forecasts has occurred in conjunction with improvement in both satellite and radar technology, which enables forecasters to have a better sense of what is occurring "upstream" of any given location. Satellite technology has also vastly improved forecasts over the oceans and in the tropics, where accurate weather observations used to be rather sparse. This is especially true in tropical cyclone forecasting. The use of satellites now enables forecasters to track tropical cyclones for several days or weeks, and has improved the tropical storm forecasting system in the US, Caribbean Sea and Southeast Asia.

Short-term and medium-term forecast models (models that go out in time 3–15 days) are mainly compiled by government agencies, and the model output is generally available to forecasters, and often to the general public. In the US, the short-term and medium-term forecast models are compiled by the National Centers for Environmental Prediction (NCEP). Other agencies that run short-term and medium-term forecast models are Environment Canada, the European Centre for Medium Range Weather Forecasting (ECMWF) and the UK Met Office.

The major short-term models that are analysed for weather prediction are the North American Mesoscale model (NAM), the Nested Grid model (NGM) and the Global Forecast System (GFS). The NAM and NGM are both short-term regional models that predict weather patterns 84 and 48 hours into the future, respectively. The GFS is a longer-range model that predicts global weather patterns 384 hours into the future. This model is also used in short-term forecasting since it provides coverage of the same time periods as the NAM and NGM. The other longer-term forecast models, the Canadian and the ECMWF, produce forecasts out to 240 hours (10 days), and the UKMET model produces forecasts out to 144 hours. Since the GFS, Canadian, ECMWF and UKMET models are global forecast models, meaning they produce forecasts for all data points around the globe, they can be used to predict weather patterns for North America, Europe and Asia, regardless of which agency runs the model. Within the energy trading and marketing sector, however, the terms "American model", "European model" and "Canadian model" are commonly used to differentiate between the individual long-range forecast models. These terms usually refer to which country or agency runs the model and not the individual forecast regions, since all of these models are global forecast models. For example, the American model usually refers to the longer-range GFS model, although the model forecasts atmospheric variables for the entire globe.

Numerical weather models are compiled using a plethora of data points from around the earth. In terms of input, models use a number of weather variables taken from several thousand data points on the earth and taken from several layers of the atmosphere. Model inputs include, but are not limited to, surface temperature and pressure (gathered by the US National Weather Service, ship reports and government agencies through the world), upper air temperature and pressure (gathered by the use of weather balloons, aircraft, etc) and winds at the surface and aloft. These data are then used as input variables for several physical equations that govern how fluids and air behave in time and space, and the model output shows the predicted output in a simplified and idealised atmospheric flow pattern. In general, forecast models tend to be more accurate if they are compiled using a higher spatial resolution. However, there are limitations to how many data can be obtained, and since more data input usually leads to longer compilation (that is model output generation) times, it is important to balance the need for more data input with the need to produce forecast model data in a timely manner.

Similar to the model input, the forecast model output includes predictions of several weather variables over the surface of the earth and aloft. The raw data are then used to generate forecast maps for different regions of the earth. Among the variables that the models predict and that can easily be calculated with the raw data are: surface temperatures, surface pressure, surface winds, surface dew point, precipitation, precipitation type, cloud cover, temperature departures from normal and a number of upper-atmospheric variables. The major users of the model output are the national weather-forecasting agencies, university and private researchers, and several airline/aviation interests.

The NAM and GFS numerical weather forecast models that are generated by NCEP are generated every six hours at 00.00 GMT, 06.00 GMT, 12.00 GMT and 18.00 GMT. The NGM is run only two times per day, at 00.00 and 12.00 GMT. The Canadian, ECMWF, and UKMET models are also run daily at 00.00 and 12.00 GMT. It then takes hours for the supercomputers to run these models, compile the data and disseminate the data to government agencies and the general public.

As with any forecast model or methodology, accuracy is higher for shorter time periods and declines considerably with each day of the forecast period. Temperature and precipitation forecasts are generally accurate to about three days since there is less uncertainty in the weather forecast variables and nearly every regional and global forecast model covers the time period, so comparisons can be made between the models. After the third day, forecast model skill begins to decline considerably, and beyond about 192 hours (8 days), the forecast accuracy is considerably lower since small errors in the forecast models early in the period tend to create much larger errors later in the forecast period.

The three major models that produce forecasts beyond 144 hours go through cycles in which their accuracy is higher than the other models, but it is difficult to pinpoint one certain model that is constantly superior to all the others. Thus, it is important for forecasters to constantly use all available model output and not to rely heavily on a certain model or model run when making a forecast. Another important forecast tool is the medium-term weather forecast model ensembles. An ensemble is a set of forecast models (of the same model, meaning the GFS, ECMWF, Canadian, etc) that are all valid at the same time but whose starting conditions differ by small amounts. The GFS ensemble, for example, has 12 members per model run, and the average of all the ensembles is called the ensemble mean. Using the model ensemble forecast mean provides forecasters with a general idea of how a forecast weather pattern is trending in the medium term, and it can sometimes prevent forecast errors that are the result of an erroneous forecast model run (for instance, bad data or initial conditions). The model ensemble means are usually available one to two hours after each model's operation run.

In the US, all model output data are made available to all agencies within the National Oceanic and Atmospheric Administration (NOAA), including the National Weather Service, National Hurricane Center and the Climate Prediction Center (CPC). All model data are also available to the general public and to the media, although it is likely that only a small percentage of the public looks at the raw model data. Since the late 1990s, many weather products have been made available on the internet, which has drastically improved the sharing of model data with the general public and nongovernment weather-forecasting entities. The advent of the internet has also made it much easier to obtain historical weather data for specific locations throughout the US, as many NWS offices make data for their local areas available on their Web site. This is not the case in many other countries, unfortunately. Many European countries charge a fee for weather data and do not make their forecast products readily available for the global population. Therefore, obtaining historical weather and forecast data from some countries outside of the US can be an expensive process.

There are a number of different forecasting agencies and companies that use the raw model output to generate forecasts. Many of these forecasting entities have different goals, which impacts how and why they make their forecasts. In the US, for example, the stated mission of the National Weather Service is as follows:

> The National Weather Service (NWS) provides weather, hydrologic, and climate forecasts and warnings for the United States, its territories, adjacent waters and ocean areas, for the protection of life and property and the enhancement of the national economy. NWS data products form a national information database and infrastructure which can be used by other governmental agencies, the private sector, the public, and the global community.

Thus, it is the goal of the NWS to mainly protect life and property, although other private entities such as private forecasting companies, the media and commodity marketing and trading operations often use their data. It is primarily for the last two entities that private forecasting companies have become established in both North America and Europe. As stated above, the US NWS has a stated purpose and most of its forecast products are created to meet its goals. Private forecast companies, on the other hand, tailor their forecast products to meet the needs of their clients. For example, if a television station or publication is a client of a private forecasting company, that forecasting company may tailor its forecasts and products to a very specific region of the country, and may make its graphics simple to read and aesthetically pleasing. On the other hand, if it is an energy-marketing or -trading operation that is the end-user of a private forecasting company's product, it may tailor its forecast product to show forecast temperatures and forecast temperature extremes over the key population centres of a country or an entire continent. There are literally hundreds of forecast products available to weather-sensitive entities in North America,

Europe, and Asia, and there seem to be just as many private forecasters and consultants available to meet these demands. Government agencies, on the other hand, tend to be very rigid with the forecast products that they offer, although their forecast products tend have a very high quality to them.

WEATHER PATTERNS, FORECASTS, AND THE ENERGY COMMODITY MARKETS

Since global weather patterns have a profound impact on energy demand, weather patterns and forecasts have a major impact on energy commodity prices worldwide. Commodities such as same-day power and next-day gas are impacted mostly by real-time weather and temperature patterns, while futures prices tend to be more sensitive to weather forecasts. There are several weather-forecasting models and services across North America and Europe, and the energy commodities markets have evolved to the point where trading and hedging decisions are made daily based on short-term and long-term weather forecasts.

The energy commodity markets react to both long-term seasonal forecasts and to medium-term forecasts that are issued daily, and in some cases, more than once per day. There are two key times of the year when energy markets tend to react to seasonal forecasts. Seasonal forecasts for the summer months are usually released in April and May, and most of these forecasts cover the five-month period from May to September. Included in these forecasts are usually temperature departures from average for the summer months, forecast precipitation patterns for the summer months and an indication of how much activity there will be in the tropics during the hurricane season, which lasts from June to November in the Atlantic Ocean basin.

The other time of the year when energy commodities markets react to seasonal forecasts is the period from early September to about the end of October, when public and private forecasters alike disseminate their seasonal forecasts for the following heating season (November–March). These forecasts mainly concentrate on temperature departures from average for the key population centres of North America and Europe. Expectations of widespread colder-than-normal temperatures usually provide strength to the energy futures markets, especially to natural gas and heating oil markets. Likewise, expectations for above-normal summertime temperatures support higher long-term power prices.

During the spring and autumn (the "shoulder months"), when energy demand falls considerably, it is usually the long-term forecasts and/or the potential for hurricanes (in the autumn months) that drive large price movements in the natural gas and power markets to the extent that the market moves based on weather. As the cooling and heating seasons progress, however, less emphasis is put on long-term seasonal forecasts and more of an emphasis is put on to short-term and medium-term weather forecasts. This is especially true early in each season, when large temperature deviations from normal can have a profound impact on end-season natural gas storage levels, or stoke fears that widespread anomalous warm or cold conditions will persist for the majority of the season. Market participants usually analyse the forecast models and forecasts for the medium-range period (1–15-day time frame) for an indication of whether or not the seasonal forecasts are coming to fruition. If the shorter-term forecasts support the seasonal forecasts, then the price action that preceded the season may continue well into the season. For example, if the seasonal forecasts indicated widespread cold anomalies for the heating season and the heating season begins colder than normal across the major population centres of North America or Europe, energy commodities' price strength will likely persist well into the winter season.

On the other hand, if early-winter temperature forecasts are contradictory to cold heating season forecasts, energy commodity price support may erode rapidly early in the heating season, especially if there is a sufficient natural gas or oil storage "cushion". By the final weeks of the cooling and heating seasons, short-term and medium-term weather forecasts tend to lose their importance in the natural gas and petroleum markets if storage levels are sufficiently high, and it is widely perceived that there will be enough end-season gas in storage to accommodate the most extreme weather patterns possible. However, this is not always the case with the electricity markets since late-winter cold spells and late-summer warm spells can still lead to strong electricity demand, and this demand still needs to be met with real-time electricity supply. In other words, electricity cannot be "stored" the way natural gas and petroleum products are stored, so electricity markets still exhibit strong weather-related volatility through the end of the heating and cooling seasons. Power prices tend to be less volatile during the off-season months of April, May, September and October, unless there are supply problems such as generation unit maintenance or unit failure.

CONCLUDING REMARKS

The weather patterns across the world are ever changing and very chaotic. Chaos in the atmosphere stems from the fact that the earth is tilted 23.5 degrees on its axis, which results in unequal heating from north to south. Chaotic patterns are also aided by the fact that the earth is constantly spinning, so surface heating for even one specific location is unequal throughout the course of a day. Thus, the earth is constantly trying to reach an energy equilibrium, which can never occur since the earth is always moving and different areas of the earth are being heated at different rates.

It is for aforementioned reasons that we have different weather from day to day, week to week and season to season. Large air masses, fronts, and cyclones are in constant motion across the earth, as are the ocean currents. The chaotic nature of the earth's atmosphere and oceans is a major reason that the accurate prediction of long-term weather patterns remains elusive. There are simply too many variables over the entire earth and throughout every layer of the atmosphere to be able to measure, quantify and model. Science and technology continue to advance, and the prediction of weather patterns several days into the future has improved considerably in the last one to two decades alone. Successful modelling of the atmosphere and the major forecast variables (temperature and precipitation) will no doubt improve over the next several decades as more capital is invested in atmospheric research and on the supercomputers that generate the weather-forecast models, resulting in major advances in the science of meteorology. Until that time, however, large errors in weather forecasts beyond only a few days will likely continue, and long-term forecasting will remain mostly a probabilistic tool for predicting mean weather patterns for a given season.

Despite the shortcomings of weather forecasting, the energy commodities markets still rely heavily on weather forecasts, both in the short term and the long term. Energy commodity traders and analysts are constantly evaluating weather data and forecasts, and timeliness of weather updates is a key component in being able to take advantage of weather-driven market fluctuations. It is a well-known fact in the energy and agriculture industries that supply and demand for their respective commodities is strongly related to the weather, so traders and analysts alike will continue to use weather forecasts as an input variable into their demand, supply and pricing models. The natural uncertainty associated with weather forecasting means that weather forecasts will constantly be changing in both the short term and long term, and, as a result, commodity futures that trade based on weather forecasts will constantly be undergoing considerable weather-driven and forecast-driven price movements.

> **PANEL 1 USEFUL RESOURCES OF WEATHER INFORMATION**
>
> The United States Climate Diagnostics Center has a Web site available for analysing past temperature anomalies by month or season. This site also enables the user to correlate climate indexes to seasonal temperature and precipitation anomalies for the United States. http://www.cdc.noaa.gov/USclimate/USclimdivs.html.
>
> The US National Hurricane Center has a Web site that includes all tropical cyclone advisories for the North Atlantic Ocean and the eastern tropical Pacific basin. This site also includes historical tropical cyclone data as well as several tropical cyclone educational tools. http://www.nhc.noaa.gov/index.shtml.
>
> The US Energy Information Administration maintains a Web site for all supply and demand statistics for energy products. This site also includes several reports that the EIA publishes. http://www.eia.doe.gov/.
>
> The US Climate Prediction Center has a detailed explanation of the ENSO cycle on its Web site. On this site, it also includes current Pacific water temperatures, water temperature anomalies and its latest ENSO discussions and forecasts. http://www.cpc.ncep.noaa.gov/products/ analysis_monitoring/lanina/.
>
> The Climate Prediction Center has a Web site that explains all of the weather and climate terms that are commonly used in its discussions and forecasts. http://www.cpc.ncep.noaa.gov/products/predictions/610day/glossary.html.
>
> The National Centres for Environmental Prediction (NCEP) maintain a Web site for GFS and NAM model output. http://www.nco.ncep.noaa.gov/pmb/nwprod/analysis/.

The US National Weather Service homepage, with links to every local NWS forecast office in the US, is at http://www.nws.noaa.gov/.

The US National Climatic Data Center homepage: http://www.ncdc.noaa.gov/oa/ncdc.html.

The US National Weather Service Aviation Weather Center maintains a Web site that includes satellite pictures, radar imagery, and upper-air weather data. http://adds.aviationweather.gov/.

Information from the National Climatic Data Center on such climate indexes as the SOI, PDO, NAO, and AO: http://lwf.ncdc.noaa.gov/oa/climate/research/teleconnect/teleconnect.html.

US Regional Climate Centres. http://www.wrcc.dri.edu/rcc.html.

The US Climate Division Center maintains a map room with more weather forecast model output. http://www.cdc.noaa.gov/map/.

10

Full-Requirement Contracts

Yan Gao; Harald Ullrich; Krzysztof Wolyniec

Progress Energy; Constellation Energy Commodities Group; Sempra Commodities

Some of the most popular contracts in energy markets are full-requirement supply contracts. Under the provisions of full-requirement contracts, the supplier is obligated to meet all the supply requirements of the buyer. Full-requirement deals involving gas are relatively straightforward; but in the case of electricity the contracts involve delivery of several related commodities/services. Those services are commodities in their own right with their own markets. They arise chiefly because of the need for the real-time balancing of supply and demand. They are called *ancillary services* and include operational reserves, forward reserves, schedule 2&3, AGC (automated generation control), NITS (network integration transmission service), regulation up/down and blackstart capability. Additionally, in some pool markets (NEPOOL, NYISO, PJM), the provision of energy involves satisfying the regulatory, environmental and long-term security of supply requirements, such as: RPS (renewable portfolio standard), ICAP/UCAP (installed capacity markets), or REC (renewable energy credits).

In this chapter, we will look closely at the pricing, hedging and risk management of energy. However, the topic itself is vast and we cannot hope to do justice to its complexity given the scope of this text. Instead we will also refer the reader to other sources for additional analysis. Before we proceed we need to note that the full-requirement contracts, by and large, do not involve any optionality on the part of the contracting parties. More specifically, the buyer of the full-requirement service does not have the option to choose the source of supply. This is of utmost importance in pricing and managing those types of contract.

LOAD AUCTIONS AND OTHER WAYS OF PROCUREMENT: REGIONAL DIFFERENCES

The full-requirement contracts are very diverse geographically. The variability is partially driven by the structure of the markets and partially by the nature of available supply. What follows is a description of the main regional methods of procurement, including drivers and motivations.

Northeast (NEISO, NYISO)

The wholesale market is dominated by the local utilities procuring their needs under the default retail programme. The default/standard service is the power service given to all those who do not choose a competitive supplier. The presence of competitive retailers/aggregators is quite limited. Consequently, the majority of energy needs for end-users is supplied through the default/standard offer programmes. The local utilities source power through repeated blind-price auctions, whereby the supply is procured for a term between three months and a year. The contracts are fixed price arrangements: where the bidder specifies a sequence of monthly prices.

Mid-Atlantic (PJM)

The market sees a significant presence of competitive suppliers. Default service programmes are still present; however, customers are relatively active in switching between the competitive suppliers and default programmes. As in the Northeast, the utilities contract-in the wholesale market for supply of the default programmes through load auctions. The supply agreements have tenors ranging from a couple of months to several years.

Texas (ERCOT)

When the demand side of the ERCOT market was deregulated in 1999, all load-serving entities except municipalities and co-ops — that were given the right to opt into competition at any time — had to open their service territories to retail competition. The formerly exclusive suppliers became automatically the default supplier for their former customers under a special tariff, the so-called "price to beat" (PTB). While the tariff was designed to foster customer switching to competitive suppliers, only roughly 10–20% of residential customers had switched to another provider market-wide at the end of 2004. In contrast, and similar to other regions, 50% or more of the commercial and industrial customer load had left the PTB service. Today, the ERCOT market is one of the most competitive load-serving markets in the US. Most of the demand for full-requirement service comes from municipalities and co-ops, with peak demands in the range of 10–700 MW. PTB providers and competitive retail providers (the latter including deregulated affiliates of PTB providers) tend to arrange for their own supply, which occasionally includes full-requirement service, fixed-shape products or weather derivatives. The single most important method of procurement in ERCOT is a one-stage or multi-stage bidding process initiated via an RFP (request for proposal) and concluded with a bilateral supply agreement. While the introduction of load auctions similar to the ones in the Northeast and East regions has been under discussion in ERCOT, at this point only congestion revenue rights and generation capacity are sold via auctions.

SERC

The Southeast markets are remarkably different from the RTO markets such as PJM. They are generally dominated by big regulated utilities. Retail competition almost doesn't exist. Most of the competitive activities are on the wholesale side, with customers ranging from local utilities to cooperatives. Most of the transactions are done bilaterally and are usually long-term (some of deals go as long as 10 years) and physically oriented.

West (WSCC)

The Western part of the US is still slowly recovering from the failed deregulation attempt in its biggest market, California (CAISO). The causes for the infamous market blow-up in 2001–2 are manifold and complex and hence beyond the scope of this chapter. But two of the most relevant and lasting consequences of this disastrous event are that retail access has been suspended in California and that many market participants in the West have impaired credit or are involved in litigation. Furthermore, other deregulation efforts outside California have been slowed down. Today, the WSCC load-serving markets are reminiscent of the ERCOT market in three aspects: first, most demand for full-requirement service comes from munis and co-ops; second, RFPs ending in bilateral service agreements are the most common form of procurement; third, the introduction of locational marginal pricing (LMP) is planned for the next couple of years. A peculiarity of the WSCC market worth noting is that, despite all of the problems mentioned above, it is probably the most liquid electricity forward market in the US. This may be partly explained by the fact that many load customers are accustomed to buying supplies for periods of up to 10 years.

Canada

The only two Canadian provinces that deregulated their load markets are Alberta and Ontario. In Alberta, retail competition began in 2001. Since then, the progress of deregulation has been slower than initially hoped, partly because of a lack of a liquid wholesale forward market. At this point, only a handful of competitive suppliers have entered the retail market to compete with the deregulated affiliates of the incumbent utilities. In Ontario, retail competition began in Spring 2002. However, the deregulation progress has been slow because of continued discussions and uncertainty about the deregulation of the generation side of the electricity market.

DEMAND STRUCTURE

It is common practice to differentiate customers and their associated load based on their consumption patterns into three main categories: commercial customers, residential customers and industrial customers. In many cases a load-serving contract will involve a combination of these categories. The exact definitions of those categories can differ from region to region. From the pricing and risk management standpoint, the three types differ mainly in two dimensions: volume and correlation of price and volume (covariance, or load convexity).

Volume. Industrial loads tend to be less sensitive to changes in weather, while suffering from higher rates of migration/attrition.

Residential loads are very sensitive to changing weather conditions (a/c demand) while exhibiting much less sensitivity to changing competitive environment (little migration). Commercial loads fall in between the two extreme types.

Price/load covariance. Residential loads exhibit the strongest covariance between prices and volume. Industrial loads tend to have a relatively weak covariance with prices, while commercial loads fall somewhere in between.

CONTRACT STRUCTURE

In the following, we will briefly look at a couple of commonly seen load contracts currently available in the market.

Flat load contract

The flat load contract clearly is the simplest load contract. It is essentially a forward contract. The supplier of such a contract only has to serve the customer a fixed amount of power in all the on-peak and/or off-peak hours of a given month. Typically, the fixed amounts vary by season or calendar month and reflect anticipated load growth over time. Obviously, it is trivial to value such a contract. The supplier has only to charge the volume-weighted average forward price plus some premium to serve this load. The most common reason that customers are willing to pay more than the forward price for this product is that the block size traded in the market is 25 or 50 MW while the customers' needs might be "odd lots" that fall, in terms of size, between multiples of the standard block size.

Fixed-shape load contract

The fixed-shape load contract is a contract that requires the supplier to provide power to the customer according to a predefined fixed shape at certain contract price. In contrast to a flat load contract the amount of the load obligation can vary by hour or even metering interval, but it is still fixed at the time of the contract execution.

Often, the predefined fixed shape will be the same for every day of a month, or the same for the weekdays or weekends separately. Unlike the full-requirement load deal, the fixed-shape load deal exposes no volumetric risks to the supplier because the shape it has to serve is predetermined. Still, as in almost every other load deal, the supplier faces the price risks.

Full-requirement contract

The most commonly seen load contract in the marketplace is the full-requirement load contract. By entering a full-requirement deal, the supplier takes on the obligation to serve energy and ancillary services, and gets paid in return. The payment schedule the supplier receives from the customer can vary from deal to deal. For instance, the payment can be a fixed dollar amount per MWh (fixed-payment full-requirement load deal, or simply full-requirement load deal) or some kind of links to the fuel prices such as gas prices (fuel-linked full-requirement deal). In the fixed-payment full-requirement load deal, the customer essentially eliminates all the price risks it faces in its day-to-day operations.

In practice, some of the full-requirement load contracts have certain bound structures embedded in the contracts. One example of such a contract is a full-requirement load contract with a cap. The supplier is not responsible for the power above the cap. Therefore, from the supplier's point of view, this contract eliminates all the potential volumetric risks associated with high loads. It follows that the price the supplier can charge the customer should be lower compared with the full-requirement load deal. One such example is shown in Figure 1.

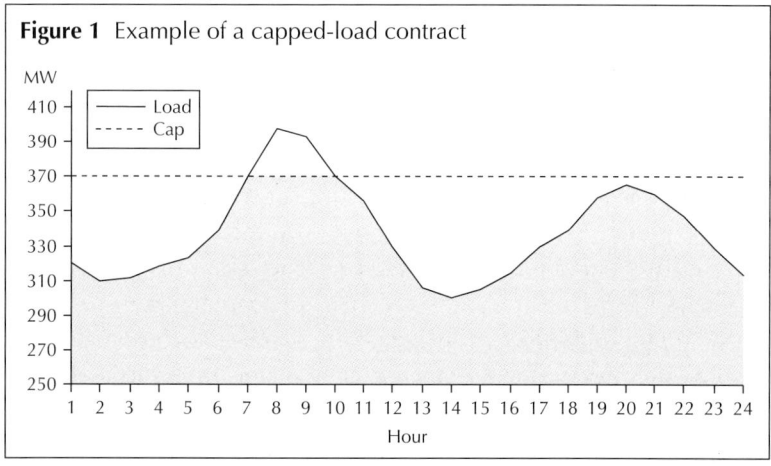

Figure 1 Example of a capped-load contract

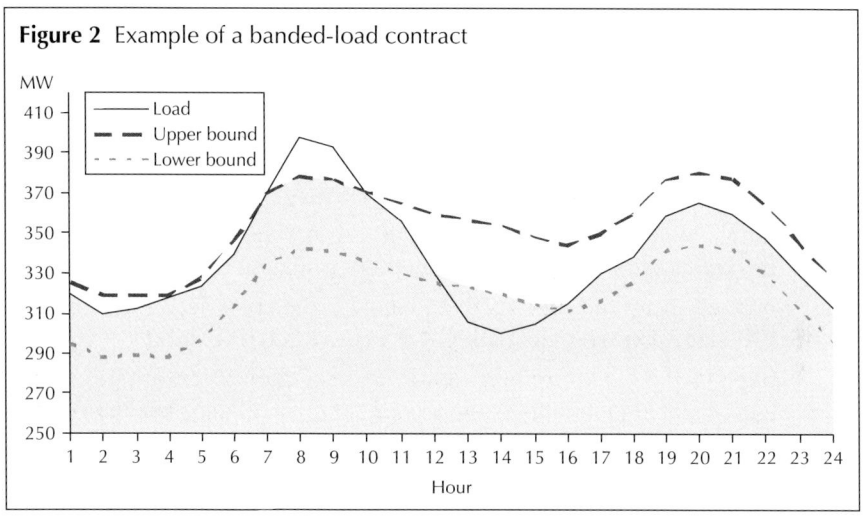

Figure 2 Example of a banded-load contract

Banded-load contract

Unlike the fixed-shape load contract, the banded-load contract gives the customer a static or partially flexible band around a predetermined load shape. In other words, the customer can still enjoy the contract price even though the realised load deviates from the predefined fixed shape, as long as the deviation is within certain band of the fixed shape (usually about 5% to 10% of the fixed shape or certain constant MW). One such example is shown in Figure 2.

Load contract with look-back features

The load contract with look-back features is a little bit more complicated than the previously discussed contracts. The essential idea of such a contract is to charge the customer different prices based on the different realised load factors.

Usually there is a predefined cap in the load contract with look-back features. As with other capped load contracts, the supplier of such a contract is responsible to serve only the amount below the cap. Anything above the cap is the customer's own responsibility. In pool markets, the volumes exceeding the cap are simply settled at the index. In bilateral markets, the customer is responsible for the physical procurement of energy. The major difference between this type of the contract *versus* the regular capped contract is that the supplier will charge the customer different levels of prices based on what the realised load factor for that particular month is. This price matrix (different price for different load factor) is predetermined at the signing of the contract.

Other hybrids of load contract
It is obvious that the list of the load contracts above cannot be complete. As in the options market, one can construct virtually any type of exotic load contract as long as there is a need from the customer.

VALUATION AND HEDGING OF FULL-REQUIREMENT CONTRACTS

In this section we discuss the pricing and hedging of load contracts. Since the perspective of this book is driven by risk management considerations, all the pricing approaches presented here take into account (and often are based on) the hedging instruments available in the market. In the US power markets, the following products tend to offer the most liquidity: on-peak and off-peak forward contracts, on-peak monthly options and strips of on-peak daily options. In some situations, this collection of tools can be enhanced by various types of weather derivatives that are helpful in managing weather-driven volumetric exposure. We will not systematically analyse pricing in terms of weather products since the market lacks any liquidity and uniformity. Consequently, we will always try to express the value of any contract in terms of the liquid products: forwards and options. For the sake of simplicity, we will assume that we are always able to transact at the "mid"-market prices; but the reader should keep in mind that the products listed above are often traded bilaterally and that liquidity and transaction costs can vary widely by region, product and term. Therefore, any deal price derived from the pricing approaches presented below needs to be adjusted for the actual costs of executing the associated hedge strategy.

Notice that we are interested in analysing our contract with a monthly resolution. This procedure is motivated by the typical resolution of the forward markets. Since traded contracts usually have monthly resolution (for example, forwards, daily options), the risks of finer resolution cannot be managed. Additionally, we usually do not have any conditional information of finer resolution either. For example, unless we are very close to the month in question, we cannot separately transact or even generate forecasts for days early or late in the month. The foregoing does not mean that we should not be concerned about, for example, hourly shapes of loads and prices. The above argument merely indicates that we are interested only in the *monthly sums* of hourly shapes, since these are the only things we can do something about. Risks with finer resolutions cannot be managed on a forward basis but should be analysed in terms of the distribution of residual deal risk. The shape of this distribution will be an important driver of the risk premium level required by the party acquiring the risk (the supplier).

In some cases intramonth risks can be mitigated contractually. For example, if we can charge a customer two different fixed prices for the first and second halves of the month, then the resolution of our analysis will have to drill down to match the resolution of the contractual terms. In practice, however, the resolution of hedging contracts is higher than that of the contractual terms of full-requirement contracts. Hence, the resolution of the valuation is driven by the hedging contracts.

To sum up, we follow one principle in the valuation process: *express all the prices in terms of available hedging instruments and charge a risk premium for unhedgeable risk.*

MATHEMATICAL REPRESENTATION OF THE PRICING PROBLEM IN TERMS OF TRADED CONTRACTS

For any specific month, the total cost to serve is given by the following expression:

$$CS = \sum_{d,h} L_d^h P_d^h \qquad (1)$$

d, h - daily and hourly indices
L - load; P - Price

For the purpose of valuation of full-requirement load contracts, we are interested in finding the expectation of the above expression conditioned on all the information available today (time t):

$$E_t[CS] = \sum_{d,h} E_t[L_d^h P_d^h] \qquad (2)$$

The expectation can be decomposed into the shape and covariance components:

$$E_t[CS] = \sum_{d,h} E_t[L_d^h P_d^h] = \sum_{d,h} \{E_t[L_d^h] E_t[P_d^h] + \text{cov}_t[L_d^h, P_d^h]\} \quad (3)$$

Since the above covariance term is conditioned on today's information, it can be called *daily covariance* in analogy to the volatility markets. In contrast to its daily volatility counterpart, however, the daily covariance cannot be observed in the forward markets. Consequently, it has to be estimated from historical or fundamental data.

Since historical prices in any given month are obviously observed for any given realisation of the forward price, any estimation based on monthly resolution in historical data will yield only *cash* statistics (such as correlations, volatilities). Since we cannot directly observe the daily covariance we need in (3), we have to decompose the daily covariance into the part we can observe in historical data (cash covariance) and the part that usually cannot be reliably estimated in historical data (monthly covariance).

We can rewrite the daily covariance by conditioning on the forward price at expirat ion in the following way:

$$\text{cov}_t[L_d^h, P_d^h] = E_t[\text{cov}_M[L_d^h, P_d^h]] + \text{cov}_t[E_M[L_d^h], E_M[P_d^h]] \quad (4)$$

The first term is today's expectation of the cash covariance, while the second term is the monthly covariance. The subscript M indicates conditioning on all the information available at the expiry of the forward contract.

We can rewrite (4) the following way:

$$\text{cov}_t[L_d^h, P_d^h] = E_t[\rho_{d,h}^C \sigma_{L_d^h}^C \sigma_{P_d^h}^C] + \rho_{d,h}^M \sigma_{L_d^h}^M \sigma_{P_d^h}^M \quad (5)$$

An aside: The above assumption implies that the expectations we are using are risk-adjusted. In historical data, we do not observe risk-adjusted quantities. We will ignore this difference here, although the adjustment can be done relatively easily.

Now, let us express the hourly loads and prices as a function of daily loads and prices and an hourly shape coefficient:

$$L_d^h = a_d^h L_d,$$
$$P_d^h = b_d^h P_d \tag{6}$$

where

$$L_d = \sum_h L_d^h$$

$$P_d = \sum_h P_d^h$$

Assuming that the shape coefficients are constant numbers we get the following:

$$E_t[\rho_{d,h}^C \sigma_{L_d^h}^C \sigma_{P_d^h}^C] = E_t[a_d^h b_d^h \rho_{d,h}^C \sigma_L^C \sigma_F^C] \tag{7}$$

Assuming also that the shape expectations are identical for all the days, we have:

$$E_t[a_d^h b_d^h \rho_{d,h}^C \sigma_{L_d}^C \sigma_{P_d}^C] = a^h b^h E_t[\rho_d^C \sigma_{L_d}^C \sigma_{P_d}^C]$$
$$\rho_{d,h}^M \sigma_{L_d^h}^M \sigma_{P_d^h}^M = a^h b^h \rho^C \sigma_F^M \sigma_L^M \tag{8}$$

Using all the above, we can rewrite (2) as follows:

$$E_t[CS] = \hat{S}(F\hat{L} + E_t[\rho^C \sigma_L^C \sigma_P^C] + \rho^M \sigma_F \sigma_L^M)$$
$$\hat{S} = \sum_h a^h b^h \tag{9}$$

We managed to express the cost-to-serve in terms of forward prices, monthly and daily volatility. The simple expression required many simplifying assumptions along the way. The relationship between the expected cost-to-serve and the available trading contracts is, in general, more complicated. In actual estimation of pricing models, we will use more complicated relationships, even though we might not always know them explicitly. Either way, the general picture is independent of a model, as it is driven by the resolution of the traded contracts. We always need to split the valuation into three components:

- ❏ flat block component (corresponding to forward prices);
- ❏ monthly covariance (corresponding to monthly volatility); and
- ❏ cash covariance (corresponding to cash volatility, which itself is a function of monthly and daily volatility).

MODELLING TECHNIQUES

There are a couple of different approaches to the valuation of the load-serving deals. As in any valuation we can use reduced-form and fundamental models. Fundamental models derive the relationship between prices and load through the direct simulation of the power price formation process. Reduced-form models make explicit assumptions about certain relevant statistics describing the relationship between loads and prices. The statistics are then estimated in various ways based on historical data or are implied from market quotes. We do not specify exactly which statistics we are concerned about, partially because the statistics of interest are determined by whether or not we make explicit distributional assumptions about load, price and their joint structure. In fact, the classification into reduced-form and fundamental models is not the most fruitful. As we can see in (9), we are ultimately concerned about the correlation and load volatility. We can model those quantities *directly* or *indirectly*. Fundamental and many reduced-form models are examples of indirect modelling. We can also consider a direct modelling approach, where we estimate the quantities in question directly from historical and forward-looking data. In general *direct* modelling methodologies do not involve making any assumptions about the underlying distributions. *Indirect* methodologies are, in contrast, built around making explicit assumptions about the underlying distributions.

Indirect methods and direct methods: a comparison

In indirect methods, we often simulate the hourly price and hourly load. Once we have the simulated hourly information, we can calculate the expected cost of serving the load as well as the revenue very easily. The key issue here is to get the joint distribution of hourly price and hourly load right. This can be done in the following two ways: either through certain reduced-form processes, or through certain fundamental and hybrid models. The major difference between the reduced-form approach and hybrid approach lies in the model of the underlying power prices. For the reduced-form approach, we first specify the underlying power prices explicitly. This involves the choosing of the proper process as well as the estimation of the model parameters. For the hybrid approach, we model the power prices by combining the market-tradable information (such as forward prices and options) with the fundamental supply-and-demand information.

In the direct approach, as we mentioned above, we estimate the required statistics (for example, price-load correlation, load volatility) either through historical simulation or by calculating correlations and volatilities directly on historical data. For non-linear load products (such as banded load), the statistics of interest might be more complicated.

Both of these approaches have their own pros and cons. The direct method does not require any distributional assumptions, hence the method tends to be much more robust, if done correctly. Alternatively, the integration of forward-looking information is definitely harder and requires much care, especially for non-linear load structures (such as banded products). In contrast, the indirect method can easily handle complex structures and incorporate forward-looking information. However, the indirect methods tend to be much more computationally intensive and also usually require a significant amount of structural assumptions, making them, in general, less robust.

Estimation
Estimation strategies are closely related to the model's structure. There are clear differences in approaches between indirect fundamental methods and indirect parametric and direct methods on the other hand.

Indirect fundamental methods: Fundamental models rely on the explicit modelling of the power price formation in terms of the underlying fundamentals: system load, fuels, outages. Naturally, the relationship between system load and power price is implicit in the modelling technique. Hence, the estimation of the model involves estimating a load process along with the mapping between load and power price (ie, the generation stack). Although those models usually rely on the system loads modelling, the model can be extended (not always fruitfully) to other loads. Finally, for some loads the relationship between the system load and load of interest can be quite high, so we can still use the basic fundamental models.

It is worth noting that, in general, we can calculate the model implied forward covariance between price and load relatively easily in fundamental models. Note that in order to get the forward covariance (or any other forward parameters), all we need is the joint distribution between monthly average price and monthly average load (or any other similar statistics). Hence we don't even have to worry about simulating the hourly load and hourly price, which is a much harder task to accomplish. This estimation methodology can be applied to almost all the forward parameters mentioned in this paper. On the other hand, we need some structural assumptions on hourly load and hourly prices in order to estimate the cash parameters.

Details on fundamental models and their application to full-requirement contract can be found in Eydeland (2002).

Indirect parametric methods: The indirect models make explicit assumptions about the joint distribution of loads and prices. The usual estimation strategy involves the application of some standard econometric estimation techniques: MLE or GMM, given the parametric specification. In general, if the specification is simple enough, the methods are passably robust. However, if one tries to impose a lot of structure on data, the resulting estimates become very unstable and inefficient.

Direct methods: The direct model is built around the estimation technique. We estimate the parameters given the representation of cost-to-serve. For the simplest case of the full-requirement contract, we estimate price and load shapes, load forward and cash volatilities and price-load forward and cash correlations from historical data. The biggest challenge with this method is the effective estimation of correlations and forward load volatilities. Correlations in general tend to be quite unstable. Forward load volatilities, on the other hand, are very hard to estimate since the available historical data are quite short, yielding highly inefficient estimates of the volatility.

HEDGING OF LOAD CONTRACTS

In Gao and Wolyniec (2004), we analysed extensively the issues of hedging of load contracts. In this section we present the highlights of that treatment as well as some practical consideration involved in managing the risks of full-requirement contracts.

When we talk about managing risk we need to be explicit about what the definition of risk is. Depending on the choice, our hedging strategies can be quite different (see Eydeland, Wolyniec, 2002). In this chapter, we will concentrate on minimising the local variance, which is roughly equivalent to minimising the terminal variance of cost-to-serve.

In general, management of full-requirement contracts can be split into two distinct time periods, which correspond to the available hedging products:

❑ monthly, when we use mainly forward contracts and monthly options to manage the exposure; and
❑ cash, when we use daily options and balance-of-the-month contracts.

Using the standard typology of risk components, we have the following considerations:

Delta: As we can see in (9), the delta of the contract (the dynamic position in the appropriate forward price contract minimising risk) will have three contributions:

❑ expected load;
❑ delta of the monthly covariance; and
❑ delta of the cash covariance.

The first term always dominates. Depending on the type of load, the two remaining terms usually contribute between 3% and 10% of the expected load to the size of the delta position.

Gamma: As we can see from (9), the delta of the full-requirement deal can change either due to underlying price moves or volatility moves. As long as the monthly and daily price volatilities are *linear* in the forward price we will not have any gamma (for example, under the Black–Scholes model, the forward price volatility is a linear function of the forward price). That is, our delta position will not change with price moves. Now, the assumption is a very good approximation of reality for the majority of the life of a forward contract.

However, close to expiration of the forward contract, we can have a contribution to the delta arising from the fact that changes in forward prices are driven by changes in expected load. We will not go into details of how to determine the exact adjustment to the position and how significant a consideration those issues are. The interested reader is referred to Gao and Wolyniec (2004) for a detailed analysis.

Vega: Equation (9) shows that we will have (potentially significant) exposure to daily and monthly volatility. We can use daily and monthly options to control the exposure. Under certain circumstances the monthly and daily Vegas can be of the opposite sign.

The above analysis in terms of the contract "greeks" may be a bit misleading. Since the volumetric risk is only partially hedgeable, we have to account for the unhedged part by charging an appropriate risk premium. We will address the issues associated with volumetric exposure in the next section.

VOLUMETRIC RISK

The subject of this chapter forms a proper subset of pricing and modelling, which we considered in the previous section. We consider the topic in a separate chapter to underscore a set of unique issues associated with pricing and managing volumetric risk.

Overview

The most distinctive feature of full-requirement deals is without a doubt volumetric uncertainty. In general, the supplier will assume all or at least a substantial amount of the volumetric risk of the customer. Therefore, analysing, pricing, and managing this risk is key for the supplier. There are a variety of factors causing volumetric uncertainty:

❑ weather;
❑ population growth and migration (customers moving in or out of the geographic service area);
❑ changes in lifestyle or technology causing changes in consumption patterns;
❑ changes in economic conditions in the service area;
❑ customers leaving service or signing up for service for economic reasons; and
❑ customers changing consumption patterns for economic reasons.

Obviously, volumetric risk is only one side of the story. Of utmost importance for pricing is the relationship between volume and price. Conceptually, it is often useful to analyse the factors above by assigning them to one of the following categories: (i) volume changes affecting market prices; (ii) market prices triggering volume changes; and (iii) volume changes that are not (or only weakly) correlated with price changes. For example, volumetric risk related to weather typically falls into the first category. Here, the relationship between volume and price is caused by the fundamentals of physical demand and supply.

Weather risk
Weather is probably the most appreciated and best-understood driver of volumetric risk, especially in electricity markets. A supplier caught short on a very hot or very cold day could be exposed to substantial financial losses. In addition, there could be other repercussions – commercial, political or regulatory – should there be any shortage of physical supply or any degradation in service.

Fortunately, analysing and modelling the weather-sensitivity of loads is usually a fairly straightforward application of statistical methods. In most cases, loads are primarily driven by temperature; therefore, quite decent models of load response to weather can be built by fitting – using regression analysis or other tools – historical hourly loads to polynomial or piecewise linear functions of temperature. The relationship between load and temperature is generally very stable: one year of load and temperature data is usually sufficient to arrive at a very robust description of the basic load response function. Obviously, a good model should take into account load response variations across seasons, hours of the day and days of the week. And if multiple years of data are available then the estimation of the load response should factor in adjustments for load growth as well (this will be discussed later). Finally, particular attention needs to be paid when analysing load data in areas with customer choice: the reason is that sudden changes in the customer base stemming from customers leaving or joining service for economic reasons can bias the results of the load response function significantly.

Once equipped with a model of load response to temperature, it is fairly easy to calculate expected loads or to characterise the full distribution of loads – all we need is a distribution of the temperatures underlying our estimation of the load response. Statistical models of temperature are easy to build. Alternatively, if no such models are readily available, one can resort to using historical distributions of temperature as a first-order approximation.

Some of the more subtle issues in modelling loads involve capturing the short-term and long-term serial correlation in loads. While understanding short-term serial correlation is crucial for short-term load forecasting, it is not an issue that is relevant for pricing a load. In contrast, a good model of long-term serial correlation – usually caused by population growth and migration – can be quite useful in assessing the unhedgeable volumetric risk involved in a long-termed full-requirement deal. Unfortunately, such models are not easy to estimate.

Population growth and migration

The terms "population growth" and "migration" are commonly used to characterise natural, gradual changes in a larger customer base. While such changes may be partly driven by economic factors – for example, by people moving to areas providing better employment prospects or better quality of life – one generally assumes that such changes are unrelated to the cost of serving when pricing a full-requirement load.[1] Consequently, by assumption, the two main issues related to population growth are quantifying and managing/mitigating the volumetric risk.

Analysing historical population growth requires access to a dataset of historical loads that stretches across more than just a couple of years. Unfortunately, full-requirement customers provide potential suppliers often only with a very limited amount of load history (one to three years seems to be a typical time window). In cases like this, it may be best to resort to analysing load growth based on neighbouring loads or regional loads for which longer consumption histories are available. The data provided by the customer can then be used to assess peculiarities or to pick up recent trends in the target service area. One should always keep in mind that models based on history may be poor predictors of the future. This is particularly true in the case of load growth. When looking at deals involving significant growth risk, suppliers should therefore consider spending time on fundamental research. This could involve sending someone to the service territory to assess first-hand current economic activity and future growth opportunities.

In cases where a longer load history is available, a stab at estimating historical load growth can be made by postulating that the load response of the "average customer" to weather remains unchanged over time. This allows us to express the total load at time t as the product of individual customer load at time t times the number of customers at time t:

$$L^{Total}(t) = L^{Cust}(t) \cdot N(t)$$

An approach for estimating the customer load response has already been discussed. A natural choice for modelling the customer count process $N(t)$ is to assume that populations grow exponentially:

$$N(t) = N(0) \cdot e^{\mu t + \sigma W(t)}$$

The term $W(t)$ denotes a Brownian Motion, $N(0)$ denotes the initial customer count, and μ and σ are the drift and volatility parameters of the process. The load response function and the parameters of the customer count process can be estimated in one step; alternatively, one could pursue an iterative, two-step estimation whereby the parameters of both processes are estimated sequentially. For smaller sets of load data we may have to resort to an even simpler, deterministic model of growth by assuming that the volatility of the growth process is zero. Conversely, if a longer load history is available and we need to model growth over a long period of time then we may be able to build a more sophisticated model of growth – for instance, by using a mean-reverting process instead of a Brownian Motion. (Mean reversion in growth rates can be observed in many market loads.) Note that we do not need to know at all the actual, physical number of customers in the service area: we can simply operate in terms of customer counts implied by the actual loads and the load response function. An important assumption underlying the approach proposed here is that the load response of the average customer is fairly stationary. This assumption is violated if there are structural changes in the customer base – if, for example, the share of residential *versus* non-residential customers is changing. In such a case it is advisable to perform the analysis separately for each customer class.

Hedging future changes in cost of service caused by population growth is practically impossible. While there exist hedges for long-term price risk (eg, forwards and options) and possibly fundamental hedges for the volumetric risk (eg, real estate in the service area), there are no instruments or assets whose value is sufficiently correlated with the product of long-term volume and price changes. This leaves a full-requirement supplier with the following options. First, he or she can simply take on the growth risk and charge a premium for it. Second, he or she can attempt to propose deal structures with shorter terms to reduce the growth risk (reasonably assuming that forecast uncertainty increases with term). Third, he can attempt to mitigate the price risk on excess loads and load shortfalls contractually by charging an index price that is correlated with the cost of serving the load. The obvious benefit of such a variable tariff is that the supplier's revenues on any volume deviations from expected loads will be close to the incremental costs. A variant of this, a look-back that sets a fixed price as a function of the realised load factor, was discussed above. Fourth, the supplier can attempt to mitigate the volume risk contractually by including caps and floors on the total load. The most common flavours are caps and floors on hourly loads and/or on energy consumption over longer periods (such as months, quarters, summer months, on-peak hours per calendar year). Usually, such bounds will reflect consumption patterns, seasonality and anticipated load growth. In most cases, loads or energy falling outside the predefined bandwidth will still be served by the supplier, but either at market prices or at some index tied to production cost. Consumption bounds do not have to be static: for example, bounds indexed on temperature are a more efficient way of separating out growth risk than static bounds.

When developing a deal proposal, a supplier will usually deploy a combination of the strategies outlined above that takes into account the customer's preferences. Depending on the framework for negotiation (auction, RFP, bilateral negotiation and so forth), a supplier may be able to submit multiple-deal structures such that the customer is able to self-select the combination of price and risk allocation that best fits his or her needs.

Customers leaving or signing up for service for economic reasons

The fundamental objective behind the deregulation of retail energy markets is to lower the cost of service market-wide through competition at retail level. A key prerequisite for competition is the customers' right to choose a supplier. In most deregulated energy markets in the US, especially in the power markets in the East and Northeast, deregulation laws have granted residential customers substantial flexibility. Customers can leave or sign up for service with very little advance notice. The situation can be compared with the long-distance telephone market, whereby customers can switch service almost instantly. In contrast, while commercial and industrial customers were initially given the same right of choice, these customers often have to commit to a certain service term when switching to a new supplier. Independent of the specific market rules – which vary by region – the common feature of deregulated markets is that retail customers have an economic option: they have the right to switch to another supplier and will generally exercise this right when savings are significant enough to justify the effort. Conversely, suppliers do not have the right to terminate written or unwritten service agreements unless customers stop paying their bills.

This means that any full-requirement supplier in such a market is short a portfolio of (American-style) put options, resulting in substantial volumetric risk that once again is adversely correlated with price. If market prices fall, customers will switch to other suppliers who can price new service at current market levels. This leaves the current supplier with long hedge positions that have to be liquidated below their initial purchase price. The risk related to customer choice is even higher for a special category of suppliers, the so-called providers of "standard service". These suppliers are required to offer full-requirement service to all qualified retail customers in the service territory at a regulated price. Therefore, customers currently served by other suppliers have the right to sign up for service with this supplier any time and can exercise this right when they save money relative to their existing contract. As a result, such a supplier is short a portfolio of put options (held by the current customers) and call options (held by customers currently served by other suppliers). In fact, given that customers can repeatedly leave and re-enter service, each customer holds a portfolio of put and call options; the supplier has the corresponding short position. In essentially every deregulated energy market, there is at least one provider of standard service. In the East and Northeast regions of the US, subsets and even slices of standard service load obligations are sold at auction.

When pricing a full-requirement deal involving customer choice, one has to model all the factors affecting directly or indirectly the retail customers' decision process. Besides the market rules for customer choice, these factors include:

- the customers' hurdle rates for switching to another supplier;
- the horizon used by customers in their decision-making process;
- the customers' switching history and related experiences;
- competitive pressure in the retail market;
- a supplier's cost to proactively approach and "acquire" a new customer; and
- customers' familiarity with the market design and associated options.

The relevance of these factors will vary significantly across customer classes. For example, a chain of retail stores spending thousands of dollars a month on air conditioning should be expected to pursue savings opportunities more aggressively than a residential household.

CONCLUSION

In this chapter we have briefly reviewed the issues involved in risk management and pricing of full-requirement contracts. We have not touched on several pertinent topics. To fully consider a realistic example of pricing and risk management, we would have to analyse the issues of bidding, relation between forecasting and hedging, performance of various hedging tools, liquidity in traded contracts, details of tariff structures, market structure and many, many others.[2]

---

1   This may be an incorrect assumption in fast-growing markets with tight supply.
2   Again, the scope of this chapter does not allow us to delve deeper into the above consideration. For a more thorough treatment of hedging issues, see Gao, Wolyniec (2004). For more extensive analysis of pricing and risk management of load contracts, see Eydeland, Wolyniec (2006).

REFERENCES

Eydeland and K. Wolyniec, 2002, *Energy and Power Risk Management*, 1st edn (New York: John Wiley & Sons).

Eydeland and K. Wolyniec, 2006, *Energy and Power Risk Management*, 2nd edn (New York: John Wiley & Sons).

Gao, Y. and K. Wolyniec, 2004, *Hedging Load Contracts* (available from the authors on request).

11

Heat Rate Options

Boris Chibisov, Alexander Eydeland; Krzysztof Wolyniec

Morgan Stanley; Sempra Commodities

In this chapter we intend to present the overview of heat rate options and their modifications. Exact definitions will be given below. Now we need to mention only that the heat rate of a power plant is a measure of its efficiency determining the amount of fuel necessary to produce 1MWh of power. In financial terms, heat rate is used to calculate the cost of fuel required to produce 1MWh of power. Similarly, the market or implied heat rate is just the ratio of power price to the primary fuel price (often natural gas). It describes the relationship between power price and primary fuel in any given market. In most general terms the heat rate option is an option on the market heat rate, and therefore it is an option on the spread (and sometimes the ratio) between the power and fuel prices. The difference between heat rate options and options in the financial markets is that heat rate options often have a physical flavour related to their origin, especially if they are used to replicate a power plant. Some of the physical characteristics of heat rate options will be discussed in this chapter.

The uses of heat rate options are numerous. The power asset operator frequently uses these options to hedge power plants against adverse moves of power and fuel prices by selling heat-rate-option-like structures most closely replicating the power plant economics. Another frequently used application is to sell a medium- to long-term strip of heat rate options for a fixed stream of option premiums to an institution with good credit. This transaction would allow the power asset operator to eliminate the market risk, would guarantee stable cashflows for a long time and leave it only to handle the operational risk, something that the operator knows how to do well. Having locked steady cashflows for a long period of time from the institution with solid credit, the operator has now significantly improved its potential for accessing capital, if needed, and achieving financing on much more favourable terms than before.

Power marketers use heat rate options as a virtual power plant. That is, they buy these options to achieve financial replication of the power plant without taking on operational and other risks and responsibilities associated with running the plant.

Heat rates, spark spreads and tolling deals
We shall start with definitions.

Heat rate
DEFINITION: Number of Btu needed to make one kilowatt-hour (kWh) of electricity.

Heat rate is a measure of how efficiently the generating unit converts the energy content of the primary fuel into power. Ideally, in the absence of any inefficiency, it would take 3,412 Btu to produce one kWh of electricity. Table 1 contains heat rate data for some frequently encountered power plants.

It is important to note that constant heat rate is just an approximation. In reality the heat rate varies with a number of parameters, particularly the ambient temperature and the plant generation level.

One use of the heat rate, particularly important in financial applications, is to determine the cost of fuel needed to generate one

Table 1 Heat rate data for frequently encountered power plants

	Heat rate Btu/kWh	Efficiency (%)
"Ideal" power plant	3,412	100
Combined-cycle combustion turbine	6,250–7,200	55
Base load unit	10,000–12,000	30–35
Single-cycle gas turbine	12,000–19,000	20–30

HEAT RATE OPTIONS

unit of power. Indeed, if it takes HR Btu/kWh to produce one 1kWh of power and the price of one 1Btu of fuel is $P_{fuel}\left[\dfrac{US\$}{Btu}\right]$, then the fuel cost of generation of 1kWh of power is

$$Fuel\_Cost_{power}\left[\dfrac{US\$}{KWh}\right] = HR\left[\dfrac{Btu}{KWh}\right] \cdot P_{fuel}\left[\dfrac{US\$}{Btu}\right]$$

or,

$$Fuel\_Cost_{power}\left[\dfrac{US\$}{MWh}\right] = \dfrac{1,000\left[\dfrac{MWh}{KWh}\right]}{1,000,000\left[\dfrac{MMBtu}{Btu}\right]} \cdot HR\left[\dfrac{Btu}{KWh}\right]$$

$$\times P_{fuel}\left[\dfrac{US\$}{MMBtu}\right]$$

$$= \dfrac{1}{1,000} \cdot HR\left[\dfrac{Btu}{KWh}\right] \cdot P_{fuel}\left[\dfrac{US\$}{MMBtu}\right]$$

Example 1

Assume that the heat rate of a gas-fired power plant is 7,000Btu/kWh, and that currently the cost of 1MMBtu of natural gas is US$6/MMBtu. Then, the cost of fuel needed to generate 1MWh of power is

$$\dfrac{7,000}{1,000} \cdot US\$6/MMBtu = US\$42/MWh$$

Table 2 contains examples of cost of generation of 1 MWh of electricity for different types of fuel and corresponding generic plants.

Table 2 Cost of power generation for different fuel types

Fuel	Fuel units	Fuel cost (US$/Unit)	Approximate heat content (MMBtu/unit)	Fuel cost (US$/MMBtu)	Heat rate (Btu/kWh)	Fuel cost (US$/MWh)
Coal	Tonnes	40	20	2	9,500	19
Gas	1,000 CF	6	1	6	7,000	42
#6 oil	Barrel	48	6	8	12,000	96
Nuclear	KG	32,000	64,000	.5	10,000	5

Thus, the significance of the heat rate is that it provides a conversion factor between fuels used to generate power and the power itself. Furthermore, it allows us to compare fuel and power prices, expressing them in the same units, typically in US$/MWh.

We have defined heat rate for a particular generating unit. However, the concept of heat rate can be generalised to be applied to a group of units constituting a particular power market.

Market or implied heat rate at a given time is the heat rate of the marginal generating unit, that is, the unit that is setting the market price at that time. An alternative definition of the market heat rate, which is frequently used since it can be easily calculated from the market data, is as follows:

$$IHR = \frac{P_{power}}{P_{fuel}}$$

where *IHR* is the implied heat rate, and P_{fuel} is the market price of the fuel used by the generating unit that sets the market power price, that is, by the marginal generating unit.

From this definition it is clear that the implied heat rate is a random quantity changing with power and fuel prices.

Spark spread
DEFINITION: Spark_Spread = Power_Price − Heat_Rate · Fuel_Price.

In this definition spark spread, power and fuel prices are random market variables and heat rate is a fixed constant. As a rule, electricity is quoted in US$/MWh, gas is quoted in US$/MMBtu, and heat rate is quoted in Btu/kWh. Hence,

$$Spread\left[\frac{US\$}{MWh}\right] = P_{power}\left[\frac{US\$}{MWh}\right] - \frac{1,000\frac{KWh}{MWh}}{1,000,000\frac{Btu}{MMBtu}} \cdot HR\left[\frac{Btu}{KWh}\right]$$
$$\times P_{fuel}\left[\frac{US\$}{MMBtu}\right]$$

or, in a more simple form,

$$Spread\left[\frac{US\$}{MWh}\right] = P_{power}\left[\frac{US\$}{MWh}\right] - \frac{1}{1,000} \cdot HR\left[\frac{Btu}{KWh}\right] \cdot P_{fuel}\left[\frac{US\$}{MMBtu}\right]$$

Example 2

Assume that at a given time P_{power} = US$70/MWh, P_{fuel} = US$6.5/MMBtu. Then the value of the spark spread with heat rate equal to 10,000 [Btu/kWh] is

$$70\left[\frac{US\$}{MWh}\right] - 10\left[\frac{MMBtu}{MWh}\right] \cdot 6.5\left[\frac{US\$}{MMBtu}\right] = 5\left[\frac{US\$}{MWh}\right]$$

HEAT RATE OPTIONS

There are many variations of heat rate options, some of which we will present in this chapter. Originally heat rate options were just that – options, calls and puts, on the implied, or market, heat rate. We can categorise heat rate options in several ways. There are in general two types of heat rate option with respect to the ownership of the output and input:

❑ physical – in those structures the owner of the option arranges the supply of fuel at the contractual location and takes the physical ownership of the output; and
❑ financial – in those structures, the owner of the option decides only on the exercise of the option but does not physically deliver fuel or controls the output; the options are cash-settled.

It might seem that the two structures differ only in some operational details (which can be significant). However, this is not the case. Those two structures can potentially offer very different payoffs and very different values. In addition, they pose different challenges from the risk management point of view.

Financial heat rate options can be further subdivided into *standard heat rate options* and *look-backs*. The classification centres on the mode of exercise. As we will see below, heat rate options are usually exercised before the relevant prices are known. This is caused by the way power plants operate. However, one can always structure a financial product in such a way that the exercise happens when the price becomes known. Those heat rate options are called look-back heat rate options. They have little to do with standard look-back options seen in the financial world. The name owes its existence to the way the contracts are settled. At the end of a contractual month, the parties look back at the realised fuel and power price and settle the payoff accordingly.

Given the extreme conditional volatility of power prices, the difference in value, risk and hedging of standard and look-back options can be significant.

Physical options always require some lead time in exercise. This is to say, the decision on the level of generation needs to be made from a few to several hours before the actual generation period.

Another classification of heat rate options is centred on the firmness of generation (supply). Both physical and financial options can be:

❏ firm – the delivery of the power is always guaranteed;
❏ unit contingent – the delivery of the power is tied to the physical availability of a particular unit; and
❏ system-contingent – the delivery of power is tied to the physical availability of a portfolio of units.

Plain-vanilla heat rate options
The payoff structure of heat rate call and put are defined in the usual manner:

$$Payoff_{call} = \max\{IHR - HR_0, 0\}$$
$$Payoff_{put} = \max\{HR_0 - IHR, 0\}$$

In this expression, HR_0 is the fixed heat rate strike; the random quantity IHR is the implied heat rate (see the definition above). The main objective of these options was to hedge risk exposure of the power plants. However, their success in reaching these objectives was quite limited for the following reasons. First, their approximation of a generating plant was far from perfect. And second, although the option payoff depends only on one random variable, IHR, one still needs two traded instruments, power and fuel futures, to hedge the option. On the positive side, studying the distribution of implied heat rates needed for valuation of this option can be useful in a number of applications. Nevertheless, the attractiveness of these options decayed with time and a different class of heat rate options with much better plant representation characteristics has emerged and achieved a greater popularity as a risk management tool.

SPARK SPREAD OPTIONS

The reason why spread options are so popular is that in the energy markets they are everywhere. Practically every energy asset, from refinery, to storage, to pipeline, to power plant, and every structured deal has a spread option embedded in it. By definition, a spread option is an option on a spread – that is, an option holder has the right but not the obligation to enter into a forward or spot spread contract. Typically, it is a regular call or put option with the exception that the underlying is now a two-commodity portfolio, instead of a single contract.

A spark spread option is the option on a spread between power and fuel prices. Most frequently, it is an option on a spread between power and natural gas prices. This option is one of the modifications of the heat rate options and is primarily used for managing risk of power assets.

Example 3
Consider the example of an option on the spread between forward prices of power and natural gas. Let

F_{power} = forward price of electricity [US$/MWh]
F_{gas} = forward price of natural gas times heat rate [US$/MMBtu]

Then, by the definition, the payoff at option expiry T_{ex} of the call option on the spread is

$$\Pi_{call} = \max\left\{\left(F_{power}(T_{ex}) - F_{gas}(T_{ex})\right) - X, 0\right\}$$

and the payoff of the put on the spread is

$$\Pi_{put} = \max\left\{X - \left(F_{power}(T_{ex}) - F_{gas}(T_{ex})\right), 0\right\}$$

with X being the strike price.

We now consider several examples of spark spread options that are frequently encountered in power markets.

Tolling agreement

One of the most popular power products, the *tolling agreement*, comes in different shapes and sizes. The simplest way to represent this agreement is to view it as a call option on power with a floating strike linked to fuel prices. In real-life applications, the tolling contracts can be interpreted as leasing contracts on a plant wherein the "toller", the buyer of the call option, has the right to the plant output at their discretion. A typical tolling agreement has the following characteristics:

- The length of the contract is typically short-to-medium (up to several years).
- The toller has the right (but not obligation) to use the plant and to call, if they so choose, for the delivery of energy on a specified time basis (for example, on a day-ahead basis).
- Whenever the toller decides to exercise this right and to call for energy, they should pay for the energy according to the contractual arrangements. Typically, the payment is computed according to the formula

$$\text{Heat\_Rate} \times \text{Fuel\_Price} + \text{VOM}$$

where *Heat_Rate* and *VOM* are contractually specified constants. The first term in the formula represents the fuel costs while the second term represents additional costs, frequently (but not always) variable costs of running the plant.

The payment is called *Energy Payment*. In physical tolling agreements, instead the energy payment, the toller is obligated to deliver the physical fuel to the plant.

- The toller has to pay regular (monthly, quarterly and so forth) premium to the plant owner for the right to the plant output. This option premium is frequently called a *capacity payment*.

Viewing the tolling agreement as a strip of daily power call options with a fuel-linked strike price allows us to write the payoff of these options as

$$\Pi_{tolling} = \max\{P - HR \cdot W - VOM, 0\}$$

where P is the spot (day ahead) price of power, W is a spot price of fuel (for example, *Gas Daily* Index), and *VOM* denotes variable and other contractual costs. This payoff is identical to a payoff of an option on a spread between power and fuel prices with *VOM* being the strike. The value of this option is what determines the premium, or capacity payment, paid by the toller to the plant operator.

HEAT RATE OPTIONS

In general, there are many variations of the standard tolling contract whose specifications are aimed at better representation of the physical realities of power plant operator. Various contracts provide for the inclusion of start-up costs and run-time operational constraints. Needless to say, these additional features result in much more complicated structures requiring rather sophisticated models to deal with complex added optionality. To illustrate this increased modelling complexity we consider the following frequently encountered modification of the regular tolling deal.

Although the tolling contract described above is the most common one, there are numerous variations on this structure, designed to make it more attractive to the buyer or to the seller. For example, the following tolling structure is created to give additional incentives to the seller, allowing them to take advantage of high natural gas prices in addition to the usual benefits of steady cashflows provided by a standard tolling arrangements. Moreover, this deal, being unit-contingent, better reflects the physical reality of the plant operator.

Unit-contingent toll with call-back on high gas

This deal consists of the following components.

- *Normal tolling structure*: Buyer has the right to call for power. Whenever the right is exercised the buyer pays the contractual energy cost: *Number MWh × Price of 1MMBtu of NG × Heat rate* + specified costs.
- *Additional features*: Seller has the right not to deliver power during not more than 10% of all hours of the year if a specified unit is forced out. Seller has the right not to deliver if a specified gas index is (for example, *Gas Daily* Index for TETCO M3) above US$9/MMBtu for not more than 50 hours/year.

Heat rate options with limited number of start-ups

This structure consists of a standard tolling arrangement with additional constraints on a number of yearly exercises of the tolling options. It is clear that in this deal the modelling challenge is due to the significant path-dependence of the corresponding heat rate option. In general it is not very fruitful to analyse those structures as a unified category. Even though, it might seem that an option with 200 annual starts and one with only 30 differ in degree, they are actually radically different. The first structure comes down to making judgement about the trade-off between start-up and cycling down; the second structure relies on our ability to catch the peaks.

In the next section we will discuss the methods and challenges of valuation heat rate options.

VALUATION
Spread option valuation – standard model

The simplest setup for the valuation of a spark spread option assumes a simple Brownian motion process for the logarithms of the power and gas forward prices. Then price evolution is described by the following set of stochastic differential equations:

$$\frac{dP_t}{P_t} = \sigma^P dW_t^P$$

$$\frac{dG_t}{G_t} = \sigma^G dW_t^G$$

$$\left\langle dW_t^G dW_t^P \right\rangle = \rho dt \qquad (1)$$

The above equations are just standard correlated geometric Brownian motion processes. If one assumes that forward and option markets are liquid enough to allow for dynamic hedging with underlying commodity contracts, then volatility of power and gas forward prices can be equated with the implied ATM volatility.

In this case the only unknown parameter, correlation, is estimated on historical price returns.

Once correlation is estimated, spread option can be priced analytically using one of the standard spread option valuation models (such as Margrabe or Pearson or a number of semi-analytical models (see Haug, 1997, and references therein)). Consider the case of a spread option between power and gas prices with some fixed heat rate HR_0 and

$$Payoff = \max\left(P - HR_0 G - VOM, 0\right) \qquad (2)$$

If the option strike VOM is zero, then the exact value of the spread option could be obtained by switching to gas numéraire.

$$V = G_0 B\left(\frac{P_0}{G_0}, \sigma^{IHR}, HR_0, T, r\right) \qquad (3)$$

where
$V = G_0 B$ (S, σ, K, T, r) is the Black value of a simple vanilla call option with current underlying S, strike $K = HR_0$ and volatility σ. Implied heat rate volatility σ^{IHR} is given by

$$\sigma^{IHR} = \sqrt{\sigma_P^2 + \sigma_G^2 - 2\sigma_P \sigma_G \rho} \qquad (4)$$

If the strike VOM of the option payoff is non-zero, we can use Kirk approximation; also the Pearson model provides a very good semi-analytical approximation to the value of the spread option. Recent work by Carmona and Durrleman (2003) gives an efficient lower bound, too.

Application and performance of the standard approach

The model is generally used in the following way. We collect spot fuel and power prices of the relevant resolution (say, on-peak power prices and *Gas Daily* Index prices for the relevant gas index). We form a time series of daily log returns, and calculate the return correlation.

However, we have to be careful about how we use the model. First thing to notice is that the correlation is only as useful as the volatility inputs we use in the valuation of the option. As we have seen above, the standard model requires several inputs: forward prices, volatilities and correlation. The choice of the forward prices is (usually) straightforward. It is not clear, however, what volatility we should use. There are two types of volatility trading in US power markets: daily and monthly. Given the trading resolution of volatility, we need to decompose the pricing of heat rate options into the cash and monthly (forward) covariance. Following Eydeland and Wolyniec (2002), we get the following representation:

$$\sigma_{Daily} = \sqrt{\sigma_{Forward}^2 \frac{(T-t)}{T-t+15} + \sigma_{Cash}^2 \frac{15}{T-t+15}}$$

$$\rho_{Daily} = \frac{\rho_{Forward}\sigma_{Forward}^P \sigma_{Forward}^G \frac{(T-t)}{T-t+15} + \rho_{Cash}\sigma_{Cash}^P \sigma_{Cash}^G \frac{15}{T-t+15}}{\sigma_{Daily}^P \sigma_{Daily}^G}$$

In power markets, we have quotes for implied daily volatility and implied forward volatility (from daily and monthly options respectively). In fuel markets, we have quotes for implied cash volatility and forward volatility (from index and monthly options respectively).

As we can see the daily correlation is affected by the ratio of cash and forward volatilities. This suggests that we should not try to estimate daily correlation on historical data since it will be affected by the historical ratio of cash and forward volatilities. Those ratios can be different from the market-quoted ones. As long as we intend to use power and gas options to hedge the volatility exposure, we cannot ignore the market quotes. What's more, even if we do not intend to use the volatility hedges, the market quotes may contain useful conditional information about future volatility. Apart from the above considerations, the direct estimation of daily correlation is greatly complicated (although not made impossible) by seasonality of prices and volatilities. The alternative solution is to use hybrid fundamental models (see Eydeland and Wolyniec, 2002). They tend to produce the correct estimates of heat rate option values, implicitly accounting for all the volatility effects (as well as scaling effects, as we will see shortly).

In general, given the resolution of the volatility market, it is much easier to estimate forward and cash correlation separately. If we have the corresponding volatilities quoted, we need to know the cash and monthly correlations, to determine the total (daily) covariance. Once we have determined the daily covariance we enter the daily volatilities for gas and power as well as the daily correlation calculated above into the formula (4).

As we can see, we need to know at least two correlation inputs to be able to calculate the appropriate daily correlation: cash and forward correlation. The first quantity is the correlation between forward contracts returns; the second is the correlation between spot prices returns inside the contract month (spot prices conditioned on the forward price at expiration: ie, cash prices). The standard procedure is to estimate the two quantities on the one-day returns of respectively forward and spot prices. We underscore the importance of the use of one-day returns. As we will shortly see, that assumption is of crucial importance.

Table 3 Daily heat rate option valuation

	Standard model	"True" value
Mass Hub – Algonquine	US$25	US$8
PJM West Hub – Transco Z6	US$17	US$6

A natural question is: how does the model perform?

We follow the procedure described in the section using historical data to estimate cash and forward correlation and using market volatility quotes and forward prices on 3rd March 2005 to price a couple of ATM daily heat rate options in two regions (PJM and NEPOOL) for January 2005 (see Table 3).

We can see that there is an immense difference in the estimate. At this stage it is not clear how we know what the true value is. The rest of the chapter will explain how we arrived at this number and why the standard model is so deeply erroneous.

Since we derived our estimates of the value in the standard model using market prices and volatilities, the only possibility is that we used the wrong correlations.[1]

The only free parameter that is not set by the market is the correlation. Therefore, the wild mispricing tells us that our estimate of the daily correlation was incorrect. This, in turn, suggests that our estimates of cash and forward correlations are incorrect.

For our specific case the daily correlation estimated on the sample (and using implied daily and monthly vols) and the "true" daily correlations are as in Table 4.

We can see that there is an immense difference in the estimate, corresponding to the significant difference in value reported above.

To understand why we see such an enormous divergence we have to look closer at the assumptions behind the model.

Table 4 Daily correlations

	Standard estimate (%)	"True" correlation (%)
Spread		
Mass Hub – Algonquine	10	70
PJM West Hub – Transco Z6	5	60

Limitations of the standard approach

The following assumptions are implicit in the standard construction.

1. Distribution of power and gas forward prices conditional on current level of forward prices is lognormal.
2. Market-implied heat rate (IHR) follows a random walk process

$$\frac{dIHR_t}{IHR_t} = \sigma^{IHR} dW_t$$

$$IHR_t = \frac{P_t}{G_t} \quad (5)$$

For the valuation of spread options the second assumption is the critical one: it implies that the variance of market-implied heat rate is increasing as a linear function of time. If the actual process for the implied heat rate evolution has bounded variance, the spread option value can be overestimated in the situation above.

To illustrate this point, let us consider the following relationship between power and gas prices, where power price, implied heat rate and gas prices are all random variables

$$P_t = IHR_t \times G_t \quad (6)$$

The value of the spark spread option with zero strike is given by the following expectation under the standard (ie, cash-bond) risk-neutral measure:

$$V = E^Q \{\max(P - HR_0 G, 0)\} \quad (7)$$

where heat rate HR_0 is a constant specified in the option contract.

By switching to the gas as a numéraire the spread option value is

$$V = G_0 E^{Q'} \{\max(IHR - HR_0, 0)\} \quad (8)$$

where G_0 is the current forward gas price. In the above equations Q stands for the risk-neutral measure with the cash bond as the numéraire, while Q' is the risk-neutral measure with the forward price of gas as the numéraire. For details on the change of numéraire techniques, see, for example, Bjork (2004).

From this formula it follows that it is the distribution of the market-implied heat rate that determines the spread options value.

Power and gas volatility, correlation and, most importantly, the specification of the implied heat rate process in Equation (5) determine the width of the heat rate distribution and, thus, the option value. Note that Equation (5) implies that the cumulative variance of implied heat rate is a linear function of time

$$\text{Var}\{IHR_T\} = \left(\sigma^{IHR}\right)^2 T \qquad (9)$$

For sufficiently long horizons, where variance becomes large, a substantial probability is assigned to the unreasonably large and unreasonably small values of implied heat rates. Eventually, one starts sampling from the implied heat rate values that do not make sense for many power markets. This leads to the overestimation of the heat rate option value and to incorrect hedging.

Below, we present a qualitative argument that for some power markets, heat rate variance, in contrast with the equation (9), is essentially constant, for time horizons over several months.

Uncertainty in implied market heat rate: a toy model of the supply curve

We have so far considered the evolution of forward prices only. To get an insight into the dynamics of implied market heat rate, let us consider the process of power price formation in the spot market.

Since power is not storable, spot power prices are determined by supply–demand balance on the market.

The cost of energy production for each power plant is determined by plant heat rate and various fixed and variable costs. Sorting all the power plants in the market, according to the cost of power production, we obtain the generation stack – the power supply curve. The demand for power (load) is assumed to be inelastic – that is to say it does not depend on the level of power prices. At equilibrium, supply equals demand. All the units with lowest cost of production that are necessary to satisfy demand for power will be producing energy. This condition allows one to determine the marginal generation unit – the unit that sets power price on the market equal to its cost of production. If in addition we assume that the part of the generation stack that is on the margin is essentially a single-fuel one, then the relative cost of production among various power plants is determined by plant heat rate. Thus, one can think of supply curve as a mapping from the load (demand) into marginal heat rate.

Assume that the shape of the supply curve could be roughly approximated as exponential in load

$$MHR_t = \exp\{\lambda \tilde{L}\}$$

where *MHR* is the heat rate of the marginal unit. While *ad hoc*, this functional form is capturing the main feature of supply curve – its convexity.

Note that market-implied heat rate introduced above is essentially a risk-adjusted expectation of the average marginal heat rates for all the tradable days inside the month. Indeed, deterministic deviations between market-implied heat rate and expected average monthly heat rate could be arbitraged away by taking forward power and gas positions to delivery.

Let us assume that load is distributed normally. This implies that, for our exponential stack, market-implied heat rate has lognormal distribution. Note that, although distribution is lognormal, the stochastic process for implied market heat rate is not necessarily a random walk. A usual random-walk assumption would imply that variance accumulates linearly with time. This clearly does not work for load forecast. Load forecast (at least long-term load forecast outside of the prompt month) is done based on unconditional load and temperature distribution (ignoring overall load growth, which contributes a couple of percentage points annually). In other words, uncertainty in load forecast essentially does not accumulate with time beyond a couple of months. Ignoring for a moment overall load growth, we see that load and as a consequence marginal heat rate distributions (in this toy model) are stationary for horizons in excess of a couple of months.

In reality, for time horizons in excess of one or two years, uncertainty in the reserve margins will become dominant. This argument holds for intermediate time horizons or for shoulder months, where uncertainty about reserve margins is much less important. The argument also fails if the generation stack on the margin has several fuels, with volatile spreads between fuel prices. This is the case for the PJM stack in winter, where both coal and gas are on the margin.

Thus, in this simple model of a generation stack, distribution of market-implied heat rate is stationary and, as a result, heat rate variance does not grow with time. However, the standard approach implies linear-in-time growth of the variance of heat rate. One has to resort to the analysis of the empirical data to determine which of the two assumptions correctly describes the behaviour of observed prices.

Next, we look at some empirical evidence that heat rate variance indeed accumulates more slowly than suggested by Equation (9).

Empirical analysis of implied heat rate volatility

To check how well a random-walk assumption holds for market-implied heat rate, let us consider a set of implied heat rate returns over various horizons:

$$r_{t,\tau}^{IHR} = \frac{1}{\sqrt{\tau}} \ln\left(\frac{IHR_{t+\tau}}{HR_t}\right) \qquad (10)$$

where τ is the return horizon and $\tau = 1$ corresponds to daily returns.

Note that, under the random-walk process in Equation (5),

$$r_{t,\tau}^{IHR} = \frac{1}{\sqrt{\tau}} \ln\left(\frac{IHR_{t+\tau}}{HR_t}\right) = \varepsilon \propto N\left(\mu, \sigma^{IHR}\right) \qquad (11)$$

Under the random-walk hypothesis standard deviation of returns $r_{t,\tau}^{IHR}$ will be the same for all horizons τ.

Figure 1 shows the estimation results for the SERC region. For each forward contract in 2004, 2005 and 2006, returns are computed using equation (10), and return horizon τ being equal to one day, two days, five days and one month. Standard deviations are computed for returns of each horizon and each contract. Resulting standard deviations are averaged across all 36 forward contracts and then divided by standard deviation of the one-day returns.

It is clear that implied heat rate volatility displays significant scaling with time. Under random walk, standard deviation *versus* time graph should be a straight horizontal line. On this figure, the standard deviation of the implied heat rate returns over the monthly horizon is less than 70% of what it should be, if random walk, implied in Equation (5), holds.

The example in Figure 1 is from SERC region. However, identical effects obtain in all power markets. In fact, similar effects will show up in the majority of energy commodities. See Chibisov and Wolyniec (2004) for details.

ENERGY MODELLING

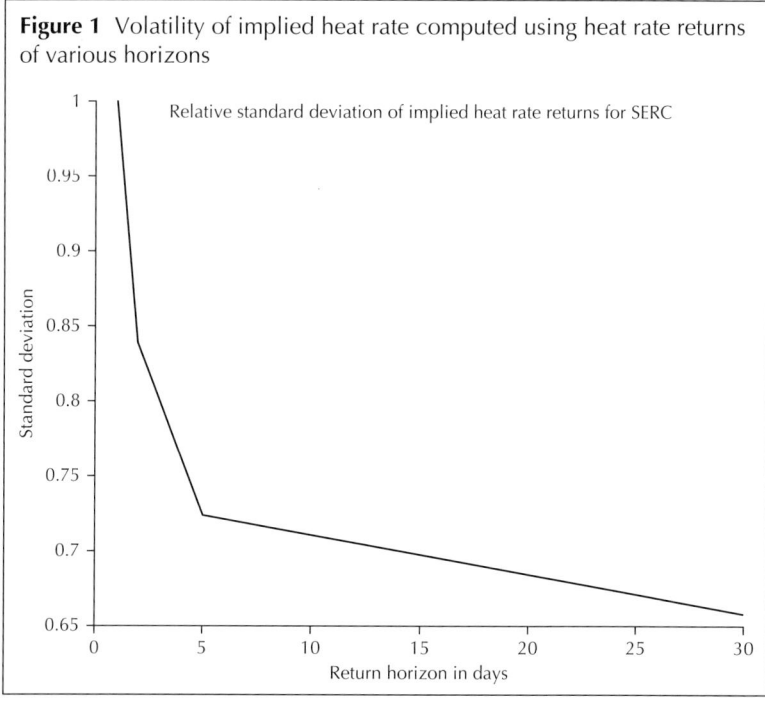

Figure 1 Volatility of implied heat rate computed using heat rate returns of various horizons

The above analysis suggests that empirical data exhibit behaviour that is not consistent with the process specification given in Equation (1), which leads to random walk in an implied heat rate process (Equation (5)).

To be consistent with the observed dynamics of the implied heat rate, it is necessary to introduce a scaling law into the correlation in Equation (1). The relationship between accumulation of the heat rate variance and correlation scaling is described below.

Correlation scaling and implied heat rate variance

Let's introduce the following notations for the correlation:

$$\rho^t_{\{P,G\}} = \frac{E\left\{\log\left(P_t/P_0\right), \log\left(G_t/G_0\right)\right\}}{\sqrt{\text{Var}\left\{\log\left(P_t/P_0\right)\right\} \text{Var}\left\{\log\left(G_t/G_0\right)\right\}}} \qquad (12)$$

Also, $E\{P_t\} = P_0$; $E\{G_t\} = G_0$, where P_0, G_0 are the current forward power and gas prices for the given month. Using (5), correlation between power and gas could be found as follows:

$$\rho^t_{\{P,G\}} = \frac{\mathrm{Cov}\left\{\log\left(IHR_t/IHR_0\right) + \log\left(G_t/G_0\right), \log\left(G_t/G_0\right)\right\}}{\sqrt{\mathrm{Var}\left\{\log\left(P_t/P_0\right)\right\}\mathrm{Var}\left\{\log\left(G_t/G_0\right)\right\}}}$$

$$= \frac{\mathrm{Var}\left\{\log\left(G_t/G_0\right)\right\} + \mathrm{Cov}\left\{\log\left(IHR_t/IHR_0\right), \log\left(G_t/G_0\right)\right\}}{\sqrt{\mathrm{Var}\left\{\log\left(P_t/P_0\right)\right\}\mathrm{Var}\left\{\log\left(G_t/G_0\right)\right\}}} \quad (13)$$

Ignoring for the moment the covariance between the implied heat rate and gas price we get

$$\mathrm{Var}\left\{\log\left(P_t/P_0\right)\right\} = \sqrt{\sigma_G^2 t + \sigma_{IHR}^2(t)} \quad (14)$$

and for correlation

$$\rho^t_{\{P,G\}} = \frac{\sigma_G^2 t}{\sqrt{\sigma_G^2 t + \sigma_{IHR}^2(t)}\sqrt{\sigma_G^2 t}} = \frac{\sigma_G \sqrt{t}}{\sqrt{\sigma_G^2 t + \sigma_{IHR}^2(t)}} \quad (15)$$

where $\sigma_G \sqrt{t}$ is the standard deviation of gas log-prices and $\sigma_{IHR}(t)$ is an arbitrary function of time.

It then follows that, if variance of the implied heat rate grows linearly with time, then the correlation is time-independent:

$$\rho^t_{\{P,G\}} = \frac{\sigma_G \sqrt{t}}{\sqrt{\sigma_G^2 t + \sigma_{IHR}^2 t}} = \frac{\sigma_G}{\sqrt{\sigma_G^2 + \sigma_{IHR}^2}} = \mathrm{const} \quad (16)$$

If, on the other hand, the implied heat rate distribution is stationary, then

$$\rho^t_{\{P,G\}} = \frac{\sigma_G \sqrt{t}}{\sqrt{\sigma_G^2 t + \sigma_{IHR}^2}} \quad (17)$$

with correlation eventually approaching one for longer time horizons.

The two cases above – stationary distribution of implied heat rate and random walk – are extreme ones. In reality, depending on the region, implied heat rate will accumulate variance at some rate. But, as long as variance in heat rate accumulates more slowly than variance in gas prices, correlation will be an increasing function of time approaching (the value of) one for longer horizons.

Empirical correlation estimates are usually carried out on the returns of some small horizon, typically from one day to one week. In the presence of correlation scaling, short-term correlation will differ significantly from the long-term correlations. This partially explains the disconnection between the high values of correlations implied from the heat rate option quotes and the lower values of the correlations observed empirically.

To price spread options in a manner consistent with the heat rate dynamics one has to infer the correlation scaling law for all the applicable horizons. This is a difficult proposition given the scarcity of data. Another possibility is to use the fundamental model, which creates joint power and fuel distribution without assuming any specific process for power prices.

Simulation analysis of correlation scaling

The analysis of data presented above shows considerable correlation scaling for horizons up to one month. To extend this analysis to longer horizons where the data are not available we perform the simulation of energy prices using a fundamental generation stack model (see Eydeland and Wolyniec, 2002, for details of the modelling approach).

The difference between the correlation estimates is illustrated in Figure 2.

Price paths from the fundamental model are used to compute three different correlation estimates. The solid line shows the correlation computed on daily power and gas returns. Power returns are constructed as log-differences of power prices in two consecutive super-peak-time blocks. The dotted line represents cumulative correlation, ie, the one computed on log-return over the horizon that starts at the beginning of the valuation period. Finally, implied correlation is computed based on the variance of the implied heat rate by reverting Equation (4). Each statistic is computed separately for every time step in the valuation period.

The model assumes that the market price for power is set through balancing supply and demand for energy. The first correlation measure is computed on daily price returns. It is not far from empirical correlation estimates. In summer months, correlation

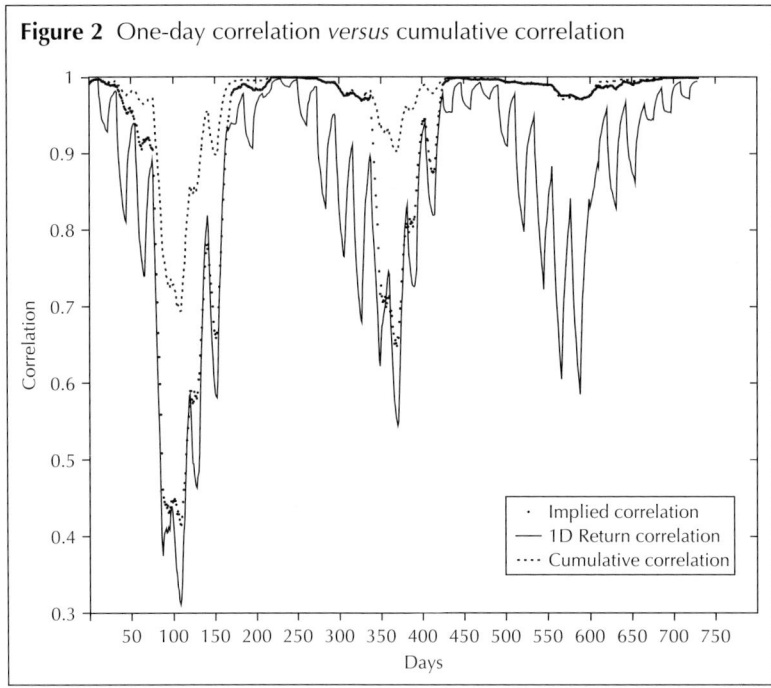

Figure 2 One-day correlation *versus* cumulative correlation

between daily returns can go as low as 20%. Note that, if one uses time series of prices instead of a price sample (which is the only thing that could be done empirically), one would mix regimes with high (shoulder months) and low (summer months) correlation. Long-term correlations (computed on price returns of longer horizon), however, are much higher. Even for summer months they do not go below 70%. For shoulder months the correlation is very close to one. Thus if one were to price summer spread option maturing in two years in an ideal world, described by the fundamental model, the correct long-term correlation would be around 95% rather than 60%, as would follow from analysis of daily returns.

These results are model-dependent. Market imperfections and hourly market balancing most certainly de-correlate power and gas and increase expected heat rate volatility even for longer terms. Other effects, such as severe transmission congestion, will reduce power price dependence on heat rate. In markets that are essentially multi-fuel markets (for instance, PJM in winter months has periods when coal generation is on the margin), power prices will be affected by the spread between fuel prices, which is potentially non-stationary. The non-stationary nature of the fuel spread can lead to the increase in the rate with which variance of heat rate accumulates with time, thus reducing the correlation scaling effect.

Econometrics of correlation scaling
The above attempts to give one intuition for the mechanism that induces the observed high correlation. However, it turns out that the scaling effect can be even stronger than it is suggested by the simple assumptions in Equation (15). The above example relied on the strong time-to-maturity effects to induce the correlation scaling. That suggests that we should see a strong scaling effect in forward correlations. Undoubtedly, we find a strong evidence for this effect. However, somewhat surprisingly, the mechanism of correlation scaling appears on even shorter time scales. Specifically, we will shortly see that the true cash correlation tends to be much higher than simple one-day return analysis would suggest. Here, we will not concentrate on what induces this behaviour. We will only review the econometric mechanism of correlation scaling and propose some estimators of correlation, which enable one to use the standard Margrabe model to properly price heat rate options.

We saw earlier that implied correlations tend to be much higher than the estimated correlations. The estimation procedure involved forming time series of one-day price returns of the component legs (power prices and primary fuel), be it for forward or spot prices. As we will now see, the choice of the return horizon is not trivial.

Non-Brownian extensions
The standard analysis crucially relies on the assumption that the increments of the underlying prices are independent. However, the empirical data suggest the presence of significant (although low) negative autocorrelation across the increments. In general, if the autocorrelation in returns is driven by the serial correlation in the uncertainty increment of the process, it can be easily shown that the presence of autocorrelation changes the effective correlation (for pricing purposes). However, if the serial correlation is induced by the dependence of the drift of the process on the price under the physical measure, the serial correlation has no impact on pricing

(ie, the effective correlation) as long as we can dynamically hedge our exposure.[2]

For example, if the serial correlation in forward prices is induced by mean reversion, the effective correlation for pricing forward monthly heat rate options does not change. In other words, we do not see any scaling effects in pricing. The picture is clear as long as we can dynamically hedge. However, intramonth (cash) hedging is not usually possible. Consequently, serial correlation induced by the drift or the diffusive mechanism will have impact on pricing. In other words, we will see correlation scaling. The scaling phenomenon in cash heat rate is just a simple consequence of the fact that, under the risk-neutral (pricing) measure, cash heat rate does not have to be a martingale. In other words, the cash heat rate can exhibit mean-reversion.

Obviously, in practice, we deal with daily options that see contribution to value from both monthly and cash volatility and covariance. To correctly price daily heat rate options, we need to consider the source of the serial correlation.

Whatever the source and relevance for pricing and proxy hedging, the presence of serial correlation has the following effect on the observed cumulative correlation.

Correlation and autocorrelation
Consider the following situation. We have two series of returns $\{x_i\}$ and $\{y_i\}$. Let's assume that the individual series exhibit serial correlation. This is to say:

$$\text{Correlation}(x_i, x_{i-1}) = \Delta\rho^x \neq 0$$
$$\text{Correlation}(y_i, y_{i-1}) = \Delta\rho^y \neq 0$$

Additionally, that the series exhibit auto cross-correlation:

$$\text{Correlation}(x_i, y_{i-1}) = \Delta\rho^{xy} \neq 0$$

When we measure the correlation between daily returns we find the following:

$$\text{Correlation}(x_i, y_i) = \rho_1^{xy} \neq 0$$

The returns correspond to one-day returns of (say) power and gas prices. Consequently the correlation above is the one-day correlation.

It would seem that, no matter what time horizon of returns we are interested in, the correlation is always the same. However, in the presence of autocorrelation, this is not the case. For example, the correlation for two-day returns will be given by the following formula:

$$\rho_2^{xy} = \frac{\rho_1^{xy}(1 + \Delta\rho^{xy}/\rho_1^{xy})}{\sqrt{(1 + \Delta\rho^x)(1 + \Delta\rho^y)}}$$

We can see that, as long as the autocorrelations are negative and the cross-autocorrelation is positive, the two-day correlation will be greater than daily correlation. This is exactly what we observe in almost all commodity markets. Details about estimation techniques and the results for a wealth of commodity markets can be found in Chibisov and Wolyniec (2004).

The analysis can be extended to long-term returns with similar conclusions that negative serial correlation induces higher long-term correlation. The effect is compounded with the increasing length of the returns.

Correlation estimation

The previous section suggests that in order to find the correct correlation to use we need to consider estimating correlations on returns of horizons more than one day. The correct horizon is a function of the term to expiry of the structure in question. For daily options at the expiry of the forward contract (cash options), the maximum return horizon is 30 days (corresponding to the length of the exercise period). In truth the "effective" length will be shorter since the daily exercise of options gives us a collection of options with the average maturity of half the trade period (ie, 15 days). We can measure the correlation on overlapping intervals (the autocorrelation induced in this manner reduces efficiency, but it does not bias our estimates). For example for January 2005, the correlation estimates between Mass Hub power price return (NEISO) and Algonquine gas index returns would yield the results seen in Table 5.

As we can see, even the 15-day returns do not fully capture the scaling effects. We need to employ more sophisticated techniques. We do not have space here to cover the details, but the interested reader is referred to Eydeland and Wolyniec (2006).

Table 5 Scaling of correlation

	One-day returns	2-day returns	15-day returns	True correlation
Correlation	69%	75%	80%	94%

Conclusion
It is worth noting again that the scaling mechanism described here is relevant for pricing as long as the true forward process is non-Brownian. In a diffusive setting, any effects due to mean-reversion or, equivalently, co-integration (equilibrating mechanisms), will have *no impact* on pricing as long as we have the ability to dynamically hedge both legs of the option. The situation changes dramatically if one leg (for example, power) is not very liquid, and dynamic hedging is not feasible. In such a situation, we can use proxy hedges (hedging power with natural gas, for example). The effectiveness of the proxy hedge is greatly affected by the equilibrating mechanisms (co-integration) and consequently the correlation scaling effects become important no matter what their source.

Given all the above, a natural question is how we can identify whether we are dealing with a diffusive or non-diffusive (non-Brownian) environment.[3]

Valuation strategies

The arguments presented above suggest that consistent estimation of spread option values in the absence of liquid marked quotes is not an easy task. There seem to be three major valuation approaches consistent with the observed dynamics of market heat rates. The first one is to build a fundamental model of the electricity markets. It avoids all the problems of estimating correct correlation altogether, since it produces joint distribution of power and fuels from the fundamental market drivers. The second approach is to model the heat rate distribution directly, incorporating all the heat rate constraints directly into the model. The joint structure of market-implied heat rates and fuel prices can be ignored in the first approximation. Finally, the third approach is to continue with the specification given in equation (1), but make correlation dependent on the maturity of the option.

As we explained earlier, we need to be careful to make sure that the restrictions on the heat rate distribution (or correlation scaling) are relevant for pricing. For example, if both power and natural gas forwards follow a mean-reverting process with the same parameters (under the physical measure), the heat rate process will also follow a mean-reverting process. However, the resulting distributional "constraints" have no impact on pricing of heat rate options, since under the risk-neutral measure mean-reversion (drift) has no impact (for details see Rebonato, 2004). In this specific case, even though we can observe correlation scaling in empirical data, it is of no interest to us as long as we are able to dynamically hedge with both legs of the spread.

In general, however, we need to be concerned about the distributional "constraints" or, equivalently, correlation scaling.

The time dependency can be estimated on the empirical data (at least for shorter time horizons) and then extrapolated in a manner suitable for the task at hand. (For example one can assume that there is no further scaling in the heat rate volatility beyond the largest horizon for which empirical estimates exist. This will place an upper bound on the spread option value.) Under certain distributional assumptions, the three approaches are fully equivalent. The choice of the methodology depends on the availability of market and non-market information. The fundamental approach has the advantage of incorporating non-market information (see Eydeland and Wolyniec, 2002), but it is cumbersome and requires extensive maintenance. The other two approaches are pretty much equivalent, and the choice is an issue of convenience and the structure of traded underlying contracts.

CONCLUSION

In this chapter, we have presented an overview of heat rate options and several issues in estimating the value of those structures. We have been able only to touch on the pertinent issues. The topic is vast and requires more extensive treatment.

---

1   An aside here. We might be also using the wrong model. However, we can always use the standard model for valuation as long as we use the correct free parameter: ie, correlation. This is fully analogous to standard options and the concept of implied volatility.
2   For accessible exposition see, for example, Rebonato, 2004.

3 Unfortunately, we have only space here to signal the issues without resolving them fully. The full treatment will be available in Eydeland and Wolyniec (2006).
4 A more thorough analysis of the impact of equilibrium conditions on the joint distributions of commodities (not limited to power and fuels only) can be found in Chibisov and Wolyniec (2004). Detailed analysis of the pricing, hedging and structuring issues can be found in Eydeland and Wolyniec (2002). An extensive update of that analysis, incorporating all the latest research, will be available in Eydeland and Wolyniec (2006).

REFERENCES

Bjork, T., 2004, *Arbitrage Theory in Continuous Time*, 2nd edn (Oxford University Press).

Carmona, R. and V. Durrleman, 2003, "Pricing and Hedging Spread Options", *SIAM Review* **45**(4), pp. 627–85.

Chibisov, B. and K. Wolyniec, 2004, *Correlation Scaling in Commodity Markets* (available on request from authors).

Eydeland, A. and K. Wolyniec, 2002, *Energy and Power Risk Management*, 1st edn (New York: John Wiley & Sons).

Eydeland, A. and K. Wolyniec, 2006, *Energy and Power Risk Management*, 2nd edn (New York: John Wiley & Sons).

Haug, E., 1997, *The Complete Guide to Option Pricing Formulas* (New York: McGraw-Hill).

Rebonato, R., 2004, *Volatility and Correlation* (New York: John Wiley & Sons).

12

Credit Risk Management for the Energy Industry – Some Perspectives

Vincent Kaminski; Vasant Shanbhogue

Citigroup; AIG Financial Products Corp*

The US merchant industry went into a tailspin in 2002, following the Enron bankruptcy and discovery of various questionable practices related to energy trading. The most popular explanation for this is that the crisis was caused by the credit rating agencies that suddenly started applying much stricter and often possibly unreasonable standards to the energy merchants. Our view is different: the crisis of energy trading and marketing was due to the fundamental flaws in business strategy common to many companies in the industry, such as highly leveraged balance sheets, low profitability of traditional market-making business and inflated organisations that overexpanded because abuses of mark-to-market (MTM) accounting masked poor financial results and justified excessive investments in physical and human trading infrastructure. It is true that credit rating agencies aggressively downgraded the parent companies of many energy trading and marketing entities, precipitating the onset of the crisis of the entire industry. These actions, in our view, represented the delayed recognition of the existing problems that were being ignored for too long. The downgrades alone did not cause the damage that the industry had inflicted upon itself.

One of the major flaws of the business model of the merchant energy industry was the approach to management of credit risk that, in combination with other factors, undermined their long-term

---

*The statements and opinions in this chapter are the authors' own, and such statements and views are not attributable to the authors' employers, which did not participate in the preparation of the chapter.

viability. This chapter contains a review of the industry approach to management of credit risk and discusses some of the analytical tools that can be useful in assessment and mitigation of credit exposure.

CREDIT RISK: DEFINITION

Credit risk is defined as a potential loss resulting from the failure of a counterparty to perform under a contract. The critical aspect of credit risk is its relationship to market risk and its asymmetric character. Credit exposure is a mirror image of a MTM gain or loss. One can lose through the counterparty's non-performance only what is owed as a receivable or what represents a MTM gain. In bankruptcy, a defaulting party can reject losing contracts and retain contracts that are to its advantage. This explains a well-known graphical representation of credit risk that looks like a hockey stick. Of course, this picture is somewhat simplified, as a non-defaulting party has certain safeguards against cherry picking through the ability to net similar exposure and be at risk only with respect to the sum of negative and positive exposures.

Credit risk is a subject of extensive research both by academics and practitioners who specialise in development of financial instruments that can be used for transfer of credit risk to another party. Most academic research applies to the events of default on corporate debentures or bank loans, resulting from bankruptcy or insolvency of a public or private company. The event of default on a commodity contract is not as precisely defined as a default on a corporate debenture or a financial loan and this represents a unique challenge to a practitioner in the energy markets.

There are many shades of non-performance that may assume such forms as contract frustration, demands to renegotiate a transaction and legal actions exploiting the imprecise and vague language of legal documents.

Contract frustration may consist, for example, in insisting on very strict compliance with even the most minute contract provisions where, by tradition and industry conventions, some flexibility is left to each counterparty. One firm in the past tried to exit a highly unprofitable gas purchase contract by complaining about the quality of natural gas being delivered. In another case, a public agency was restarting and shutting down a combined-cycle power plant from which it had contracted electricity under a highly unprofitable long-term transaction, causing major disruptions in operations and an excessive wear and tear of the turbines. Such practices may become very expensive to a firm with a credit exposure, and typically they represent an alternative, or the first step, to a legal action. Often a counterparty facing a high loss will insist on renegotiation of a contract, threatening to discontinue future business relationships. Renegotiations often lead to a new transaction with modified terms that formally leave the firm whole from the point of view of MTM valuation but expose it to more risk over a longer time period. For example, a losing firm may renegotiate the contract price and agree to the extension of the tenor of the transaction.

The further complicating factor is that two counterparties are often dependent on each other and they don't want to create a situation leading to a disappearance of the other side. A producer or marketer of natural gas needs an industrial customer as much as the industrial customer needs physical fuel input. In many cases, contracts involve physical arrangements for commodity transfers that require long-term cooperation and involvement of personnel with highly specific skills, and cannot be recreated overnight if one side disappears. This is why managing credit in the energy business is often similar to extending and/or renegotiating international debt. The credit problems are addressed not through bankruptcy and reorganisation but through long and complex negotiations, resulting in mutual concessions. This is why mechanical transplantation of the analytical tools borrowed from the financial markets to manage energy risk is often a bad practice.

Also, another complication arises from the fact that some transactions are subject to regulatory review and can be invalidated through regulatory oversight. A counterparty to an energy transaction is often a public or semi-public (or heavily regulated) entity. The danger one faces is that the counterparty may enter into a contract that is incompatible with its charter. Such a contract may be determined by the courts to be *ultra vires*.[1] One can invoke the case of two British municipalities transacting in the financial markets in the early 1990's.

> The United Kingdom's House of Lords determined that the London Borough of Hammersmith and Fulham lacked capacity to transact in derivatives linked to interest rates. Not only were contracts dating back to the mid-1980s with that borough declared void, but contracts

with over 130 other councils were effectively invalidated. A number of derivatives dealers suffered losses.[2]

In the energy markets, especially in the international transactions, this exposure is very high.

The challenge a practitioner faces is to use models developed for credit risk defined for a different purpose in a creative way to address the problem at hand: management of credit risk in the portfolios of energy-related transactions. We hope that this chapter will contribute to the effort in this area.

CREDIT RISK MANAGEMENT: CURRENT PRACTICES

The credit risk management practices that evolved during the 1990s were defined around management of bilateral exposures and, as the subsequent developments demonstrated, contained the seeds of future problems that transpired to be very serious for the health of the entire industry.

The credit risk management process starts with the signing of one or more documents that define the rules for processes such as margining, netting, close-outs, etc. The industry developed a number of standardised contracts that contain credit language, either in the body of the document or in a special annexe. The most widely used documents are:

- International Swap Derivatives Association Inc (ISDA) – Master Agreement
- Edison Electric Institute (EEI) – Master Purchase & Sale Agreement;
- Western Systems Power Pool (WSPP) – Western Systems Power Pool Agreement;
- North American Energy Standards Board (NAESB) – Base Contract for Sale and Purchase of Natural Gas; and
- Gas Industry Standards Board (GISB) – Base Contract for Short-Term Sale and Purchase of Natural Gas

The large number of overlapping credit documents can be explained by the evolution of the merchant energy business in the 1990s, when energy trading expanded from the crude and refined products markets to natural gas, and later electricity, coal, emissions and other commodities and instruments. Given the fundamental differences between the physical attributes of different commodities with respect to production and distribution, and varying levels of sophistication of the counterparties, it was natural to develop and use specialised credit documents. In many cases, the industry was rushing forward, trying to conquer new deregulated markets, with competitive pressures leaving little time to build a solid legal infrastructure for merchant operations. Over time, this approach produced many inefficiencies and increased costs of doing business and increased uncertainty with respect to the legal foundations of the industry.

These documents are typically used as templates, providing a starting point, and are typically modified through a process of long and laborious negotiations. The modified documents can differ for each counterparty and this creates the first obvious problem: one has to maintain the expensive machinery in place to manage credit exposures that may be defined differently for different firms. Note: in case trading with a counterparty starts before the documents are negotiated and signed, one has to use a long-form confirmation that contains detailed credit language. The negotiated documents cover basic issues critical to credit risk management, which are discussed below.

Netting
Netting means offsetting positive credit exposure on one set of transactions with a given counterparty with negative exposure on another set of transactions. In energy trading, the netting rules may become very complicated as netting may apply differently, depending on the language used and the current law, to:

a. transactions related to different physical commodities (for example, natural gas and power);
b. physical and financial transactions;
c. transactions with different affiliates of a given counterparty; and
d. transactions across different jurisdictions.

The industry is slowly evolving towards more comprehensive arrangements that will allow for more extensive netting across all the dimensions mentioned above. This can be accomplished by negotiating a master netting agreement that will bridge and cross-reference different credit documents and will eliminate potential discrepancies and differences in the legal language.

This chapter should not be interpreted as the legal guide to the rules that apply to netting under bankruptcy. The legal profession has not yet agreed on the interpretation of the US laws in this area and the current practice varies from firm to firm. Some companies assume that netting will apply across all the dimensions. Other firms take a more conservative approach and act under more restrictive assumptions, given that the laws have not been extensively tested under bankruptcy conditions. A credit officer should be aware of the differences in interpretation of the law and work with the company lawyers to arrive at acceptable practical guidelines. They should be based on the review of specific credit agreements executed by his or her firm.

Margining
A company vulnerable from the credit point of view seeks protection in the form of collateral (cash, letters of credit, marketable securities, guarantees) from the counterparty. The level of collateral depends on a number of factors such as netting of credit exposures and thresholds, that is, the minimum levels of exposure above which the margining begins. The thresholds may be dynamic: their level may depend on predefined factors, such as credit quality of the counterparty, defined in terms of credit ratings or certain financial ratios that may evolve over time. The documents define also the rules for calculating exposures, the way of resolving differences in valuations, the frequency of exposure monitoring and frequency of, and time allowed for, collateral transfer.

Dynamic thresholds can trigger demand for additional collateral in two different ways. Under one approach, the collateral threshold levels depend on the credit ratings by one or more recognised credit rating agencies. The thresholds would be adjusted downward in case of a negative credit rating adjustment and would go to zero if the credit rating dropped below investment grade level. In case of disparity between credit ratings of two counterparties, the thresholds could be significantly higher for the higher-credit-quality counterparty. Under the second approach, known as adequate assurance, one counterparty can ask, at least in principle, for full collateralisation in case of an adverse development related to the counterparty, with the definition of an adverse development left to the discretion of the creditor.

One of the consequences of using dynamic thresholds was a chain reaction of credit difficulties, once the process of downgrades of energy merchants started. Lower credit ratings translated into lower thresholds and this in turn translated into higher working capital requirements related to additional collateral that had to be posted. Impaired credit ratings made it more difficult for the affected companies to borrow additional funds, leading to emergency measures such as forced liquidation of assets or trading books.

The credit agreements negotiated by the counterparties were often administered in a somewhat haphazard way. For example, in some cases one counterparty would waive its rights to receive collateral for commercial reasons: margining would preclude a weak counterparty from transacting. Of course, this practice was quite myopic, since it produced asymmetry in margin collections and postings. In consequence a portfolio with a very low market risk could contain a huge credit risk or a cashflow risk, since cash would have to be posted on one side and collateral was not collected for the offsetting (hedging) transaction.

The practice of managing credit through bilateral contracts and the characteristics of the energy markets explains why working capital needs in the energy trading are significantly higher than in the case of financial trading. Financial institutions, and financial hedgers, can rely to a larger degree on organised exchanges that offer the benefits of multilateral netting and clearing. Bilateral arrangements allow for limited netting. Many energy transactions have long tenors and significant changes in the levels of forward prices and volatilities can produce drastic changes in the level of MTM valuations and credit exposures. Many energy companies have exacerbated the problem through inefficient and irrational management practices. The most common mistakes, signalled above, include the following.

- ❏ The failure to insist on the rights to collateral collection and to negotiate credit agreements with many counterparties. These practices resulted in asymmetric collateral collection and increased the need for working capital.
- ❏ Some counterparties refuse to post collateral though they have rigid requirements regarding collateral that has to be posted by the counterparties. For example, the wholesale power marketers participating in the Basic Generation Service (BGS) auctions have to post collateral with the New Jersey utilities they supply, but the margining requirements are unidirectional. A utility does not have to post collateral with a supplier.

- In many cases, counterparties with stronger credit ratings can substitute letters of credit or corporate guarantees in place of cash or cash equivalents. This means that a weaker counterparty that has to post cash does not receive offsetting cashflows, even if it is running a balanced book of business. This problem may be exacerbated in the case of a liquidity crisis and a confidence crisis. A counterparty with a rapidly deteriorating credit quality may face calls to post collateral, facing at the same time substitution of letters of credit for cash and a deteriorating liquidity position. Of course, the behaviour of stronger counterparties is fully justified, given the current practice of collateral management. The funds posted as collateral are generally not kept in a segregated account but are commingled with other company funds and can be used for general corporate activities, such as meeting the payroll (this means that in many cases the funds can be lost through mismanagement and fraud). This is known as the so-called rehypothecation risk. Posting letters of credit as collateral offers an additional layer of protection in the case of potential bankruptcy proceedings.
- Trading was often carried through many different subsidiaries operating under different rules and in different jurisdictions. For example, a regulated utility could engage in merchant energy business under one umbrella and buy fuel for the regulated business through a different unit. The existing credit agreements sometimes could fail to guarantee netting across different corporate structures.

EVALUATING A COUNTERPARTY

Credit risk analysis in the energy industry poses unique challenges related to the heterogeneous character of the institutions functioning in these markets. Potential counterparties in energy trading are entities functioning under different laws, varying with respect to corporate governance, subjected to different laws, regulatory oversight, and reporting requirements. A potential trading partner can be one of the following.

- A hedge fund operating under limited regulatory scrutiny, with none or only limited information available about its strategies and finances. It is estimated that there are currently 300 hedge funds operating in the US that have energy-related strategies. It is worth noting that hedge funds are getting active in equity investments, which have their own kind of risk different from trading. This blurring between hedge funds and private equity firms only makes credit management more complicated, and highlights the point that the whole company needs to analysed and not just a component operation.

- A regulated utility that operates in the markets to procure fuel, improve efficiency of its physical assets, such as natural gas storage and generation units, sell excess supply or acquire energy commodities to cover its short position.
- An unregulated affiliate of a regulated utility, engaging in proprietary, speculative trading, providing merchant energy services in the wholesale and retail markets, and offering risk management tools to the end users and producers of energy commodities.
- A producer of energy commodities: an independent exploration and production company (natural gas, crude, coal) or an oil and natural gas major.
- A merchant electricity producer operating in the markets by managing its generation assets and also offering merchant energy services, and trading on its own account.
- A big industrial consumer of energy or a transportation company.
- A big financial institution (a commercial bank, an investment bank) operating an energy-trading desk.
- A standalone merchant energy company.
- A stand-alone physical asset developed under a project-finance venture.

The challenge a credit analyst is facing is that their company may be transacting with many different entities operating under the same corporate umbrella but under different credit agreements and with different rules applying to netting of credit exposures and different levels of explicit and implicit credit support from the parent. In many cases, the credit support of the parent may take the form of an explicit guarantee but in many cases it is implied by the corporate structure and may be withdrawn under the conditions of financial distress. One of the critical preconditions of effective credit risk management practice is a customer database that captures the information about different counterparties, the negotiated credit agreements, the rules that apply to netting of credit exposures, credit guarantees in place, credit insurance and credit derivative transactions, approved credit lines and the rules for collecting and posting collateral. Given the complexity of the corporate structures of potential counterparties, and complicated parent–child relationships, the design and administration of such a database is a very challenging task. Many credit-risk-related trading disasters happened due to making excessively optimistic assumptions regarding the credit support that can be extended by a parent to an affiliate entity.[3]

MULTILATERAL NETTING AND CLEARING

The crisis of energy trading that started in 2002 in the US triggered an industry-wide review of credit management practices. The bilateral approach to credit management was correctly identified as one of the root causes of the problem. It is not surprising that the industry started searching for ways to utilise advantages of multilateral netting and clearing.

Multilateral clearing and netting is usually identified with transacting on an organised exchange, although technically it is not necessary. The essence of clearing is that a bilateral transaction is broken up into two separate transactions and a central institution is substituted as counterparty to each original leg. This process is called *novation* and offers a number of benefits to the participants. A typical clearing house has direct members who participate in the clearing process, and can submit, in addition, trades on behalf of their customers. It is important to note that the safeguards that a clearing house offers apply directly to clearing members and not to their customers, who benefit only indirectly. Once a central counterparty is substituted for the counterparty, netting across offsetting positions becomes possible (see Davidson, 2003).

The clearing house requires that all the cleared transactions be margined. Margining takes the form of initial margin that represents only a small portion of the value of the underlying contract and it gives the participants the benefits of leverage. Variation margin reflects daily changes in the MTM value of the position and has to be posted typically within a day. The variation margin posted by

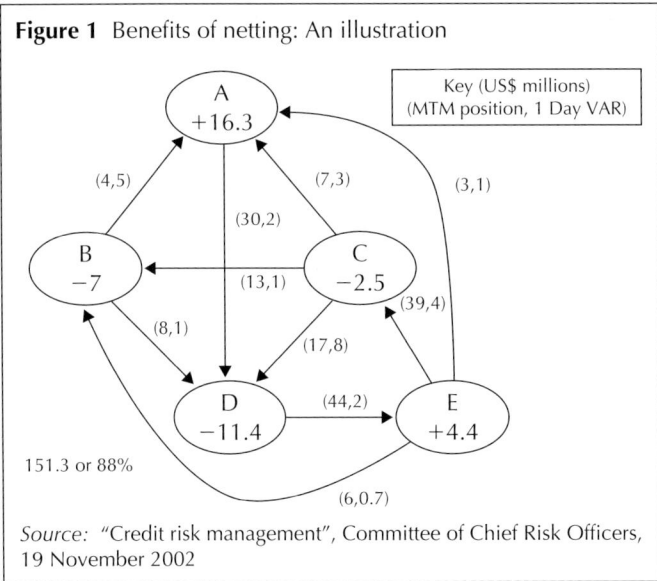

Figure 1 Benefits of netting: An illustration

Source: "Credit risk management", Committee of Chief Risk Officers, 19 November 2002

a counterparty that incurs a loss is credited to the account of a counterparty that has a MTM gain.

A clearing house additionally collects a special fee from direct members, which provides additional protection against direct exchange members.

The benefits of netting may be illustrated using a well-known example included in the White Paper on Credit published by the Committee of Chief Risk Officers in 2002. The example assumes that five companies (A through E) engage in interlocking transactions in the energy markets. Figure 1 shows the magnitude and direction of credit exposure. An arrow going from B to A shows the direction of credit exposure, in this case US$4 million (the second number shows daily value-at-risk – or VAR) and this number means that A stands to lose in case of default by B (B represents a credit risk from the point of view of A). In other words, A has entered into transactions with B that have a net MTM gain to A of US$4 million. The information contained in Figure 1 is summarised in Table 1 for easier aggregation. The rows show credit exposure of companies listed at the left-hand side with respect to the companies listed at the top. The diagonal is always 0, as a company has no credit exposure to itself. One can see that the total

Table 1 Bilateral credit exposures

		With respect to					
		A	B	C	D	E	Total
Credit exposure of:	A	0	4	7	–	3	14
	B	–	0	13	–	6	19
	C	–	–	0	–	39	39
	D	30	8	17	0	–	55
	E	–	–	–	44	0	44
Column total		30	12	37	44	48	171
Row total		14	19	39	55	44	–
Difference		16	–7	–2	–11	4	0

Source: "Credit risk management", Committee of Chief Risk Officers, 18 November 2002.

credit exposure is equal to US$171 million. The sum of the rows shows the total credit exposure of each company. The sum of the columns shows the total credit risk a company imposes on others. One can calculate the difference between the sum of columns and the sum of rows. A negative number in this case shows that a company (B, C and D) is a net creditor. B, for example, has a total credit exposure of US$19 million, but imposes credit risk on others of only US$12 million. The total of net credit exposures is zero, as one could expect to be the case in a closed system. The negative numbers in the bottom row add up to –20; the positive numbers add up to US$20 million.

If a system based on multilateral clearing and netting is introduced, the amount that is required for collateral drops from US$171 to US$20 million, a very significant reduction. This results from the fact that all exposures are transferred to, and assumed by, a central counterparty, and the offsetting exposures are offset and extinguished.

Given obvious advantages of this approach, many industry observers were surprised that netting and clearing was not used on a larger scale. The slow pace of acceptance of the proposed clearing platforms was, however, expected by many practitioners.[4] We agree in principle with the general conclusion of the example produced in the White Paper. We believe, however, that the Paper paints an excessively optimistic picture of the potential for netting. The example includes companies with credit exposures going in both directions and all of them willing to "play in the same sandbox" to address the credit exposure problem. This is why an overall level of exposure in this simple system was relatively small.

One omission in this example is that it does not take into account any adjustments for default probabilities and recoveries. Such adjustments may very well make the net exposure significantly larger and different from zero. Differing credit qualities of different firms impose different levels of costs on firms participating in such a joint programme – so they may become unwilling to participate. Moreover, in many cases a company's credit exposure may show to be very high or very low due to missing pieces in the matrix of transactions. For example, if an airline hedges its fuel risk and the prices run up or drop a lot subsequently, the credit exposure may become very high. There is no offset to the credit exposure from this fuel purchase – the opposite exposure is related to the future revenues from ticket sales that will materialise over time and that don't show up as an offsetting claim or a credit exposure.

So one of the main issues here is that clearing and netting apply only to the transactions that can be marked to market daily, in a non-controversial way. This can be easily done in the case of the short-term, standardised transactions. Unfortunately, the biggest credit exposures arise typically in the long-term, highly structured and unique transactions. Given the current level of market transparency and the absence of a generally accepted valuation procedures in the energy industry, it is more difficult to use clearing in this case, compared with the financial markets. The industry could, of course, agree on a valuation by a third party, but other obstacles would remain. Many long-term transactions contain confidentiality provisions that may make some counterparties reluctant to submit the contracts to the clearing house.

A second impediment to wider reliance on netting is the physical aspects of most energy transactions. Many energy transactions result in physical delivery that requires access to the transportation, transmission and storage infrastructure. The choice of counterparty is often motivated by its control of specific physical assets and experience in managing the supply chain in a specific region and even for specific grades of a given commodity. A clearing institution does not control physical assets, and this means it cannot offer guarantees of physical performance, as opposed to financial performance. The latter is very important but one cannot overlook the fact that many end users of energy attach overriding importance to having uninterruptible and secure supply. A fuel manager of a utility or an industrial plant is relatively less sensitive to cost (that can often be transferred to the ratepayers or represents a relatively small percentage of overall production costs) but risks his or her career if a physical shortage develops.

The reliance on multilateral clearing and netting is promoted as a solution that lowers the trading costs and working capital needed for collateral, but the true picture is more complicated even here. Many energy companies negotiated arrangements with their counterparties that allowed them to avoid posting collateral below certain exposure levels, on conditions of reciprocity. Collateral may, of course, be required in the case of an unusual market move that increases credit exposure or in the case of an adverse credit rating event. Using solutions based on multilateral clearing requires posting initial and variation margins even for small positions. Given that a novation of a bilateral transaction onto an exchange requires consent of two counterparties, this creates an obstacle to clearing.

The comments made above indicate that credit risk will continue to be a critical constraint in the merchant energy business. This is why a system for estimation of the overall credit exposure of a portfolio of energy-related transactions is critical to any company in this business. The remaining part of the chapter reviews the basic architecture of such a system.

CREDIT MODELS FOR THE FINANCIAL MARKETS – A REVIEW

There are three classes of model to analyse credit risk. The first, called *structural models*, are most popularly represented by the Moody's KMV approach. This is based upon the Merton framework of analysing the assets and liabilities of a firm. Equity is viewed as a call option on asset value, and risky debt is viewed as a combination of risk-free debt and a short position in a put option on asset value. The strike for these options is taken to be the amount of debt. The advantage of the Moody's KMV approach is that the concept is intuitively defensible, and the methodology is widely available as a commercial package for companies to use in their risk management operations. In general, the structural models try to weave a theory of *why* a firm defaults.

The second class, called *reduced-form models*, are popular among practitioners for their tractability in coming up with pricing models for varieties of credit derivatives. In this class of models, the focus is not on why a firm defaults, but on using a parameter (called default intensity) to represent the term structure of default probabilities. Parameter values are fitted from observable data such as bond prices and credit default swap prices, and then used in pricing models. One example of such fitting is the estimation of the default probabilities for different time periods from the prices of bonds of different tenors, as suggested by Fons (1994).

Both these classes of model have shortcomings that practitioners tend to overlook for lack of any convenient fix. To address these shortcomings, a third class of model, called *incomplete information models*, have recently been studied, for example, by Duffie and Lando (2001), and by Giesecke and Goldberg (2004), that combine the tractability of the reduced-form models with the attempt to explain the cause of defaults as in the structural models.

In this chapter, we will avoid going into detailed descriptions of these three classes of model, as there are already excellent papers doing this (see Giesecke, 2004). We will, however, give a brief overview of varieties of structural models and reduced-form models, and how they relate to the incomplete information models. In particular, we will mention the applicability and usability of these models for energy companies.

STRUCTURAL MODELS FOR CREDIT RISK
Structural models try to explain why a firm defaults on its debt by explicitly modelling the value of the firm assets.

We elaborate a little here about what constitutes assets and liabilities. Consider a firm with current assets, property, plant and equipment, and other securities that have a certain market value. For this analysis, we consider market value as opposed to book value, because market value is more relevant to the value that a firm may receive for its assets from an independent buyer. Most models treat asset value as a stochastic variable (and we do the same here) but it is well worth noting that composition of asset value would depend on whether the expected scenario to pay-off liabilities would involve a sale of the firm as a whole or involve a liquidation. Line items such as goodwill on a balance sheet may not be relevant in a liquidation scenario, even though these line items may arise naturally over the course of a business, such as during acquisitions. This should be taken into account when an asset is modelled.

The liability side of the balance sheet would include debt and equity, and we must be sure to include the consideration of off-balance-sheet liabilities as well. This is another potential stumbling block for credit models, as estimation of off-balance-sheet liabilities can be difficult for an analyst. Analysts must be sure to study the history of a company and make allowances for potential off-balance-sheet liabilities in their credit models. This may be accomplished by allowing for the debt and accounts-payable liabilities to be in a range and examine their impact.

Merton's framework takes a simplified view of taking the debt and accounts-payable liabilities as a single known number, or barrier. The asset value (more specifically, the market value of the assets of the firm either in liquidation or as a block for resale) is then modelled and examined relative to this barrier.

Equity holders own the net assets of the firm, or assets less debt and accounts-payable liabilities, if this net amount is positive. So equity is viewed as a call option on asset value.

If we refer to the cumulative amount of debt and accounts payable as risky debt, then this is viewed as a combination of risk-free debt (or a promise of certain payment for the debt holder) and a put option on asset value. This approach can be taken many steps further by looking at different tranches of debt. The senior most tranche could be viewed as a combination of risk-free debt and a put option on asset value. The next tranche could be viewed as involving a put option on a portfolio consisting of the firm assets and the senior most tranche debt obligations, and so forth. For now, we stick to the simple view of debt as a fixed amount of zero coupon debt.

The simplest way to look at this is to consider a fixed amount of zero-coupon debt that is due on a fixed date T in the future. Then the firm is considered to default if the firm asset value at time T is less than the amount to be repaid to the debt holders at time T. However this simple approach fails to take into account the behaviour of firm asset value prior to time T.

The extension to this approach that looks at the entire path of firm asset value is called the *first-passage time model*. In this approach, the firm is considered to default on its debt if the firm asset value drops below a fixed barrier D any time before the final payment date, *or* if the firm asset value at time T is less than the amount to be repaid to the debt holders at time T. The barrier D is assumed to be known in advance, and Giesecke (2004) discusses the implications of D's being greater or less than the debt amount.

Since the structural models focus on modelling firm asset value, calibration of these models naturally looks to equity prices and volatilities.

REDUCED-FORM MODELS FOR CREDIT RISK

Reduced-form models do not try to explain why a default occurs. Rather they model the default rate as an entity with its own dynamics through time. No attention is paid to the capital structure of the firm, and how the value of the firm's assets evolves. In particular, since we do not look to a smooth evolution of the value of firm assets, the firm is allowed to default suddenly at any time, as is observed in practice. However, these models do depart from intuition in that the key parameter of these models is a default intensity that is exogenously specified and does not relate to how a firm operates.

Calibration of these models is more naturally done with data on bond prices and credit default swaps. Equity prices and volatilities play no role in this calibration.

INCOMPLETE INFORMATION MODELS FOR CREDIT RISK

There are two styles of incomplete information models, and both of these are based on extensions of the first-passage time structural models. In one style, the default barrier is taken to be the optimal liquidation value of the firm, as determined by the equity holders, and uncertainty is attributed to the specification of the process for firm value. This is the approach taken in Duffie–Lando. The specification uncertainty, or "imperfect information", leads to the formulation of a default intensity process with the characteristic that the implied credit spreads do not go to zero as tenor goes to zero. In another style, the process for firm value is assumed to be well specified; however, uncertainty is attributed to an explicit default barrier level, which has an expected value and a volatility. This is the approach taken by Giesecke–Goldberg.

An important characteristic of the output of these models is that they predict strictly positive credit spreads even in the short term for risky bonds. This intuitively allows for sudden jumps in the firm value leading to default, and incorporates the fact that not all of the information about a firm may be publicly available (see Duffie and Lando, 2001). In contrast, the structural models assume that the process for firm value is well specified and there are no sudden jumps possible, so that defaults may be anticipated to some extent. In this case, the bond value will smoothly transition to its recovery value, and the predicted credit spreads in the short term will be close to zero, which is against intuition.

BOND DEFAULT *VERSUS* CONTRACT DEFAULT

So far, we have described models that are used to capture bond default. For energy firms that have many transactions such as financial swaps, physical sales and purchases, and financial and physical options, they are exposed to counterparty credit risk on these transactions. The reality of such transactions is that there typically are no cross-default provisions between these transactions and the firm's debt, and, even if there are cross-default provisions, there is a grey area that leads to dispute resolution rather than cross-default.

For example, a physical sales contract of natural gas at a particular point on a particular pipeline is typically subject to force majeure clauses. This means that, for a variety of reasons including "acts of God", unforeseen or unavoidable events, physical delivery may be interrupted by the seller with no penalty. Events that allow such excuses typically include natural disasters, wars, riots and performance failures of parties not controlled by either of the contracting parties. One point of dispute on contract performance could be whether an event was a force majeure event or not.

Another example of a dispute that may come up for a physical contract is about the quality of the commodity being delivered, and whether the quality is within acceptable tolerance or not. Resolution of such disputes takes time, and the onset or resolution of such disputes may have nothing to do with the firm asset value at those times, or with the manner of evolution of firm asset value.

We consider a firm to default on its debt if the firm does not make its contractual payment of interest and/or principal to the debt holders. We consider a firm to default on a general contract if the firm does not make any required contractual payment or deliver/receive physical commodity, as defined in the deal. In considering contract defaults, one must note that a firm typically has dealings not just with a single counterparty but with multiple counterparties. For commodity transactions, the firm may buy commodity from some counterparties and sell to other counterparties – the direction of transaction (buy/sell) would depend on the natural inclination of the counterparty (producer/consumer/market maker) and the extent to which the counterparty participates in the market. As a consequence, if commodity prices go up, then the MTM value of the firm's deals with one counterparty may go up (for example, if the firm has bought commodity at a fixed price from the counterparty), and the MTM value of the firm's deals with another counterparty may go down (for example, if the firm has sold commodity at a fixed price to the counterparty). The overall financial health of the firm would depend on the extent to which the firm buys and sells and what its existing asset position is. Without extensive examination of a firm's portfolio, which is typically not publicly available even for publicly traded firms, it is then hard to judge what the correlation is between commodity price moves and overall financial health of the company – all this even before allowing that the firm could have positions in many different commodities.

The financial health of the firm impacts its ability to make interest and principal payments to debt holders, dividend payments to equity holders, and contractual payments to counterparties. It is no wonder that the presence of off-balance-sheet accounts and special-purpose vehicles to manage these makes it harder to analyse the credit of a company. This puts a much higher burden on accountants and auditors to ensure that the financial health of a company can be adequately analysed.

The standard approach to modelling counterparty default is to still look to models of bond default to come up with probabilities of default, but the discussion here suggests that the answer is not so simple. One possible way to handle this is to start from the probability of default inferred from the models for bond default. This is generally a reasonable place to start since the ability of a firm to make good on its debt should impact its ability to honour its commercial transactions. Then we can layer on an uncertainty adjustment to the probability of default, in a manner reminiscent of the reduced-form models that subjectively accounts for our expectations of how the capital structure of the firm will relate to the performance on the portfolio of transactions with a specific counterparty. Probabilities of default of a firm to each of its counterparties need not be the same but should be correlated through the common factor of overall firm performance.

ESTIMATION OF PORTFOLIO CREDIT EXPOSURE

The credit department in any company has the task of monitoring exposures to different counterparties. It utilises available market information and available quantitative models developed either internally or by third parties. Given the earlier discussion on how existing credit models fall short for the energy markets, the question may be asked as to why we should even bother with developing systems and methodologies for calculating credit exposure. The answer is that credit managers have to make consistent and reasonable decisions given the information available to them, and the best way to do this is to build a good knowledge base and avail themselves of existing technology and methodology. They can then use this understanding of the myriad issues to complement the systems, and even direct ways to improve upon the existing systems. A consistent approach must be agreed upon, and communicated to senior management. Since different companies have different levels of sophistication and data availability, one expects and sees credit management procedures with varying levels of complexity.

In this section, we will describe some of the approaches for tracking credit exposure, with emphasis on energy commodity risk. Some of these views may also be found in an earlier paper (see Buy *et al*, 1998). When considering credit exposures for portfolios, one has to ensure that the netting language, as discussed in the earlier sections of this chapter, is properly incorporated into the calculations, which must correspond with the negotiated documents. These documents can vary from deal to deal, and document changes must immediately be reflected into the calculation engine. This requirement points to the need for a good document repository that allows up-to-date information to be quickly available.

High-level steps for credit exposure calculation

The *first step* for calculating exposure is to have a system representation of all deals, with all the required forward price curves, volatility curves, correlation information, interest-rate curves, foreign exchange rate curves and any other information needed to calculate an accurate fair market value for the deals.

The *second step* is to have access to historical information as appropriate, with historical prices being the mostly commonly used. The historical information may consist not just of spot prices, but also of forward price curves, and forward volatility curves. This provides guidance as to how an existing deal may perform in the context of actual historical scenarios.

The *third step* is to decide if the existing portfolio is to be considered throughout its remaining life with no deals excluded from the portfolio and no extra deals included in the portfolio. The alternative is to decide on a methodology for adding new deals to the portfolio over time. For example, if we start with a portfolio of swaps with total notional amortising down over time, a viable business may be presumed to at least maintain the total notional over time. This would require, say, assuming that the company adds at-the-money swaps every quarter to maintain this.

The *fourth step* is to calculate a representation of distribution of deal value, and distribution of portfolio value for selected times in the future. The portfolios that are selected by the credit department would be the sets of deals with each individual counterparty, provided that the deals are allowed to be netted against each other – otherwise we can keep track of multiple subportfolios of deals with each counterparty.

There are several ways of looking at future exposure. One method is to calculate the future exposure at specific points of time in the future by simply revaluing the unexpired portions of the portfolios of deals at those points in time. The revaluation may use current price curves or shifted price curves. If shifted price curves are used, the shift may be defined in terms of some absolute value or in terms of some number of volatilities. Volatility curves may also be shifted to account for changing volatilities. One point to note here is that, since the portfolio may have exposure to multiple commodities, multiple price curves (and multiple volatility curves) need to be simultaneously shifted, and care needs to be taken that the joint shifts are not inconsistent with the perceived correlations between the commodities.

The second method, which is preferred by us, is to use Monte Carlo simulation to simulate a number of price paths, and to revalue the portfolios of deals at specific future times along the price paths. The advantage of this method is evident for portfolios containing deals other than just swaps and forwards, and for portfolios with exposure to multiple commodities, as correlation between commodities prices can be handled in a more natural way through correlation between the stochastic factors driving the price processes for these commodities.

Focusing on forward prices, there needs to be a methodology to simulate the entire forward curve – in fact, to jointly simulate a set of forward curves. There is a well-established methodology, which started with the formulation by Heath, Jarrow and Morton (1992) for interest rates, and extended to commodity forward curves by Cortazar and Schwartz (1994). The idea here is to have multiple independent factors that describe the movement of the entire forward curve – a one-factor model corresponds to model where all prices along the forward curve are perfectly correlated; an N-factor model for a forward curve with N prices corresponds to using the full $N \times N$ correlation matrix for the forward prices. Three to five factors provide a happy medium for commodity forward curves, with the first factor explaining most of the curve movement, sometimes referred to as the "parallel shift" factor. For the three-factor case, the simulation would use an $N \times 3$ matrix of so-called factor loadings, which are used to weight three independent shocks (the factors) to get the shocks for each of the N forward prices.

Although the concept of simulating the forward curve with either a full correlation matrix or a reduced matrix of factor loadings, together with a starting forward price curve and a starting volatility curve sounds simple, the devil is in the detail. One has to be especially clear about whether, and how, volatilities and correlations will evolve through time. For energy commodity futures contracts that have fixed expiry dates, it is typical for the forward price volatility to be higher when the time to expiry is shorter. Correlations between forward prices are also typically less for shorter-tenor futures contracts than for longer-tenor futures contracts.

When looking at future credit exposures, we have to model price evolution of the futures contracts, keeping the fixed expiration dates of the contracts in mind. The credit exposures estimated for each future date are then a function of the decisions made with regards to the volatility and correlation evolution.

POTENTIAL CREDIT EXPOSURE

Following the steps in the previous section for estimating credit exposure, we get an estimated credit exposure distribution for each future date for a given starting portfolio. For each such future date and credit exposure distribution, we can calculate the expected value and any desired percentile. Typically, the 95th percentile or the 99th percentile is used as an estimate for the maximum credit exposure (even though the percentile is not strictly a true maximum).

We refer to the maximum of the max credit exposures over all future dates as the *max potential exposure*, and we refer to the maximum of the expected credit exposures over all future dates as the *expected potential exposure*. These are very useful concepts in order to put a limit on how large the potential exposure could become.

AN EXAMPLE OF POTENTIAL CREDIT EXPOSURE CALCULATIONS

We now describe an example of estimating credit exposure for a single standalone commodity swap. In principle, we can consider a portfolio, and this would involve making sure that all the terms of the netting agreements as discussed in the earlier sections of this chapter are adequately considered in the calculations.

Consider a commodity swap on natural gas that is priced off the settlement of the front 12 months of natural gas Nymex futures contracts. We take 1 May 2005 as the valuation date, and refer to it as T0. The months of the swap consist of the futures contracts for the months from June 2005 to May 2006. Each futures contract is a contract for physical delivery of natural gas approximately evenly throughout the calendar month of the contract. The contract expires (or has final settlement) three Nymex business days before the first calendar day of the delivery month. For the contracts of this swap, these expiration dates start with 26 May 2005 and end with 26 April 2006. We refer to these dates as $T1, \ldots T12$.

We consider a notional volume of 10,000 MMBtu, and consider a forward price curve consisting of the prices US$6.00, US$6.10, US$6.20, US$6.10, US$6.00, US$6.00, US$6.30, US$6.50, US$6.40, US$6.10, US$6.00, US$5.70. Note that any realistic example must consider seasonality in the forward price curve (with a seasonal peak for winter gas, and a possible seasonal peak for summer gas as well). We do not distinguish between forward prices and futures prices in this example.

Typically, for this commodity swap, a single price is taken to be the fixed price for all twelve months of the swap. This would mean that the difference between the fixed and forward prices would be positive for some months and negative for some months. For ease of exposition, we consider different fixed prices for the 12 months, and, in fact, take the fixed price for each month to be the current forward price for that month. Note that there is no reason why such a swap cannot be transacted in the market. The net value of the fixed and floating legs for each month is now zero, and the MTM value on the valuation date is also zero. We assume a fixed discount rate of 4% with semiannual ACT/365 discounting (this just means that, for any time horizon of D days, the discount factor for a 4% rate will be $(1+4\%/2)^{-2*D/365}$).

For expediting the calculations, we decide to monitor the potential future MTM values at monthly intervals. This can be done at daily intervals if a finer resolution is desired. We further choose the dates for estimating the potential future MTM values to be the expiration dates for the futures contracts.

For a commodity swap between counterparties A and B, in which A pays floating price to B and B pays fixed price to A, credit exposure arises for B when prices rise. The MTM at time T0 for our example is zero. Consider a US$0.10 move up of the entire forward price curve per month. Then the MTM at time T1 will simply be the discounted value of 10,000 times US$0.10 for each monthly leg for times T1, ... T12, with the discounting done to time T1. The MTM at time T2 for the same scenario will be the discounted value of 10,000 times US$0.20 for each monthly leg for times T2, ... T12, with discounting done to time T2. The MTM at time T12 for the same scenario will be 10,000 times US$1.20. For each time Ti, the undiscounted values of the monthly legs from Ti to T12 are discounted to Ti. Figure 2 shows the graph of potential credit exposure *versus* time.

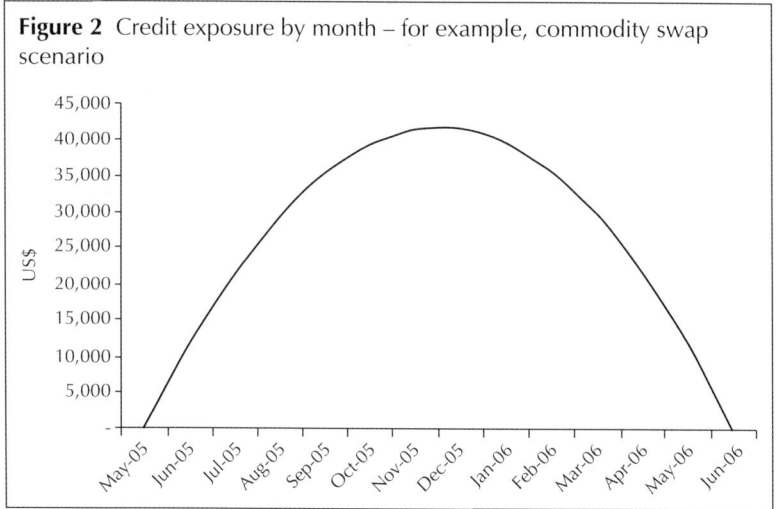

Figure 2 Credit exposure by month – for example, commodity swap scenario

The parallel shift selected in this example is arbitrary. The appropriate choice of price shift should be dictated by the standard deviation of forward prices, and also by the historical worst-case price moves. The selection should be consistent with the credit department's methodology for setting credit limits and estimating credit exposures.

SIMULATION OF DEFAULTS AND EXPECTED LOSS

Simulating defaults is a challenging task, especially given the distinction between bond defaults and contract defaults, as noted earlier. Nevertheless, once we decide on a definition of default, we need to superimpose events of default onto the estimation of credit exposure. By doing this, we can estimate the potential MTM exposures at times of default.

One possible approach that is commonly employed is to use static probabilities of default. In this methodology, a term structure of probabilities of default is assumed or estimated. We define *cumulative probability of default* over a time period as the probability of a default event happening during that time period, and define *cumulative probability of no default* over a time period as the probability of no default event happening at any time during that time period. For example, for the swap example earlier, we can estimate cumulative probabilities of default for the time periods from T0 to T1, from T0 to T2, and so on, up to the time period T0 to T12. Cumulative probabilities of no default P1, P2, ... P12 are then obtained for the same time periods. These probabilities are assumed to stay unchanged in the static approach. We can take as reasonable $P1 >= P2 >= P3 >= ... >= P12$.

Now, to estimate credit losses in a simulation approach, for each simulation path for deal/portfolio MTM value, a random number x is drawn from the uniform distribution [0,1]. If x is less than P12, then there is no default for this simulation path. Otherwise if $x >= P1$, then a default is assumed in the first period from T0 to T1. Otherwise, there must be an index I such that $x >= P\_I$ and $x < P\_(I-1)$. In this case, a default is assumed in the period from $T\_(I-1)$ to T_I.

When a default is assumed in the period $T\_(I-1)$ to T_I, then the credit loss before recovery for this simulation path is assumed to be the MTM value along this simulation path at time T_I, if such MTM value is positive.

The additional assumption required is about recovery rates, given the event of default. There is a vast body of analysis on recovery rates, with Moody's publishing historical recovery rates on bonds of different seniority. Statistical assumptions about recovery rates have also been studied, and practitioners have even settled for a point estimate of recovery rate.

The combined assumptions of default simulations and recovery-rate assumptions give the potential credit loss amounts for each simulation path.

RELATIONSHIP OF THE THREE STYLES OF CREDIT MODELS TO CREDIT EXPOSURE ESTIMATION

The three styles of credit models we have mentioned are structural models, reduced-form models and incomplete-information models. Both structural models and incomplete-information models utilise the concept of an asset process, which is critical to estimating the events of default, whereas the reduced-form models view the events of default as exogenous events, without any explicit dependence to an asset process.

The expected loss calculations are greatly simplified, from a practitioner's point of view, by the reduced-form models, as the (exogenous) events of default may then be superimposed onto the portfolio value simulations to get expected loss values. The complexity of joint estimation of multiple asset processes is avoided in these reduced-form models, and consequently the implementation is simplified.

CONCLUSIONS

In this chapter, we have tried to give a flavour of how one can think about credit risk management for the energy industry. The overriding theme is that energy companies may have diverse businesses, and may even be embedded in companies with even broader business lines. Adequate systems infrastructure and experienced credit personnel are critical to keep on top of estimating and managing credit exposures to portfolios of diverse deals following diverse regulations. Credit analysis models from the financial markets, though not perfectly suited for energy markets, do provide important signals to credit situations, and are useful to monitor and use. The different styles of credit models here may then be integrated into a company-wide credit management system.

1   According to Stephen Griffin, in "The Rise and Fall of the Ultra Vires Rule in Corporate Law", this legal concept is defined as follows: "the ultra vires rule was a regulatory device which sought to prevent a registered company from entering into any type of transaction which exceeded the scope of the company's contractual capacity; contractual capacity being determined by the contents of a company's object clause." See http://www.solent.ac.uk/law/mjls/papers/griffen.pdf.
2   http://www.riskglossary.com/articles/legal_risk.htm.
3   One example is the TXU Europe, which was denied credit support by its corporate parent domiciled in the USA.
4   This problem was a subject of discussion at the University of Houston conference organised by the Global Energy Management Institute (GEMI) in January 2003. The subsequent developments confirmed the opinions voiced by the more cautious speakers. Another excellent publication on this topic is "Clearing Natural Gas and Power: Silver Bullet or Silver Lining?", by Denise Furey, Ellen Lapson and Richard Hunter from FitchRatings.

REFERENCES

Buy, R., et al, 1998, "Actively managing corporate credit risk – new methodologies and instruments for non-financial firms", in J. Gregory (ed), *Credit Derivatives* (London: Risk Books).

Cortazar G. and E. S. Schwartz, 1994, "The valuation of commodity contingent claims", *Journal of Derivatives*, pp. 27–39, Summer.

Davidson, J. P. III, 2003, "Fundamentals of clearing", FERC/CFTC Joint Technical Conference, 5 February.

Duffie, D. and D. Lando, 2001, "Term structures of credit spreads with incomplete accounting information", *Econometrica*, **69(3)**, pp. 633–64.

Fons, J. S., 1994, "Using default rates to model the term structure of credit risk", *Financial Analysts Journal*, pp. 25–32, September – October.

Giesecke, K., 2004, "Credit risk modelling and valuation: an introduction", draft working paper, October.

Goldberg L., 2004, "How good is your information?", *Risk*, January.

Heath D. R., R. Jarrow, and A. Morton, 1992, "Bond pricing and the term structure of interest rates: a new methodology", *Econometrica*, **60(1)**, pp. 77–105.

13

Capital Adequacy for Companies Transacting in US Electric Power Markets

Laura L. Brooks

PSEG

Capital adequacy is not a new concept. However, the capital required for successful participation in today's energy markets is dramatically different from what it was 20 years ago. In order to explore the reasons for this and offer some tools for assessment it is first necessary to set a scope and context for the discussion. In discussing capital adequacy this chapter will focus on the uses of capital, not the sources. In assessing capital adequacy both sources and uses need to be considered. The Basel Committee on Banking Supervision and Standard & Poor's (S&P) each calculate ratios of potential uses and available sources.

For purposes of the discussion that follows we will use the term *capital* in a very general sense to include all possible sources of funds (including but not limited to debt, equity, cash, inventory, reserves and earnings) for the requirements described. Two points to keep in mind as we explore capital requirements are that:

❑ time frames should match for sources and uses in terms of immediacy of availability and duration of use – one cannot wait on a bond or equity issue to fund tomorrow's margin calls; and
❑ when calculating adequacy to satisfy a rating agency or regulator's requirements, you should be sure that you are including only their allowed sources in the calculation.

The energy transactors whose capital requirements we will discuss in this chapter are those who are involved in the value chain that produces electricity and provides power to end use customers

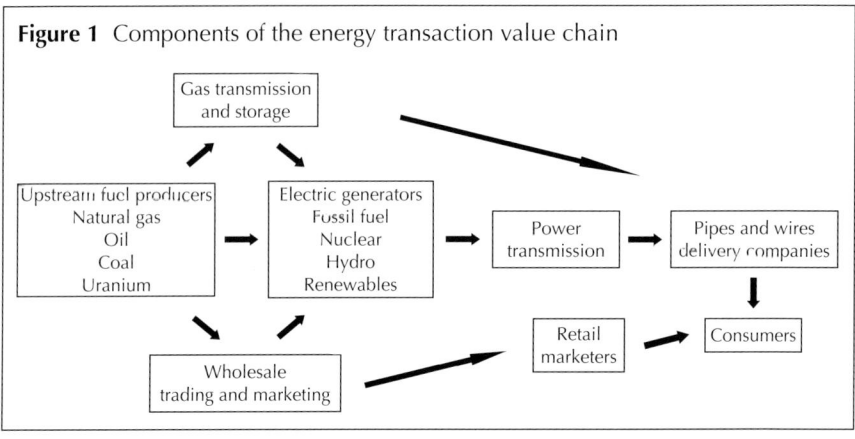

Figure 1 Components of the energy transaction value chain

(see Figure 1). Thus we include electricity distribution, power transmission, marketing and trading companies, and power generators. Upstream fuel producers, gas transmission and gas storage providers, retail marketers, and electricity consumers are not included but do serve as sources of market and credit risk to the included entities.

The capital requirements we will discuss for each of the value chain entities identified above will include the capital required to sustain profitability and absorb losses, to grow earnings and to meet daily liquidity obligations. The rest of the chapter will address the nuances of energy that complicate the assessment of capital adequacy and tools necessary to assess and mitigate the requirements.

BUSINESS MODELS OF THE ENERGY TRANSACTORS AND THE CAPITAL THEY CONSUME

The energy transactors above fall into three broad classes:

1. those whose revenues are largely derived from products and services sold under rates governed by state or federal regulators such as distribution and transmission companies;
2. those whose businesses are largely unregulated – merchant generators, and trading and marketing companies; and
3. integrated companies that combine regulated and unregulated businesses under one parent.

This is the classification we will use in discussing the business models under which capital requirements are to be addressed.

Distribution companies: Involved in the delivery of power to end-use customers including residential and commercial accounts, these entities are heavily regulated. When they are well managed and have good working relations with their local regulators, these companies enjoy good credit ratings, receive reasonable returns (for the risks incurred) on the capital they invest in delivery infrastructure and pass most of their costs, including prudently incurred commodity costs, on to their customers through regulatory mechanisms. The capital challenge for distribution companies will be the investment of sufficient capital in soft assets (billing systems, smart metering, appliance service and insurance programmes) in order to maintain or grow market share if and when retail choice programmes expand.

> So far, some 18 states have deregulated their residential electricity markets, using the same approach: they have retained control of the energy delivery, or "wires", side of the business, but have removed restrictions on who could generate and sell electricity.
>
> But stirring up competition has turned out to be much tougher than states expected, mainly because erratic wholesale electricity prices have pushed up retail rates as well. On top of that, a credit "crunch" has thinned the pool of competitors.
>
> As a result, instead of moving deeper into deregulation, some states, such as Arizona and Montana, are retreating from it. In Montana, the state now requires that residential customers buy power through a regulated utility. And then there's the case of California, whose electricity industry is still recovering from a collapse in 2000, when energy prices soared amid predatory trading by Enron Corp and others

(Smith, 2005)

Transmission companies: Regulated at the federal level by the Federal Energy Regulatory Commission (FERC), where interstate tariffs are involved, by the North American Electric Reliability Council (NERC) for reliability issues, and by State Commissions, these companies are currently struggling to balance the need to upgrade facilities and add capacity with the fact that there is little incentive to do so through granted regulated rates when compared to other investment opportunities. Three-quarters of US electricity transmission is owned by electricity utilities (distribution and integrated companies). New technologies such as real-time network management, superconductive cable, fault current limiters and broadband over power lines may change the landscape both literally and figuratively for transmission companies. An adequate investment in research and development as well as demonstration projects may play a key role for their future success.

Merchant generators: Most US markets are still characterised by an oversupply of generating capacity relative to demand growth. These companies are consequently struggling to meet the returns initially promised to their investors and many have had to relinquish ownership of facilities to the banks that financed their construction (some only partially completed). Without an affiliated marketing and trading company these companies often pay a premium for third-party risk management or for long-term fixed-rate contracts to manage volatility of earnings and lock in returns.

Traders and marketers: The highest returns have been expected from this segment thanks to the Internet boom and the buzz created by the spectacular rise of Enron, Dynegy and other trading shops. The most severe scrutiny is now being applied to every activity of these same companies due to the equally spectacular fall of Enron and other wounded participants in this sector, both the innocent and those guilty by association. When executed primarily for asset management, trading and marketing can be an enormous consumer of short-term capital to meet daily cashflow and collateral requirements. The reward, however, is the reduction in capital required by the asset side of the business (the merchant generator) to cover longer-term shortfalls in earnings due to unforeseen and unhedged market and credit events. Other capital needs for these businesses include investment in people, systems and information. Well-capitalised new entrants to the market in this sector, hedge funds and investment banks, are seriously challenging the viability of the trading and marketing entrants that were formed as subsidiaries of utilities and other energy companies in the early days of deregulation.

Integrated utility companies: Combining two or more of the above pieces and typically both regulated and unregulated businesses under one parent, the integrated utility company has requirements for all of the capital types needed by its component parts. In return,

Table 1 Capital use by activity

	Distribution or transmission utility	Merchant generator	Marketer/ trader
Assets	Pipes & wires, customers	Generating facilities	People, systems, information
Maintenance	Physical plant, customer satisfaction	Physical plant	Cash collateral
Growth	Acquisition of service territories, investment in new technologies	New facilities, investment in new technologies	New products, services, or markets
Protection	Insurance	Insurance	Insurance, VAR

it should enjoy some reduction in total capital requirements through diversification if it manages the parts as a portfolio. Table 1 summarises the types of capital required by each of these business models.

ENERGY IS DIFFERENT
Infrastructure, technology and regulation

The final barriers to wholesale competition in the electric power industry were only recently removed by the federal government in the US with the Energy Policy Act of 1992, in a process begun in 1978 with the Public Utility Regulatory Policies Act (PURPA) (see Philipson, 1999). Prior to this time, businesses involved in the production of electricity were seen as natural monopolies whose operation would result in severe inefficiencies if left to competitive market forces. When combined with the essential human-needs aspect of reliable electricity service, the economies of scale in supply and demand predicted that larger generating units serving larger distinct franchise markets could do so at lower cost. There are also practical considerations and resource conservation issues that make it more sensible to construct single infrastructure systems rather than multiple competing networks of pipelines, overhead wires, and underground cables. Taken together, these issues created a regulated market where not only was it more desirable to have a single large firm serving an entire market but also to have vertically integrated firms offering generation, transmission and distribution (see Eustache, 1998). New technologies allowing economically competitive distributed generation (on-site) are finally challenging the economy of scale assumption that created today's infrastructure.

Human needs and regulation: It is common to talk about the human needs for heat and light as drivers for regulation of the power industry, but these are not the only human-needs drivers. People also need clean air and water and desire peace, safety and security. Generating electricity can produce air and water pollutants and stray current, consume non-renewable fossil resources, and produce radioactive waste with related serious disposal concerns and military/terrorist security issues. The mitigation of these side effects of the industry, along with the additional regulations imposed as a result, requires additional capital expenditure with minimal financial return – albeit significant relative social return that may provide some competitive advantage.

Physical properties: The supply-and-demand balance for many commodities is regulated by the ability to store excess quantities for periods of time, thus smoothing out irregularities and stabilising prices to some degree. There is no way to store electricity supply. Potential energy for electric power generation can be stored as water pressure behind dams but this is possible only where nature or humankind has created the reservoirs or dammed rivers to do so. Hydroelectric power provided less than 7% of power consumed in the US in 2004 (see PennWell, 2005) and very few sites are available for new generation. In fact many existing smaller projects are being decommissioned through river restoration projects. In addition, geography plays another role in confounding the demand–supply equation for power. Unlike natural gas, which can be moved from point A to point B by lowering the pressure at point B relative to A, electricity follows a path of least resistance that is much more difficult to control.

Supply, demand and the weather: Another driver of volatility in power prices is the relationship between supply, demand and the weather. Extreme heat not only increases the demand for electricity, to power air conditioning, but also causes equipment inefficiencies and related failures that reduce available supply. Weather being only somewhat predictable means that supply and demand forecasts must include a stochastic or random variable and therefore can only be relied on to produce mean or expected results with some probability.

Age and depth of markets and the availability of risk management tools: As was discussed earlier, deregulation of the electric power industry is a recent phenomenon that has gained a foothold in the US only in the last ten years. For this reason, market mechanisms to mitigate risk in terms of financial products such as futures and other forward contracts, swaps and options are also recent phenomena. The suitability of these products is also hampered by the imperfect correlation between the pricing points of standard products and the delivery locations and conditions that need to be hedged. For generators this is further complicated by the need to hedge both the power output from generating plants and the fuel that is input. The risk management tools available for hedging fuel requirements are not much more evolved than those for the power output. A futures market was established for crude oil in the 1980s, a Nymex contract for natural gas delivered at the Henry Hub in Louisiana in 1990, a coal futures contract in 2001, and nuclear fuel is still sold mainly under long-term contracts. Add to this the complexity of non-standard contracts within and across fuel types, unilateral credit provisions dictated by the deregulation of the wholesale but not the retail side of the business, and you have many illiquid markets with little price transparency to aid in the development of models and forecasts.

Volatility: All of the above factors contribute to making power the most volatile commodity traded today. While most traded commodities, equities, currencies and interest rates exhibit high-side annualised volatilities in the 30–60% range, power can exhibit short-term spikes in prices that on an annualised basis produce volatilities in excess of 1,000%. The extreme events that create this price volatility also cause the distribution of power prices to be non-normal by having "fat tails". This means that, relative to a normal distribution, there is a higher probability of prices far from the mean.[1]

Sector Ratings and Investor Confidence: Finally, energy transacting as an industry has suffered an enormous blow to its reputation and consequently to investor confidence and sector ratings due to the scandals surrounding the behaviour of some of its largest and most respected traders, the mismanagement of deregulation in California and the overbuild of generating capacity in much of the US. These factors have resulted in credit downgrading and/or bankruptcy of much of the merchant-generating segment. In 2004, S&P assigned new business profile scores to US utility and power companies to better reflect business risk among companies in the sector. As part of this process S&P segmented the sector into sub-sectors based on dominant corporate strategy and published a new ranking list to reflect the sub-sectors. The five sub-sectors defined were:

- transmission and distribution;
- transmission only;
- integrated electricity, gas and combination utilities;
- diversified energy and diversified non-energy; and
- energy merchant/power developer/trading and marketing companies.

Not surprisingly, this last sub-sector has the highest business profile scores (highest risk) with more than 75% of included companies scoring 8 or higher on a ten point scale (see Barone, 2004a).

Energy is different: So, yes, energy is different and these differences both compound the need for capital and liquidity and complicate the assessment process, as we shall soon demonstrate.

CAPITAL FOR SUSTAINED PROFITABILITY – ECONOMIC OR REGULATORY CAPITAL
Definition

To sustain profitability, participants in the wholesale power markets need economic capital to manage the risk of:

- losses on hedge strategies employed to reduce earnings volatility;
- losses incurred due to the failure of counterparties to fulfil contractual obligations; and
- losses incurred due to operational failure.

The banking industry calls the capital required for managing these losses *regulatory capital*. Both the banks and their regulators have expended much effort to quantify the appropriate amount of regulatory capital to be maintained.

The Basel framework

In 1988 the Basel Committee on Banking Supervision issued a Capital Accord to define capital requirements for the supervision of the banking industry. The primary risk factor concentrated on in 1988 was credit. The Capital Accord was amended in 1996 to capture market risks by including "an explicit capital cushion for the price risks to which banks are exposed, particularly those arising from their trading activities" (see Basel Committee on Banking Supervision, 1996a). In June 2004, the Committee published the "International Convergence of Capital Measurement and Capital Standards, A Revised Framework", which has become known as Basel II. The combined Basel documents, the Basel framework, spell out guidelines for banking supervision and regulatory capital requirements to

> further strengthen the soundness and stability of the international banking system while maintaining sufficient consistency that capital adequacy regulation will not be a significant source of competitive inequality among internationally active banks. The Committee believes that the revised framework will promote the adoption of stronger risk management practices by the banking industry, and views this as one of its major benefits.
> (see Basel Committee on Banking Supervision, 2004, p. 2)

The Basel framework mandates a quantitative assessment of minimum capital requirements for credit, market and operational risk.

The 2004 work allows a choice between a standardised approach for calculating credit risk and an internal ratings-based (IRB) approach. In either case the end result is the multiplication of a capital requirement for unexpected loss and the calculated exposure at default. Unexpected loss for credit is defined as the potential loss at some confidence level and over some time horizon minus the expected loss over the same time horizon, as shown in Figure 2. Expected and unexpected losses can be calculated for individual transactions, for counterparties or for portfolios. The section below on credit risk gives more detail on these calculations.

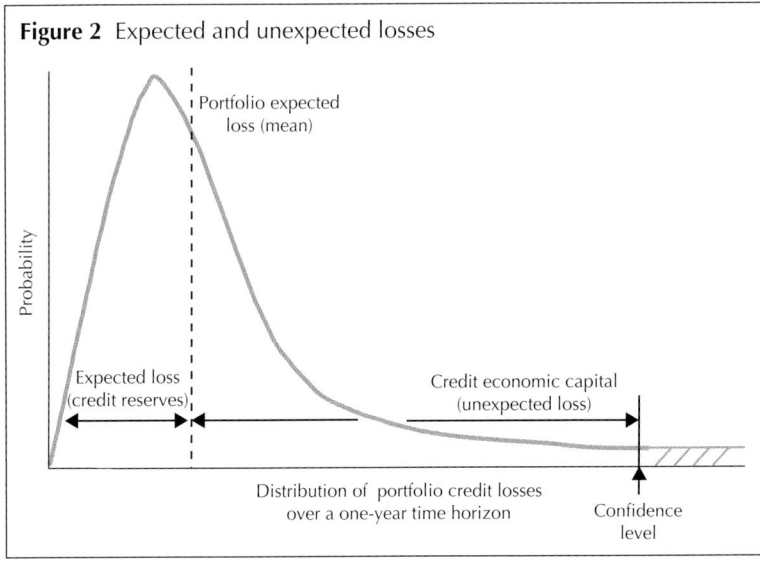

Figure 2 Expected and unexpected losses

For operational risk three approaches are allowed "in a continuum of increasing sophistication and risk sensitivity: (i) the basic indicator approach; (ii) the standardised approach; and (iii) advanced measurement approach" (see Basel Committee on Banking Supervision, 2004, p. 137). These methods are each discussed in the section below on operative risk.

The primary reference for the calculation of market risk remains the 1996 Amendment, which allows three approaches for measuring commodities risk: a maturity ladder approach, a simplified approach, and internal value-at-risk (VAR) models. (See Panel 1 for a primer on VAR.) Additional guidance is given in 2004 for market risk by redefining the trading book, establishing prudent valuation practices, discussing the treatment of credit risk in the trading book and updating the treatment of specific risk under the standardised approach. The Basel framework specifies that when using internal models to calculate VAR:

❏ VAR must be calculated daily;
❏ a 99th-percentile, one-tailed confidence interval is to be used;
❏ a ten-day holding period is to be used (one-day-holding-period VARs can be scaled up by multiplying by the square root of time);[2]

❑ the length of the sample historical observation period will be at least a year;[3]
❑ datasets should be updated at least every three months and more often during periods of increased volatility;
❑ discretion is granted for the use of correlations both within and across broad risk categories (interest rates, exchange rates, equities, commodities and related options volatilities), provided that the methodology for measuring correlations is sound and implemented with integrity;
❑ models must account for the non-linearity of options positions and include risk factors that capture the volatilities of the prices underlying options positions or vega risk; and
❑ the use of internal models must be augmented by rigorous and comprehensive stress testing.

The capital requirement calculated through the use of internal models, on any given day, will be the higher of: (i) the previous day's value-at-risk measured as above and (ii) an average of the daily VAR measures for each of the prior 60 days, multiplied by a factor determined by an assessment of the quality of the bank's risk management system. This factor will be at an absolute minimum of 3 and will have added to it a plus factor ranging from 0 to 1 based on the outcome of "backtesting" the models used (see Basel Committee on Banking Supervision, 1996b, pp 50–2). The Basel

---

PANEL 1 VALUE-AT-RISK (VAR)

Value-at-risk is a function of the current value of a portfolio, p_0, at a specified time t-zero and the distribution of future values of the same portfolio, P_1, at a later time, t-one, conditional on information available at time zero (see Holton, 2003). The common function used in most VAR implementations is the standard deviation of portfolio values at time one conditional on information available at time zero. There are two key assumptions in this seemingly simple definition. One is that the portfolio doesn't change between time zero and time one and the second is that the distribution of future values of the portfolio is dependent (conditional) on p_0 and the information available at time zero. The first assumption is easy to understand but nonetheless has large consequences for the informative value of the metric. Actively managed portfolios change constantly. Components are bought, sold or mature in value; options are exercised. It is therefore necessary to ensure that users of the VAR metric understand that the value is predicated on this important assumption.

What of the second assumption? What is the information available at time zero? This information includes things like the path of prior values that lead up to p_0, the distribution of these prior values – was it normal, uniform or some other distribution? – and also any market expectations as to future value (such as might be inferred from current, t_0, option values). To understand this concept, suppose p_0 = US$100. What other information might be available at time zero? The portfolio might consist of a US$100 dollar bill, a US$100 savings bond that has just matured, or a stock certificate from a 100-year-old company with a current value of US$100. In the absence of inflation the value of the US$100 dollar bill has not changed since it was acquired, nor will its future value change. For the stock certificate and the bond we have histories of prior values and in the case of the bond a known positive path for gains in future value, while in the case of the stock, the future value is random and could experience both gains and losses. The distribution of future values influences the variability of future values; therefore, assumptions made about the distribution are critical to the accuracy of the VAR calculation. This makes the choice of VAR calculation method very important when dealing with energy portfolios. It also means that it is helpful to know how a VAR metric was calculated in order to gauge its reliability and comparability to other VAR metrics.

A VAR metric is specified by the following:

1) the length of time that the portfolio will be held: one day, one week, one month, and so on –this is the holding period or horizon;
2) the function of p_0 and the conditional distribution of P_1 that is being measured (standard deviation of portfolio values, standard deviation of portfolio returns, expected tail loss and so forth); and
3) the base currency in which p_0 and P_1 are denominated.

We will follow the naming convention used by Holton (2003) for naming VAR metrics – horizon, function, and currency followed by VAR. Days are assumed to be trading days unless otherwise qualified (1 week = 5 trading days) and if the function is a quantile of loss, it is indicated simply as a percentage. Thus 2-week 95% US VAR indicates the value that a US dollar-denominated portfolio might lose 1 day in 20 (95%) if held for the next 10 trading days (see Holton, 2003).

One-tailed or two?

Sometimes the modifiers one-tailed or two-tailed will be associated with a VAR statistic. A 95% one-tailed VAR for a portfolio with normally distributed values indicates the use of 1.65 standard deviations while a 95% two-tailed VAR would use 1.96 standard deviations. In the first case, 5% of portfolio losses will be bigger than the VAR. In the

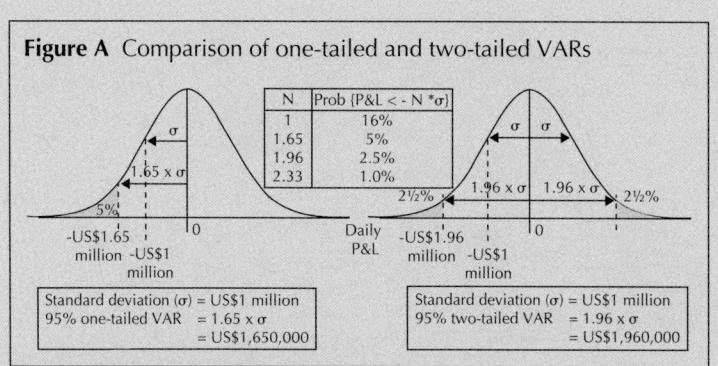

Figure A Comparison of one-tailed and two-tailed VARs

second case, 2.5% of the losses will be bigger than the VAR, as will 2.5% of the gains as shown in Figure A.

Calculation methods
Many VAR calculations are based on the simplifying assumption that the distribution of future portfolio values will be normal; others assume that the future distribution of prices will be the same as the past. In looking at the future value of a portfolio of energy commodities, we can see some of the difficulties in these assumptions. Power price distributions tend to have "fat tails" and are therefore not normal. Price distributions can change when regulatory environments change or when equipment failures or extreme weather events cause sudden shifts in the balance of supply and demand making reliance on historical distributions problematic.

Three methods are commonly used for calculating VAR:

1) Parametric – also called analytic, delta-normal, or variance-covariance: The portfolio is valued once at t_0 and then analytic methods are used to calculate local derivatives to infer potential price movements. Potential loss is then defined as:

$$\text{potential loss} = \text{sensitivities to rate changes} * \text{potential changes in rates}$$

The assumption of normality is often employed because a normal distribution is completely specified by knowing just its mean and standard deviation.

2) Historical simulation: Using the same price data as employed in the parametric VAR calculation, a structured Monte Carlo simulation of future portfolio values can be made by doing a full revaluation of the portfolio over scenarios consistent with the historical or implied

> volatilities and correlations from the available data. With full revaluation methods, potential loss is defined as:
>
> potential loss = value (at potentially changed rates)
> − value (at original rate)
>
> 3) Monte Carlo simulation: scenarios defined by a combination of selected historical periods and educated guesses are used to perform a full revaluation of the portfolio (see Guldmann, 1995).
>
> **Backtesting**
> Comparing actual profits and losses with VAR model-generated values is called *backtesting* and allows firms to gauge the quality and accuracy of their models. If a 95% loss quantile is used, for instance, one would expect to see actual losses greater than the calculated VAR for one trading day in twenty. Using higher quantiles (97% or 99%) sounds better or more accurate but in fact is often less accurate[4] and is certainly harder to backtest as it takes longer to observe losses that exceed calculated values.[5]

framework recommends the use of internal models for banks conducting more than a limited amount of commodities business (see Basel Committee on Banking Supervision, 1996b, p 31).

We will discuss these methods and others for evaluating market, credit and operational risk in the paragraphs that follow, but first we will look at the ways in which these three risks are combined to determine capital adequacy requirements.

Aggregating risk types and correlation effects
While the 1996 amendment of the Basel Accord allows for the possibility of capturing correlation effects across (as well as within) broad risk factor categories for market risk (with the use of internal VAR models), no such accommodation is allowed in the Basel framework for the interaction of market, credit and operational risks. Under the 2004 framework regulatory capital is determined by the formula:

$$12.5 * (\text{Market Risk} + \text{Operational Risk} + \text{Credit Risk}) \quad (1)$$

where market, operational and credit risks are calculated as outlined above and the multiplier of 12.5 is used to ensure that the ratio of capital to weighted risk assets is no less than 8% (12.5 is the reciprocal of 8%).

Credit, market and operational risks are correlated in energy, as is the case for financial markets – but not perfectly. The joint probability distribution of these risks is not easily constructed from theoretical frameworks. For our energy transactors, this joint distribution is also not readily determined by examining historical data due to the youth and evolving nature of many of the markets involved, as described above (see "Energy is different"). Consequently, it is easy to take an additive approach and calculate very conservative capital requirements that ignore the diversification benefits of correlated risks. S&P has also taken the Basel approach by looking at:

$$4\text{VAR} + 4(\Sigma_{\text{overall ratings categories}} \text{ (granted credit lines)} *\text{(one year default rate)}) + 2\text{VAR} \qquad (2)$$

as a proxy for regulatory capital in rating utilities that have trading and marketing affiliates (see Wolinsky, 2003). Here the 10-day 99% VAR of the trading book is used to calculate both a market and an operational risk component. Maximum utilisation of all credit lines is assumed in calculating the credit component.

The assumption of perfect correlation (the additive approach) is the most conservative way of aggregating risks. This can be seen by taking the opposite extreme and assuming there is no correlation. In this case one would square each component risk and then take the square root of the sum. An example helps:

If risk A = US$400 million, risk B = US$300 million and risk C = US$200 million then perfect correlation ($\rho = 1$) implies that the total risk capital required is:

$$400 + 300 + 200 = \text{US\$900 million}$$

while assuming no correlation ($\rho = 0$) yields:

the square root of $(400^2 + 300^2 + 200^2)$ = US$538.5 million.

When millions of dollars are at stake, this is a substantial difference. The truth lies somewhere in between but may be difficult to arrive at. Using analytic means to calculate the correlations of the risks is almost impossible due to a lack of historical data for covariance. Monte Carlo simulation holds some promise for arriving at a joint distribution of the three risks but is only as good as

> **PANEL 2 PROPERTIES OF COVARIANCE AND CORRELATION**
>
> Two quantities that measure the association between two jointly distributed random variables are covariance and correlation. These parameters can help us understand whether variables are independent or vary together. For random variables X and Y we use the standard notations of μ_X for the mean of X, σ_X^2 for the variance of X, and σ_X for the standard deviation of X. The covariance of X and Y, Cov(X,Y) is defined as:
>
> $$\text{Cov}(X,Y) = E[(X - \mu_X)(Y - \mu_Y)] \text{ where } E(Z) \text{ is the expected value of } Z$$
>
> and the correlation of X and Y, $\rho(X,Y)$ is:
>
> $$\rho(X,Y) = \text{Cov}(X,Y)/\sigma_X\sigma_Y \text{ for } \sigma_X^2, \sigma_Y^2 \text{ finite and positive.}$$
>
> The following properties of covariance and correlation are given in most texts on probability and statistics:
>
> 1) $-1 \leq \rho(X,Y) \leq 1$
>   If ρ is > 0 the variables are said to be positively correlated and move together in the same direction (if one goes up the other goes up). If $\rho < 0$ the variables are negatively correlated (if one goes up the other goes down). If $\rho = 0$ the variables are uncorrelated.
> 2) If X and Y are independent random variables with finite non-zero variance, then $\text{Cov}(X, Y) = \rho(X, Y) = 0$. The converse is not true! Two dependent variables can be uncorrelated.
> 3) $\text{Var}(X + Y) = \text{Var}(X) + \text{Var}(Y) + 2\text{Cov}(X, Y)$
>   These properties can be used to set bounds when we aggregate risk metrics. If we think of risk metrics as standard deviations of distributions of random variables then the standard deviation of the joint distribution of market risk, M and credit risk, C would be:
>
> $$\sigma(M + C) = \text{the square root of } (\sigma_M^2 + \sigma_C^2 + 2\rho\sigma_M\sigma_C)$$
> [from property 3 and definitions]
>
> This shows that if the risks are independent or dependent but uncorrelated, $\rho = 0$, then in aggregate they will have a standard deviation equal to the square root of the sum of their squares. Only if $\rho = 1$ will you have the maximal case that $\sigma(M + C) = \sigma_M + \sigma_C$ and the risks are additive since $(\sigma_M + \sigma_C)^2 = \sigma_M^2 + \sigma_C^2 + 2\sigma_M\sigma_C$. From property 1, ρ could be less than zero and the aggregate risk could be as small as the absolute value of $\sigma_M - \sigma_C$ (see DeGroot, 1986).

the numerous assumptions that are used as input to the model. In the meantime, the additive approach gives a maximum value while the square root sum of squares method can serve as a lower bound. See Panel 2 for properties of covariance and correlation.

THE COMPONENTS OF REGULATORY RISK: MARKET, CREDIT AND OPERATIVE RISKS

The Committee of Chief Risk Officers' emerging practices

For a summary overview of methods for calculating and aggregating the components of economic or regulatory risk capital for energy transactors, we look to the work of the Committee of Chief Risk Officers, CCRO.[6] In "Emerging Practices for Assessing Capital Adequacy" (see Committee of Chief Risk Officers, 2003), the CCRO addresses:

- both economic (regulatory) capital and capital required for financial liquidity;
- methods for aggregating market, credit and operative risks; and
- methods for calculating each of the component risks.

It also gives a simple example calculation using a Monte Carlo simulation to calculate and aggregate the three components (see Panel 3).

PANEL 3 A MONTE CARLO SIMULATION OF MARKET, CREDIT AND OPERATIVE RISKS

The following is part of an example that was prepared for the Committee of Chief Risk Officers' white paper on capital adequacy by the Enterprise Risk Management group at PSEG. The capital requirements for market, credit and operative risks of a small portfolio of generating assets are assessed using a structured Monte Carlo simulation. The portfolio owner is long in power in multiple power pools and short in the fuels used to generate the power. The capital requirements are assessed with and without hedges in place, to demonstrate the trade-offs between managing market and credit risk. The requirements associated with market, credit and operative risks are viewed for the portfolio of operating assets and for each asset independently to demonstrate the diversification benefits of owning a portfolio.

The portfolio of assets includes a gas-fired combined-cycle plant, a coal-fired base-load plant and a peaking facility. The portfolio owner at the outset sells all expected output from the power plant to an A-rated counterparty, purchases all expected fuel consumption from a CCC-rated counterparty and operates a small speculative trading operation that enters into deals with a BB-rated counterparty. The fact that the portfolio owner enters into forward contracts based on expected output and fuel consumption rather than tolling agreements leaves some market risk in the portfolio. A five-year time horizon was chosen for the example because it is a realistic hedging horizon for a generation portfolio in the chosen markets (ECAR, NEPOOL, and PJM).

Market, credit and operative risks are calculated for a one-year holding period at a 95% confidence interval through a Monte Carlo simulation consisting of 5,000 trials for positions with a five-year duration. Fuel and power price changes are modelled at spot for the first year and as forward prices for Years 2 through 5, with forward prices modelled as correlated Brownian motion. Credit, operations, and operational risk are simulated by an event methodology based on uniform distributions. If the simulation pull is smaller than the probability of the event (for instance, simulation pull = 0.03 and probability of default = 0.05), then there is an event. Each counterparty has an associated probability of default, each plant has a probability of an outage, and there is a probability of trader misconduct.

The interplay between the three risks is revealed by looking at alternative ways of summing the risks to arrive at a total capital requirement. Required capital and the diversification effect are reflected under three methods (square root sum of squares, Monte Carlo simulation and simple sum) of calculating combined capital. The results of the simulations, shown below, demonstrate that the joint simulation produces a capital requirement in between the square root sum of squares approach and the simple sum result. It is also shown that market, credit and operative risks can be diversified in a portfolio:

❑ *Market* – risk reduction can be achieved through fuel or geographic diversity;
❑ *Credit* – risk reduction through multilateral netting, clearing, diversity of contract durations, and counterparty ratings; and
❑ *Operative* – risk reduction through fleet diversity (fuel, geographic location, regulatory markets) and internal controls.

Simulation results:

Table A

Required capital (US$ millions)	Square root sum of squares	Monte carlo simulation	Simple sum
Required economical capital			
Market risk	6	6	6
Credit risk	16	16	16
Operative risk	22	22	22
Diversification effect – across risks	−16	−13	0
Total required economic capital	28	31	44

Market risk

In the work cited above, the CCRO discusses and compares the three methods of calculating VAR that are allowed under the internal models approach of the Basel framework:

1. analytical (based on variance-covariance matrices);
2. Monte Carlo simulation; and
3. historical simulation.

A good comparison of these three modelling approaches is given in Table 2. These methods are each described in great detail in the works of Guldimann, Jorion, Dowd, and Holton listed in the references.

The CCRO distinguishes between market risk calculations for trading portfolios similar to those held by banks and market risk calculations for portfolios that comprise physical assets and hedges that are held for long periods of time and may not receive mark-to-market accounting treatment. Asset-based portfolios often comprise long-term hedges for fuel positions and structured transactions that attempt to match the real option value and operating characteristics of different types of generation. The holding periods, modelling requirements for handling optionality, and management interventions for risk mitigation are different for these non-trading portfolios. Table 3 highlights some of these differences.

Credit risk

Market risk is often traded for credit risk when hedging the fuel requirements and output of generating assets. Measuring credit risk involves the calculation of many variables: some related to the counterparties selling fuel and buying power, some to contract terms and provisions, and the rest to market prices and general market conditions. The important components are aggregated for each counterparty and include the following.

1. Current exposure – the exposure to each counterparty at the time of evaluation. Current exposure is made up of accounts receivable (net of payables where netting provisions exist contractually), plus the value of any quantities that have been delivered but not yet billed for, plus the mark-to-market value of the remaining term of the deal, minus any collateral held.

Table 2 Comparison of modelling approaches for market risk

Modelling approaches	Price behaviour process	Market exposures	Pros/Cons	Comments
Analytical	Closed-form approach for modelling price movements	Works well for linear type exposures, but analytical solutions are available for non-linear positions (Taylor series expansion, etc)	Pros: ☐ simple and fast ☐ easy to derive ☐ easy to change as assumptions change Cons: ☐ does not capture optionality well ☐ assumes log-normal price distributions ☐ limited ability to model complexities over a longer period of time	☐ Works well for determining shorter term price moves for a trading portfolio ☐ Can be used to determine the contribution of the various portfolio components to total risks (eg, component VAR) ☐ Can be used as quick metric to help manage portfolio positions
Monte carlo simulation	5000-10000 iterations of potential price outcomes. Robust methodology for mean reversion,	Full revaluation at each price iteration better approximates non-linearity of asset/option positions. Full-blown probability distribution of financial	Pros: ☐ robust ☐ captures optionality ☐ provides a full distribution of outcomes	☐ As the time horizon is extended and the need to model certain energy price characteristics increases, simulation becomes a more suitable solution. Meanwhile,

	jumps, linking spot & forward prices	outcomes.	Cons: ☐ complex to construct the simulation model ☐ difficult to derive ☐ only as good as model input parameters the technical difficulties increase and the model needs to be modified to fit the long-term simulation purpose. ☐ Can also be used to model credit and operations/ operational risks
Historical simulation	Observed set of historical commodity prices	Revaluation of instruments based on observed market movements through price history. Easy to explain since "esoteric" models are not used and yet all the peculiarities of the market are captured (spikes, mean reversion, seasonality, etc)	Pros: ☐ robust ☐ captures optionality Cons: ☐ the simulated values are constrained to conform to history, which may be irrelevant because of market economic or regulatory changes ☐ Can be modified by appending tails to the price distributions and expanding the simulated possibilities ☐ Can be considered as a form of stress test when the historical data include extreme market episodes ☐ Can provide valuable insights and should complement analytical and Monte Carlo assessments

Table 3 Comparison of trading and non-trading activity

	Trading	Non-trading
Purpose	❏ Positions to facilitate marketing ❏ Proprietary trading positions	❏ Positions generated by asset/customer business ❏ Strategic "buy and hold" hedges
Liquidity	❏ Liquid, actively funded positions across many markets ❏ Holding period measured in days/weeks	❏ Illiquid or "buy and hold" positions ❏ Holding period measured in months/years
Optionality	❏ Price-driven exchange traded or OTC options ❏ Short holding period allows linear approximations	❏ Asset/Customer-driven embedded options ❏ Long holding period makes non-linearity material
Valuation	❏ Short-term volatilities and correlation ❏ Jump Diffusion, intra-day VAR-Analytical, Simulation	❏ Long-term volatilities and correlation ❏ Mean reversion, seasonality Simulation, Earnings at Risk
Risk Management/ Intervention	❏ VAR limit reduction, stop-loss limits, hedging with traded instruments	❏ Structured solutions, contract re-negotiations, asset sales and purchases

Contract provisions play a major role here and dictate that these calculations must be aggregated not only on a per-counterparty basis but also first on a per-contract basis. Multiple contracts with the same counterparty may have different provisions with regard to netting, margin triggers and provisions and other collateral terms. In addition, master netting agreements may exist although reliance on these requires reform to the bankruptcy code.[7]

2. Potential exposure – the change in value of current exposure over the remaining term of the deal due to changes in market prices. Potential exposure is calculated as a distribution of values at each point in time over the remaining life of the contract. Plotting the mean of this distribution at each point in time yields the expected potential exposure over time. The maximum value attained by this function can be used to evaluate the ability of a counterparty to perform under the contract. It can also be used to set credit reserves and is a component of expected loss although most energy transactors base their credit reserves on

components of current exposure. Capital adequacy requirements are set based on the unexpected (and therefore unreserved) loss at some confidence level. A similar curve to the expected potential exposure curve can be plotted using points from each distribution for a given confidence level. The maximum value attained by this curve would be a factor in the unexpected loss.
3. Probability of default (PD) – the rating agencies compile default probabilities over time by ratings categories from statistical data. Internal scoring models can be used for un-rated counterparties.
4. Loss-given default (LGD) – once a counterparty has defaulted there is still some expectation of recovery. The loss-given default percentage is calculated as 1 – the expected recovery rate.

Once all of the above components have been calculated, a loss distribution for each counterparty can be calculated from:

$$Loss(t) = PD(t) * LGD(t) * EAD(t)$$

where EAD is the exposure at default at time t.

Issues that arise in calculating the loss distribution for a single counterparty include the following.

1. Choice of time horizon: Default probabilities published for rated entities are usually in annual increments. Using a one-year default probability for short dated transactions (less than a year) will overestimate potential losses. Combining long and short dated transactions can also be problematic.
2. PD: Published probabilities of default are just that – the probability of default over a given time horizon. They give no additional information concerning default volatility (accuracy of prediction), ratings migration (path to default), or the impact of industry concentration risk and correlation effects (domino scenario).
3. LGD: There is a lack of history for industry-specific data for energy utility defaults. Large numbers of utility defaults are a fairly recent phenomena and a dramatic change from the past.[8]
4. EAD: The calculated values for exposure at default are heavily dependent on the accuracy and reliability of models of forward price behaviour.

Table 4 Overview of approaches for modelling credit risk

Feature	Approach			
	Merton-based	Econometric	Actuarial	Generalised structural model
Underlying assumptions	Default is function of asset value change – bottom-up approach	Default is function of macroeconomic variables – bottom-up approach	Default is based on buckets of similar size and characteristics – top-down approach*	Hybrid model using characteristics of the other three models – bottom-up approach
Computational technique & performance	Simulation with correlation – slow analytic with no correlation	Simulation – slow	Numerical algorithm – fast	Numerical algorithm – fast
Input format	Counterparty-level records	Counterparty-level records	Size-based cell totals	Rating/size/sub-portfolio cell totals
Expected loss	Yes	Yes	Yes	Yes
key parameters	Default rate by rating LGD "Asset" correlations	Default rate by rating LGD Macroeconomic variable coefficients	Expected loss by exposure bucket Factor loadings	Default rate by rating LGD Borrower to factor correlations
Unexpected loss	Yes	Yes	Yes	Yes
default correlation	Yes	Yes	Yes	Yes
loss severity correlation	No	No	No	Yes
Nondefault economic loss approach	Ratings migration matrix	Ratings migration matrix	None	Changes in future expected loss expectations
Default rate volatility	Yes	Yes	Yes	Yes
Economic capital	Yes	Yes	Yes	Yes

*The actuarial model is referred to as "top-down" because unlike the other three, it stipulates the shape of a distribution for the output and then uses the data to fit the distribution. The other three types allow the distribution to be formed more freely through the calculation process.

In case this isn't scary enough, the tricky part comes in creating a distribution of losses for the whole portfolio of counterparties from which an unexpected loss is calculated as the loss at the given confidence level minus the expected loss as shown in Figure 2. Table 4, from the CCRO (see Committee of Chief Risk Officers, 2004) compares various credit modelling approaches and their ability to address some of these issues.

Operative risk
The CCRO uses the term "operative risk" when referring to operations and operational risk collectively. It uses operations risk to refer to the risks associated with producing, storing or delivering physical energy products. The CCRO and the Basel Committee both use the term operational risk to refer to direct or indirect losses from inadequate or failed internal processes, people and systems, or from external events. Thus the risk of a long-term outage at a nuclear plant due to equipment failure would be operations risk while the risk of a long-term outage at the same plant due to a terrorist attack or employee sabotage of systems would be operational risk. The first thing required in measuring operative risks for inclusion in capital requirements is to create a risk taxonomy that coupled with an internal rating-based scorecard can be used to assess which risks can actually be quantified and which will require more qualitative assessment. The taxonomy should be checked for completeness by all parties within the company with "risk assessment" responsibilities including but not limited to the insurance group, operations quality control organisations, internal audit, risk management, legal compliance and the Sarbanes–Oxley team. It should also be compared to the list of qualifiers that appear in SEC statements as reliance risks to accuracy or completeness of information contained in the financial statements.

Once the list of risks is complete, the taxonomy can be fleshed out to include an assessment of impact and mitigation strategies: Is the risk HFLI (high-frequency, low-impact) or LFHI (low-frequency, high-impact)? Is the risk insurable or preventable with internal controls?

The CCRO lists five measurement techniques for operative risks for energy transactors (1–5) and briefly discusses two other approaches allowed under the Basel framework (6–7) that are more appropriate for financial institutions.

1. Scorecard approach – a qualitative, subjective assessment comparing actual practice to some benchmark or standard.
2. Scenario planning – assesses losses that would occur under a variety of scenarios and is often used to test preparedness and estimate losses if the event were to occur.
3. Advanced measurement approach – the most sophisticated of the approaches allowed by the Basel framework for quantifying operational risk and uses historical data and statistical methods to derive a distribution for expected losses. Using this approach operative risk capital is defined as:

$$OREC = \Sigma\ \gamma_I * EI_i * PE_i * LGE_i\ \text{over all events i} = 1 \text{ to n}$$

where the gamma factor determines the chosen quantile of loss, EI is an assessment of exposure (severity), PE is the probability of the event and LGE is the loss given the occurrence of the event. This additive formula assumes an unlikely perfect correlation between events such as internal or external fraud, workplace safety, damage to physical assets, business disruption, delivery process management and business practices that are the event-type categories recommended for coverage in the Basel framework. Availability of data is an obvious problem for this approach but even when available it is difficult to scale across disparate event categories.
4. Internal and external loss databases – loss databases are the way that insurance companies compile the actuarial data that they use to calculate expected losses and price their products. Collaborative loss databases such as the information compiled by industry groups such as the Institute of Nuclear Power Operators[9] and the American Gas Association create a larger base for statistical exploration and also give companies some external benchmarks. Not all information is suitable for collaborative databases – for reasons of confidentiality, or commercial or legal sensitivity.
5. Joint simulation – there is some promise in the use of joint simulation of some operative risks in conjunction with market and credit risks through the use of appropriately structured

scenarios. This is the approach taken in the example given in the CCRO white paper.
6. Basic indicator approach – the Basel framework allows banks to use 15% of average gross income averaged over the last three years (if gross income is positive for all three years, that is to say only positive years are used). This is certainly a simple approach but may, when added to correlated market and credit risk amounts, overstate total capital requirements and doesn't allow for mitigations achievable through insurance or internal controls.
7. Basel Standardised Approach – an extension of the basic indicator approach using different percentage multipliers for different banking business lines. The formula is: $K_{TSA} = \{\Sigma_{years\ 1-3} \max[\Sigma(GI_{1-8} * \beta_{1-8}), 0]\}/3$ where K_{TSA} is the capital requirement for operational risk under the standardised approach, GI_{1-8} is gross income for each of the eight specified business lines and β_{1-8} is a fixed percentage for each business line. The percentage for trading and sales as a business line is 18%.

The Basel framework acknowledges that there is a continuing evolution of analytical approaches for operational risk and does not specify the approach or distribution assumptions to be used by internal models for generating operational risk requirements. However, banks using this approach must be able to demonstrate that their models capture potentially severe tail-loss events and meet a soundness standard comparable to that of the internal ratings-based approach for credit – that is, comparable to a one-year holding period and a 99.9 percentile confidence interval (Basel Committee on Banking Supervision, 2004).

Capital for growth
In a regulated setting utilities don't compete with other users of capital in order to build facilities. Instead, they become expert in managing the regulatory process to maximise return. Today, in a deregulated setting, companies must compete for the capital required for the construction and maintenance of generating facilities and the delivery system based on forecast earnings and returns. Companies must now also develop expertise in these competitive markets and in new technologies that can create new markets and opportunities. The development of this expertise requires

human capital and investment in computer systems, information and demonstration projects, all of which must be economically justified and factored into the budgeting process.

LIQUIDITY ADEQUACY

Managing market and credit risk consumes as well as protects capital. Some capital must be maintained to meet the collateral requirements associated with risk management contracts. This is the capital required for liquidity adequacy. This area has received a lot of attention lately in the energy/utility sector as S&P began work on a refined analysis of liquidity adequacy for trading operations in early 2003 (see Hsieh, 2003) and began requesting responses to a monthly survey first issued in June of 2004. The survey requested information on primary and secondary sources of liquidity. It also required the calculation of potential draws on liquidity due to a downgrade in credit rating below investment grade both with and without a simultaneous extreme movement in the market prices of natural gas and power. Two ratios were to be calculated CELA (credit-event liquidity adequacy) and MCELA (market and credit-event liquidity adequacy), both defined as primary liquidity divided by exposure.

As a result of this initial survey, the CCRO created a working group to look at liquidity adequacy and S&P participated in the working group. The CCRO issued a position paper in November 2004 (see Committee of Chief Risk Officers, 2004) focusing on five areas where S&P's approach could be enhanced to reflect industry reality:

1. Adequate assurance – differentiating between collateral payments made due to hard and soft triggers. (Some contracts trigger a requirement to provide "assurance" in the form of additional collateral or immediate payment of amounts owing due to a specific triggering event such as a rating agency downgrade; others require the provision of "assurance" due to any material adverse change in the counterparty's circumstances. Such assurance as sometimes a pre-specified objective amount and sometimes it is a negotiated "commercially reasonable amount.)
2. Market stress scenarios – creating scenarios based on forward curve volatility rather than spot volatility. (Forward prices are used in the mark-to market valuations for margin calls and are usually less volatile than spot prices.)

3. Operating cashflows – including cashflows from operations and trading inventories as sources of liquidity.
4. Rating migration – accounting for the fact that an A-rated company will typically have more time to manage liquidity before falling below investment grade than will a company that is already on the edge at BBB–.
5. Netting of accounts receivable with accounts payable – this is often allowed for margining purposes under energy master agreements and has already been incorporated in S&P's survey (see Barone, 2004b).

A measured approach to maintaining adequate capital for liquidity that reflects the reality of contractual terms and provisions and includes the evaluation of likely management interventions in times of market or company crises must be tempered with cognisance of the very real risk of a credit death spiral caused by improper management of these scenarios.

WHO WILL REGULATE THIS?

The capital adequacy requirements of banks are supervised in the US by the Federal Reserve but using standards created by the Basel Committee on Banking Supervision. Companies with public debt offerings are monitored and rated by independent rating agencies so that lenders and creditors can gauge the risk they are taking when extending credit. Again, these agencies do look to industry, as in the case of the CCRO, to set standards. Equity investors rely on the oversight of the SEC as applied to the efforts of companies to disclose accurate information in their public filings. Thus capital adequacy for energy transactors is already subject to regulatory scrutiny from a number of perspectives. Public utility commissions and the Federal Energy Regulatory Commission are also concerned with the financial viability of energy transactors and with the health and liquidity of the energy commodity markets. In this area as in many areas of regulation and supervision there is an opportunity for the regulated and supervised entities to seek out and create the standards that they will and should be held accountable to. This is a role that needs to continue to be shouldered by industry groups such as the Committee of Chief Risk Officers, the Edison Electric Institute and others, if regulatory oversight is to be based on effective best practice.

1 A good discussion of the factors that drive volatility in energy markets and the resultant price distributions can be found in Pilipovic (1998).
2 The square root of time scalar becomes less accurate for longer periods of time as the likelihood of constant portfolio and constant volatility assumptions decreases over time.
3 If a weighting scheme or other method for the sample period is used, then the "effective" observation period must be at least one year (the weighted average time lag of the individual observations cannot be less than six months – see Basel Committee on Banking Supervision, 1996b). This requirement is problematic because energy portfolios often contain new products that do not have a full year's price history and, as previously mentioned, regulatory and other external events can cause regime shifts that move the mean for mean reverting prices.
4 See the work of Stephen Figlewski (2003) at the NYU Stern School of Business.
5 For reference material on backtesting, see Jorion (2001), Lopez (2004) and Basel Committee on Banking Supervision (1996c).
6 The Committee of Chief Risk Officers is a diverse, international coalition of energy companies developing best practices to strengthen and standardise risk management and financial management practices in the physical and financial trading and marketing of electricity and natural gas for investors, regulators, financial institutions and other energy companies. Compliance with CCRO's best-practice recommendations by its members is voluntary. More information is available at http://www.ccro.org or can be obtained by emailing the CCRO at info@ccro.org.
7 The bankruptcy code today does not clearly recognise the validity of master netting agreements in terms of the ability to terminate, close out, set off or net physical and financial forward contracts covered by such agreements. If passed in May/June 2005, the Bankruptcy Abuse Prevention and Consumer Protection Act of 2005 will solve some but not all of the issues for energy transactors.
8 The maximum percentage default rate for the utility sector between 1981 and 2001 was 0.98% and the default rate percent was 0 for 13 of these 21 years. In 2002 the rate jumped to 4.12% (see Vazza, 2005).
9 INPO was formed in 1979, following the accident at the Three Mile Island nuclear reactor, to promote the highest levels of safety and reliability – to promote excellence – in the operation of nuclear electric generating plants. One way that INPO implements its mission is by sharing commercial industry operating experience and data (lessons learned).

REFERENCES

Barone, R. M., R. Cortright, S. Smith, J. Whitlock, A. Watt, and A. Simonson, 2004a, "New business profile scores assigned for U.S. utility and power companies; financial guidelines revised", *RatingsDirect* (New York: Satndard & Poor's), June 2.

Barone, R. M., 2004b, *Private Correspondence Regarding New Quarterly Liquidity Survey*, (New York: Standard & Poor's), October 22.

Basel Committee on Banking Supervision, 1996a, "Overview of the amendment to the capital accord to incorporate market risks", Basel, January, www.bis.org/publ/bcbsca.mtm.

Basel Committee on Banking Supervision, 1996b, "Amendment to the Capital Accord to Incorporate Market Risks", Basel, January.

Basel Committee on Banking Supervision, 1996c, "Supervisory framework for the use of 'Backtesting' in conjunction with the internal models approach to market risk capital requirements", Basel, January.

Basel Committee on Banking Supervision, 2004, "International convergence of capital measurement and capital standards, a revised framework", Basel, June.

Committee of Chief Risk Officers, 2003, "Emerging practices for assessing capital adequacy", white paper published September 17, www.ccro.org.

Committee of Chief Risk Officers, 2004, "Measuring financial liquidity", position paper published November, www.ccro.org.

DeGroot, M. H., 1986, *Probability and Statistics* (Reading, Massachusetts: Addison-Wesley Publishing Company).

Dowd, K., 1998, *Beyond Value at Risk The New Science of Risk Management* (New York: John Wiley & Sons).

Eustache, A., 1998, "The evolution of US electricity markets", in P. C. Fusaro (ed) *Energy Risk Management* (New York: McGraw-Hill).

Figlewski, S., 2003, "Estimation error in the assessment of financial risk exposure", working paper version of June 29, from pages.stern.nyu.edu/~sfiglews/home.html.

Guldimannn, T., 1995, *RiskMetrics™ – Technical Document,* (New York: Morgan Guaranty Trust Company Global Research), May 26.

Holton, G., 2003, *Value-at-Risk Theory and Practice,* (New York: Academic Press).

Hsieh, T. and A. Simonson, 2003, "A Refined approach for assessing liquidity risk in energy trading operations", *Utilities & Perspectives* (New York: Standard & Poor's) March 3.

Jorion, P., 2001, *Value at Risk* (New York: McGraw-Hill).

Lopez, J., 2004, "Regulatory evaluation of value-at-risk models", in P. Jorion, *Innovations in Risk Management* (London: Risk Books).

PennWell, 2005, "Market analysis, regulation, fuel resources, key policies, and market statistics", *Global Power Review March 2005* (Essex, United Kingdom: PennWell Global Energy Group).

Philipson, L. and H. L. Willis, 1999, *Understanding Electric Utilities and De-Regulation* (New York: Marcel Dekker, Inc.).

Pilipovic, D., 1998, *Energy Risk* (New York: McGraw-Hill).

Smith, R., 2005, "A test of the results of electricity deregulation", the Wall Street Journal, March 1, 2005.

Vazza, D., D. Aurora, and R. Schneck, 2005, "Annual global corporate default study: corporate defaults poised to rise in 2005", *Global Fixed Income Research* (New York: Standard & Poor's) January 2005.

Wolinsky, J., S. Smith, and J. Kennedy, 2003, "Debt treatment of contingent capital for energy marketing and trading", *Credit Analyst* (New York: Standard & Poor's).

14

Generator Bid Strategies in Deregulated Markets: an Empirical Approach

Paul Flemming
Energy Security Analysis, Inc

Energy Security Analysis, Inc (ESAI), has conducted a study on the New York Independent System Operator (NYISO) bid-based wholesale electric power market for the summer of 2001. Contrary to the belief of many market observers in the industry – that generators manipulate their bids to inflate prices and exploit market power – one of the key findings of the study is that suppliers actually made no attempt to withhold power from the market during the potentially volatile summer season, and that in fact most New York suppliers implemented suboptimal bidding strategies for their megawatts in the summer of 2001.

Another key finding of the study is that "active" generators (that subset whose bids change) have distinct strategies or patterns depending on what area of the load demand curve their generators lie on. These strategies are influenced by the specific types of plant (base load, mid-merit or peaking) and by the characteristics of the demand curve for each of those types of plant. This is understandable: generation owners develop specialities over time; specialise in particular areas of the bid curve; and acquire assets in the sector they know best. Generators who do not understand the bidding dynamics of, for example, oil-fired peaking facilities are well advised not to buy them unless they are willing to make considerable efforts at developing bidding strategies.

PANEL 1 THE NEW YORK CONTROL AREA

The New York electrical grid is self-contained within the state of New York, as opposed to most other grids in the US Northeast, which tend to be regional, multi-state systems. Much of the low-cost nuclear and coal generation capacity is located in the western zones A, B and C.

Most of the hydro assets are located in zones D, E and F. In contrast, the majority of the load is located in New York City (Zone J) and Long Island (Zone K).

Figure A The New York control area

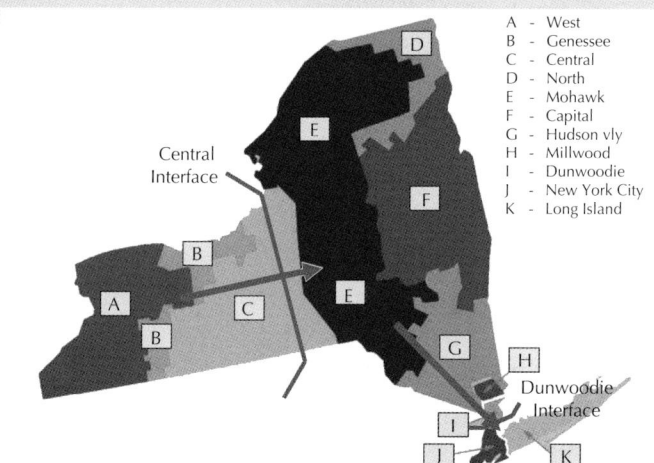

The New York grid transports lower-cost power from the west and north through zones G, H and I for delivery to New York City and Long Island. There are a number of interfaces that limit flows on the system. Two of the major interfaces are shown on Figure A.

The Central Interface limits flows from the coal and nuclear facilities in the west, generally causing prices in the west to be discounted under the locational marginal pricing scheme. The other major interface, Dunwoodie, limits flows into New York City and Long Island during peak hours.

This results in the dispatch of additional generation in New York City and Long Island at the expense of more cost-effective generation production that is constrained from entering New York City. Prices in New York City and Long Island tend to be higher during peak hours than in the other zones due to the high cost of the relatively old and inefficient oil- and gas-fired steam generation and a high reliance on peaking units during the high-load summer and winter periods.

ANALYSIS OF BIDDING BEHAVIOUR – NEW YORK CONTROL AREA

The capacity profile and dispatch stack

ESAI conducted a study of the New York Day Ahead Market generator bid data from the summer of 2001. The study was undertaken in order to understand the bid patterns of generators operating under the standard market design (SMD) paradigm. Under SMD, generators receive a clearing price that is specific to their location within the pool also known as the locational marginal clearing price (LMP). A more detailed outline of LMP is provided in Panel 3 of this chapter.

The New York Power Pool provides an excellent model for the study of generator bid behaviour because, in addition to operating under the deregulated LMP market design, it has a wide range of generators which utilise a very diverse range of fuels. As can be seen from Figure 1, New York depends upon a variety of fuels for its generator operations: Natural Gas, #6 Fuel Oil, Nuclear, Coal, Hydro and Diesel-FO2. This diversity of fuels provides a range of fuel costs that yields a wide variance in production costs. In addition to the fuel costs, the generator types or technologies dictate their efficiencies, which, in turn, impact the production costs of the generated power.

The capacity profile is commonly referred to as the dispatch stack. The generators in the dispatch stack are differentiated by fuel costs and plant types. The dispatch stack for New York is outlined in Figure 2. The least-cost units are at the bottom of the stack and the highest-cost units are at the top of the stack. As load increases, the NYISO will dispatch more capacity based upon cost. The last unit dispatched to meet the last increment of load at any given hour is the most expensive or marginal unit. This marginal unit sets the energy clearing price of the power for the pool (further adjustments are made for congestion and losses to determine the price at the generator's location). As the load increases, the marginal unit dispatched becomes increasingly expensive. As can be seen in Figure 2, hydro units are the least-cost units followed by nuclear. Coal-fired units have a significant fuel cost advantage over gas-fired units and are typically more competitive. This fuel cost advantage overcomes the greater efficiency of a state-of-the-art, combined cycle gas unit, which is 30 to 50% more efficient than a coal-fired steam unit.

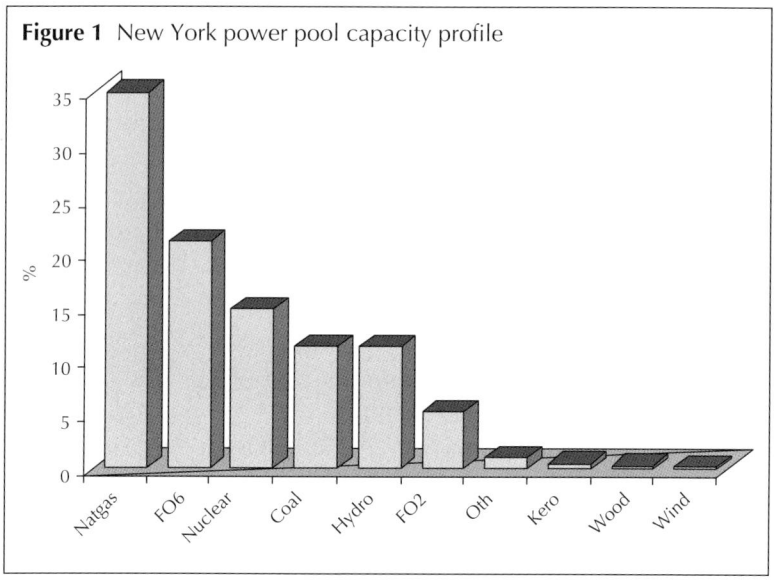

Figure 1 New York power pool capacity profile

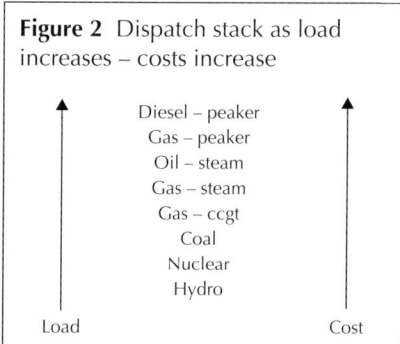

Figure 2 Dispatch stack as load increases – costs increase

New York has approximately 35% of its capacity based on natural gas as the primary fuel. There are three types of plant that are fired on gas, and each has different operating characteristics and efficiencies. Combined cycle gas turbines (CCGTs) are the most efficient and have heat rates of 7,000 to 8,000 Btu/kW (heat rate is the conversion factor from fuel to electricity, which represents the plant's efficiency). Gas-fired steam units are less efficient and have heat rates of 9,000 to 12,000 Btu/kW. Gas-fired peaking units are typically standalone turbines with heat rates of 9,000 to 13,000

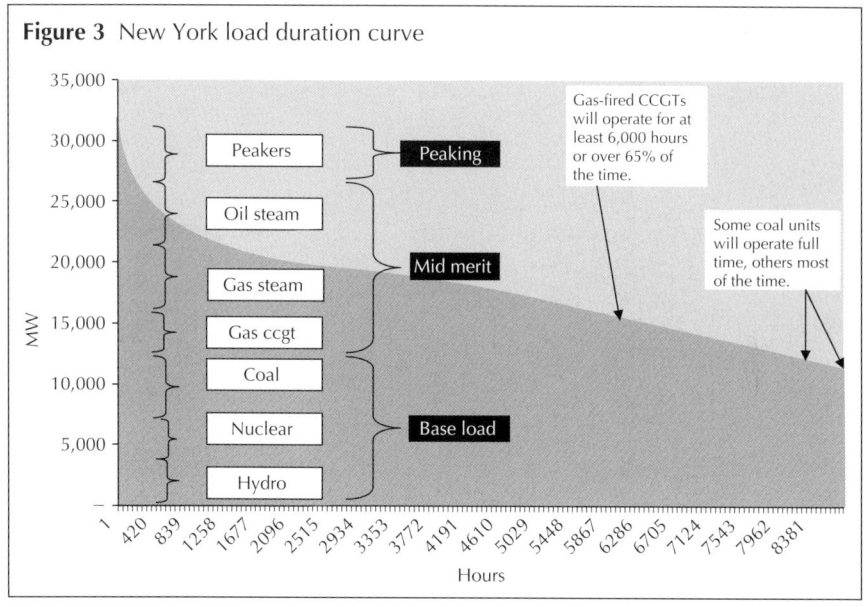

Figure 3 New York load duration curve

Btu/kW and higher, depending on age and technology. Although these units utilise the same fuel, their relative placements in the dispatch stack will dictate how often they are dispatched and where they are located on the demand curve (load duration curve).

Load duration curve

The load duration curve represents the load (electricity demand) over each of the 8,760 hours in a year, sorted from highest demand to lowest. Hour 1 represents the highest load and hour 8,760 the lowest load.

Base load
Figure 3 illustrates the load duration curve and its relationship with the dispatch stack. The lowest load seen during the 8,760 hours in this example was 11,300 MW and is referred to as the base load, or the minimum amount of power that is always being demanded by the market. New York's combined hydro and nuclear capacity totals over 9,000 MW and represents the lowest-cost capacity to the system. This combined low-cost capacity can

operate fully and still not meet the minimum demand requirement of 11,300 MW. We would therefore expect these lowest cost units to operate continuously, aside from operating and maintenance constraints.

Over 2,000 MW of base-load demand is unmet by the nuclear and hydro generation. The next band of generation in the dispatch stack is the coal-fired capacity. As mentioned above, there is a range of efficiency in the coal plants that serve New York state. The newer units are more efficient and will operate continuously to meet the additional 2,000 MW of base-load demand. Another 2,000 MW of coal capacity exists that will operate most of the time (high-capacity factors) but will be forced to run at reduced rates or not at all during the 600–700 lowest-load hours. For all other hours, coal units would be expected to be fully on-line.

Mid merit

The base-load hydro, nuclear and coal units are needed almost 100% of the time and therefore these units do not often play the role of the marginal unit. As load increases during most on-peak and much of the off-peak periods, additional generation must be called upon. The units that do not meet base-load cost competitiveness but are needed for regular service are called mid-merit. The most cost-competitive of these is the combined cycle gas turbine capacity (CCGT) and represents approximately 10% of New York's capacity or 3,000 MW. Oil and gas steam units are also included in the mid-merit category. The cost competitiveness of these units depends on the relative price of oil and gas as well as the cost of emissions credits. In some instances, oil prices could be lower than gas prices, but the additional cost of emissions credits for oil-fired units could render them less competitive than the gas-fired units.

Peaking

During the highest-demand hours, base load and mid-merit capacity cannot meet the total load. Peaking capacity is used to meet the highest-load requirements, typically seen during hot summer periods. Figure 3 shows that peaking capacity would be called upon for only about 200 hours per year. Due to the very low capacity factors of these units and the need to recover costs in a relatively short period of time, the bidding strategies of these units tend to be aggressive.

The generator bid data
New York generators are required to bid their capacity into the day-ahead market (DAM) on a daily basis. Generators formulate their bids during the morning of the preceding day and submit the bids to the NYISO by noon. The ISO then analyses the generator bids using the day-ahead pricing model, which also accounts for transmission congestion and losses. By 4 pm, the ISO releases the dispatch orders for the following day. Generators that did not get dispatched in the DAM are free to re-bid into the hour-ahead market (HAM) during the following day.

The NYISO releases the actual generator bid data on a six-month lagged basis. Because the generator IDs are masked, the data do not specify which units the bids are coming from. This makes the analysis much more complex. However, certain inferences can be made from the size of the bids. As will be summarised below, a number of bidding patterns emerge and conclusions can be drawn as to the types of unit that utilise a particular bidding pattern or strategy. More details concerning the analysis of the bid data are included at the end of this chapter.

EIGHT BIDDING PATTERNS EMERGE[1]
Most generators bid the same curve into the HAM and DAM, and did not appear to change their bids significantly over time. In the DAM, eight bidding patterns seemed to dominate, each corresponding directly to a class of generators. Bid curves for base-load generators were largely determined by must-run constraints that impose strict price discipline. Mid-merit and peaking units showed signs of bidding strategies that were more internally focused than aimed at collaboration with others, and many appeared to have attempted some price optimisation based on costs and demand price elasticity.

Cluster analysis (see Panel 2 for definition of this analytical technique) revealed eight predominant bidding patterns that are conceptually consistent with the market load duration curve and generation mix of the NYISO. Each bid cluster appears to correspond to a class of generators. Although this cannot be proven without unmasking generator bids, the creation of a supply curve

ENERGY MODELLING

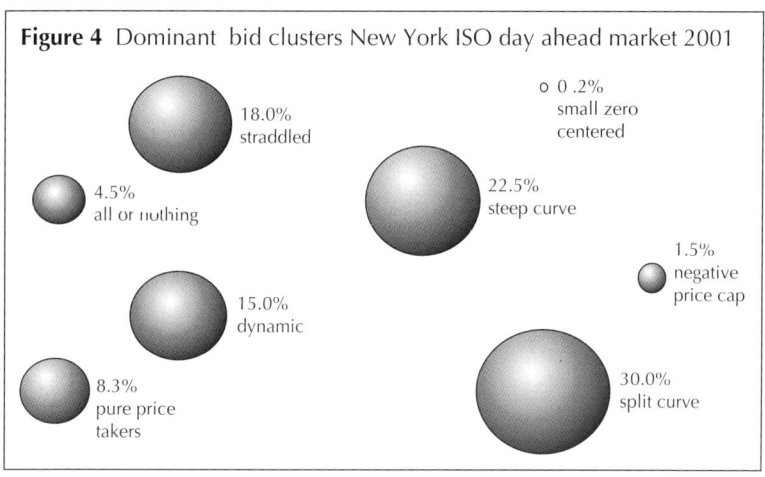

Figure 4 Dominant bid clusters New York ISO day ahead market 2001

- 18.0% straddled
- 4.5% all or nothing
- 0.2% small zero centered
- 22.5% steep curve
- 1.5% negative price cap
- 15.0% dynamic
- 8.3% pure price takers
- 30.0% split curve

PANEL 2 K-MEDOIDS CLUSTER ANALYSIS

Many mathematical techniques exist to find clusters in data. The method applied in this study was a k-medoids clustering technique. This algorithm divides the data into a predetermined number of clusters, hence the "k". To form a cluster, each record is mapped to a point that can be interpreted as a distance by the clustering algorithm, and the k-dimensions of the dataset are normalised so that a distance in one dimension is consistent with an equal change in another. In the first iteration, k points are randomly selected and midpoint boundaries are drawn between these points. In a two-cluster case, records falling to the right of the boundary are assigned to the right cluster and records that fall on the left side of the boundary are assigned to the left cluster. The equivalent sorting is performed in the k-dimensional case. The most centre point of each cluster is selected and the midpoints between this new set of medoids form the new boundaries of the clustering. This process is iterated until the medoids remain unchanged. The clusters are scored to determine whether they are well classified or poorly classified. The clusters defined in this study were well classified and eight clusters were optimum-based on up to fifty samples of 60,000 records each.

---

from the bid data supports this hypothesis. The eight bidding patterns are generalised as follows:

❑ *Split-curve bidders* – These are large, base-load generators with must-run constraints. A portion of the bid curve is priced below

cost, and the remaining capacity on a steeply sloped portion of the curve. Bids in this category are fairly static, and constitute almost 10,000 MW in the DAM.
- *Dynamic bidders* – These bidders are also large-base-load generators, but have less rigid must-run constraints. A portion of the bid curve is priced at or below cost, with the remainder on a steep slope. Bids in this category are very dynamic, and are highly correlated with the NYCA Zonal LBMP 6% of the on-peak hours in the summer of 2001 These bids represent less than 5,000 MW in the DAM.
- *Steep curve bidders* – These are mid-merit operators with minimal must-run constraints or efficient peaking units. The first bin^2 on the bid curve is approximately at cost or slightly below. The entire curve is very steeply sloped with the final bin often close to the price cap. These bids are highly correlated with the Day Ahead Zonal LBMP 71% of the on-peak summer hours in 2001, and provide 7,500 to 10,000 MW in the DAM.
- *All-or-nothing bidders* – Peaking unit that bids all of its generating capacity in a single bid block. These are the highest-priced bids, and their bid curve rarely changes. This cluster is most highly correlated to the Day Ahead Zonal LBMP about 6% of the on-peak summer hours in 2001, and represents about 1,500 MW of capacity.
- *Straddled bids* – This cluster's bid curve resembles the steep slope bidder, except that these bids usually have three bins priced below zero and three bins above zero. These generators face some must-run constraints, but try to relate the operating level of the plant to the bid slope, including the portion that is below cost. Bids in this cluster are somewhat dynamic, but less so than the dynamic bidders. This group supplies about 5,500 MW of capacity in the DAM.
- *Pure price-taking bidders* – These suppliers normally bid all their power in a single negatively valued bid price and simply take whatever price the market sets. While this may sound like a naïve, or even negligent, methodology, there are circumstances where this is actually an optimal pricing strategy. This category represents about 2,750 MW of capacity.
- *Small zero-centred bids* – This is a group of small plants that bid in two or three bins, half of their capacity at approximately zero, and

the other half at the price cap. These bids almost never change. In total this cluster represents less than 50 MW of capacity.
❑ *Negative price cap bidders* – This cluster represents a small group of bidders totalling about 500 MW of generating capacity that bid –US$999.00 or –US$1,000.00.

LOAD DURATION CURVE

The NYCA load duration or demand curve can be partitioned into three sections – that served by the base load, the mid-merit and the peaker (see Figure 3). Each partition has very different price elasticity characteristics and a distinct portion of the available generating capacity serves each partition. In a way, each can be seen as a completely different but related business environment. In other words, different classes of generator face different price elasticity, therefore follow very different bid patterns and price procedures. Base-load generators face heavy price competition. Mid-merit generators are in a less competitive setting, populating the inelastic, or flat, section of the demand curve where some level of price optimisation may be undertaken. The upward-sloping portion of the demand curve contains peaking units.

While this demand curve analysis might seem to reaffirm the obvious – demand is not homogeneous across all suppliers and firms face a different price elasticity depending upon whether they are base-load, mid-merit, or peaking – it is the slope of the individual unit's demand curve that determines its bidding strategy. Each group of generators must develop a pricing strategy that determines the "optimal" price. Evidence from the bid curves suggests most generators have evolved a pricing strategy that is consistent with the price elasticity on its partition of the demand curve.

STICKY BIDS

The study finds that a large percentage of bidders into the NYISO market are bidding suboptimally. Supplier bids are "sticky" and demonstrate lagged responses to changes in demand or fail to respond to shifts in demand below some threshold. An example of this behaviour is indicated by shifts in the supply curve during the hottest day of the year, but the bids show little, if any, deviation from average levels for the second- and third-hottest days of the year. Most generators bid the same curve into the HAM and the DAM, and do not change their bids from hour to hour or day to day. So, while we can assume many generators are trying to optimise prices, the market as a whole is fairly static.

The key component of any optimal price calculation is generator cost. Given that exogenous factors in the market are constantly changing (fuel prices, transmission constraints, competitor behaviour, interest rates, new generation, old generation derate or retirement), the optimal price calculation should change with every fluctuation of these exogenous factors. Therefore, if bids are static, they are by definition suboptimal.

STRATEGY

The very word "strategy" is laden with thousands of years of military and economic history, and holds as many interpretations as there are individuals to provide them. One particular interpretation is that to act strategically presumes two facts: that there is an imbalance of information between the players and that each player seeks to defeat the other(s). In the New York power market, the ability to act strategically has been mitigated to some degree by the design and rules of the NYISO and FERC. First, the NYISO has made enormous amounts of information available to all participants in the New York electric power market, eroding information imbalances between suppliers and helping to level the playing field for all the participants that have an appetite for complex information and analysis. Second, the New York ISO requires an 18% installed reserve margin, ensuring that "cheap" (highly inefficient) generators will not be driven out of the market. These two factors combined prevent competitively superior generators from driving less competitive generators out of the New York market. This is not peculiar to New York, of course, but a general characteristic of power markets.

On a more practical level, the analysis of the bids and demand curve implies that most generators are internally focused, and not necessarily motivated to drive other generators out of business in order to gain market share. For the most part, generators appear to be relying on internal marginal costs and price elasticity to determine their bid curves, rather than constructing bids in relation to other suppliers. This is not only intuitively appealing, but also is a consequence of the physical generating systems. Real-price competition is restricted to base-load operators and some large mid-merit plants. Any must-run constraint imposes severe price discipline on a generator; once beyond the must-run constraint, a generator can begin to exercise some strategic pricing practices.

Competition between base-load generators can have broad market-level implications. These units make up the negatively sloped portion of the demand curve in which suppliers must compete to sell power. As stated above, once we move beyond the negatively sloped portion of the demand curve, price inelasticity provides suppliers with substantial price leverage. In the neighbouring PJM power pool, extensive base-load capacity (primarily coal) helps ensure that consumers benefit from "normal" price competition and market behaviour, while explaining the comparatively low off-peak prices in the region. New York is less well endowed with such facilities and hence is likely to have more price volatility.

FERC's Standard Market Design rules, as they now stand, envision a market that would function smoothly with minimal oversight. The reality is, however, that in a market without a substantial portion of base-load generation, price elasticity interferes with "normal" price competition and market behaviour. Therefore, without either construction of generation within load pockets, or construction of transmission capacity that can transfer base load power from external sources to the constrained sinks, locational price spreads will remain. As a result, these severely constrained markets with expensive (and many times inefficient) native generation will continue to produce higher prices that will continue to fall prey to mitigation.

SUPPLIER BID CHARACTERISTICS

A cluster analysis of the DAM bids for 2001 partitioned them into eight clusters, determined to be optimal by a k-medoids scoring algorithm, the scoring details of which are provided in the statistical appendix to the study. Assigning a descriptor to each cluster can oversimplify the characteristics, as they frequently include exceptions to a general rule or pattern. Often the edges between clusters become blurred as to whether a particular bid falls into one category or another. In the study, the general pattern of each cluster appears to be conceptually consistent with the market supply curve and volume of bilateral bids. The eight clusters are:

- split-curve bidders;
- dynamic bidder(s);
- steep-curve bidders;
- all-or-nothing bidders;
- straddled bidders;
- pure-price-taking bidders;
- small zero-centred bidders; and
- negative-price-cap bidders.

Split-curve bidder

The split-curve bidder is most likely a base-load unit confronting some minimum must-run constraint. These units competitively bid in the minimum must-run generation, frequently at negative levels. With the proportion of generating capacity that is not must-run, the supplier can hope to attain near-optimal prices, which can be seen in Figure 5 in the sudden steep slope. On average, almost 10,000 MW per hour are bid into the DAM in this manner and the

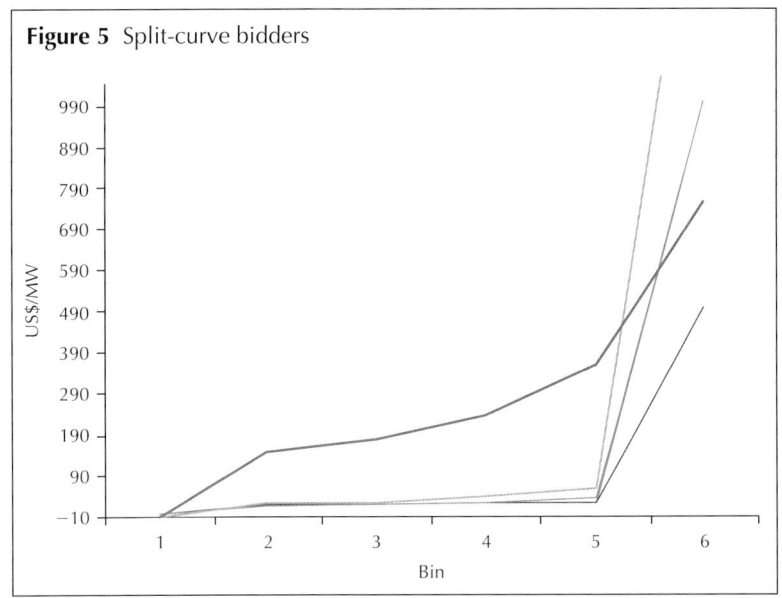

Figure 5 Split-curve bidders

average generator-operating limit is over 350 MW. These bids tend to be static and not fluctuate substantially from one day to the next or over the course of a day.

Nuclear units and many coal units will fall into this category. Under this strategy, the unit bids most of its capacity at very low levels and essentially becomes a price taker in order to guarantee dispatch. It leaves its last portion, or bin, of capacity to be bid at higher prices, which may provide some price optimisation for the unit. In the event that the generator is a base-load unit, it can bid into the hourly market or provide real-time ancillary services with the last portion of its capacity.

Dynamic bidder

The dynamic bidder is very similar to the split-curve bidder. Generators tend to be large-base-load units with some minimum must-run constraint. They share the same L-shaped curve with the split-curve bidder; however, the dynamic bidder changes bids over the course of a day and from day to day as new market information becomes available. Figure 6 displays a typical set of bid curves over the course of a single day. The minimum generating level is always bid in at near zero dollars per MW, but the slope of the curve

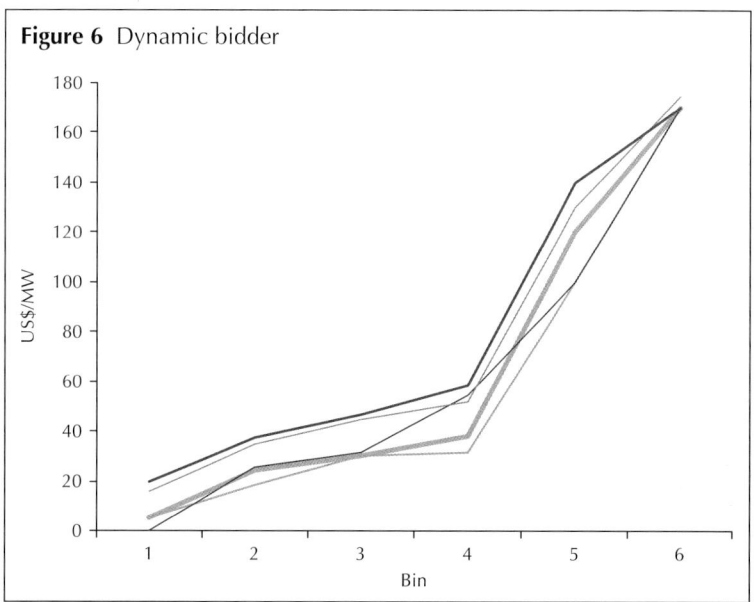

Figure 6 Dynamic bidder

changes with increasing levels of utilisation and changing levels of forecast demand. This cluster represents less than 5,000 MW of capacity, but it was the most highly correlated cluster with the New York City Zonal price 6% of the hours in 2001.

Steep-curve bidder

Another class of bid curves revealed by the cluster analysis was the steep-curve bidder. This group, more than any other, determines market prices. Its bid pattern, like that of the dynamic bidder, requires detailed knowledge of load forecasts and dispatch probabilities. Most likely this generator does not face a must-run constraint, and can withhold selling into the market until some minimum price is reached. Bid curves for this group vary more widely than other clusters, as the slopes of the bid curve tend to vary more often. The suppliers of power under this bid curve grouping offer 8,800 MW per hour to the market, and the operating capacity of these generators is, on average, between 50 and 350 MW. The steep-curve bidders are the second-largest cluster, but they are probably the most important, because the steep-curve bidders are the most highly correlated with the market price 71% of the time. Savvy operators of smaller, less efficient CCGT units may

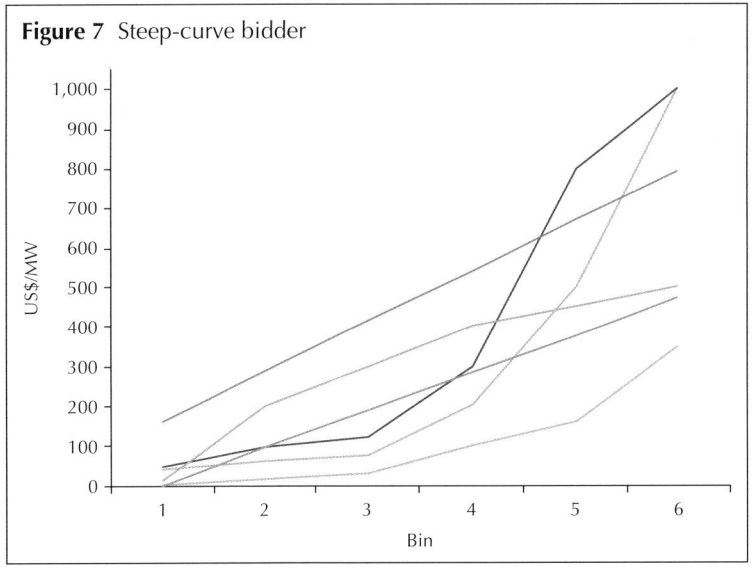

Figure 7 Steep-curve bidder

use this strategy to optimise economics when they are likely to be on the margin. By increasing their bids towards the cost of less efficient units above them in the dispatch stack, they can increase revenues. There is some risk of not receiving a dispatch order for the full capacity, but the hourly market represents another opportunity to re-bid the same capacity.

All-or-nothing bidders

The most static of the bid clusters is the "all-or-nothing" bid grouping. These generators do not face a must-run constraint. Therefore they have the option not to sell into the market until their optimal price is met. They usually bid in a single block, and the average generator operating limit is about 50 MW. These units represent approximately 1,500 to 2,000 MW per hour in the DAM. Since these generators can hold out until some optimal price is met, it is not surprising that their bids rarely change. Some of the bids in this category also include those that are at or above the current price cap, and do not necessarily represent bids to actually sell power. These bids are represented by the gas turbine and distillate fuel turbine peaking units. They will look to recover all of their costs in a relatively few days of operations. To accomplish this, the price they receive for their production needs to be quite high.

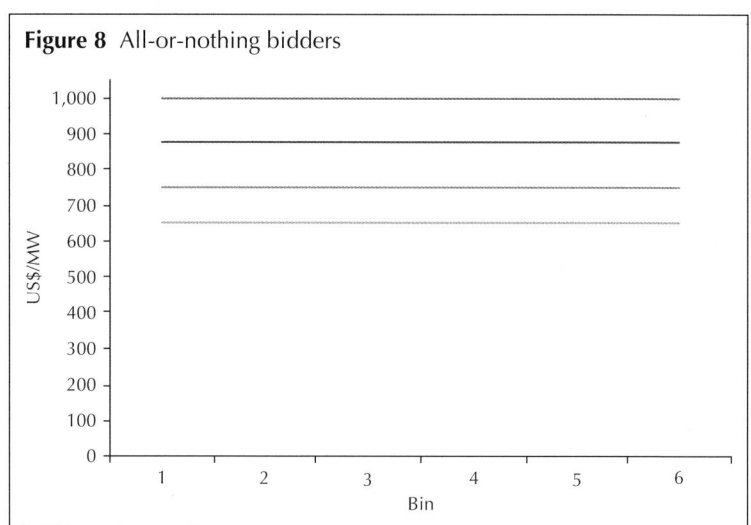

Figure 8 All-or-nothing bidders

Final three

The last three clusters contain negative bids, frequently referred to as "price takers" in the market. The first partition in the negatively valued bids is a group that straddled bids in such a way that three of the bins had negative prices and three were positive. It is possible that these straddled bidders are a variant of the split-curve bidder, where a portion of power is bid in low and, above some threshold, the price per MW increases drastically. However, this cluster of bidders partitions the minimum generating levels into separate bins, instead of a single bin representing the entire minimum operating level. The price-per-MW bids for the lower levels of utilisation are all negative to ensure they are dispatched, but monotonically increase in the hope of determining the level at which they are dispatched.

This requires very detailed knowledge about where that generator's bids are ranked within the dispatch stack. For the higher levels of utilisation, the supplier has more flexibility and no longer needs to ensure its dispatch, enabling attempts to optimise price. Once this carefully structured bid curve has been calculated, it tends to remain unchanged, likely undermining the ability of the supplier to influence the level at which it is dispatched.

The next cluster in the price-taker category comprises bidders that bid in a single block at a negative dollar-per-MW price,

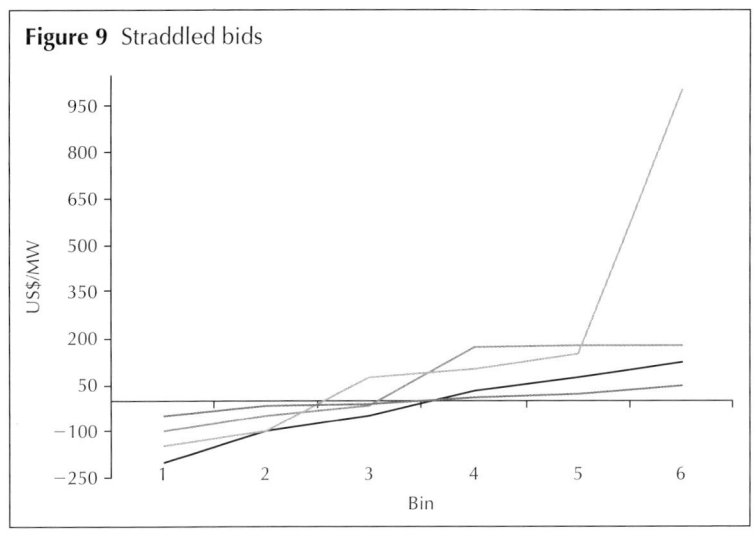

Figure 9 Straddled bids

although the values of bids vary dramatically. These are pure-price-taking bidders that may benefit from a beneficial position geographically – acting as a price taker at their bus may represent a completely rational strategy. For example, if a mid-size hydro facility were located on the same bus as a natural-gas-fired generator and both the hydro unit and the natural gas unit were consistently dispatched (or dispatched simultaneously), the natural gas unit would always set the price. Under such a scenario, the hydro unit would always be a price taker anyway, and would not sacrifice any revenue by submitting a price-taking bid. Some of the generators in this category spread their bids across three bins to influence the dispatch level, as was outlined in the straddled-bidder case.

The final two clusters represent a very small share of the market. One group bids its entire capacity at –US$998/MW to –US$1,000/MW. While this group was partitioned out as a separate cluster by the clustering algorithm, it is actually a subcategory of the pure-price-taking bidders. The final group is made up of very small units that bid a truncated version of the split-curve bidder. The units are often under 5 MW and bid half their capacity in at a zero or –US$1.00/MW, and the other half at near the US$1,000/MW price cap.

It can be deceptive that these categories appear to neatly partition different bidding patterns, because in reality these boundaries will always be blurred. For example, when a split-curve bidder submits a bid for the first several hundred megawatts at –US$100/MW, this generator is actually acting like a pure-price-taking bidder for that portion of its power and bidding as a steep-slope bidder for the remainder of its capacity. Deciding whether a particular bid is a straddled bidder or a steep-slope bidder is not necessarily a straightforward task, and we do not wish to present this taxonomy as absolute.

The bid data

Each month, between 200,000 and 250,000 DAM bids and between 170,000 and 190,000 HAM bids are submitted to the New York ISO. The summary statistics on such a dataset can be misleading without an understanding of the fuel mix and type of generating units. Nonetheless, some characteristics are worth noting. First, a sizable

PANEL 3 OPERATING IN THE DEREGULATED MARKETS

The three Northeast US power pools – New York, New England and PJM – have all completed the shift to deregulated market operations. PJM has been the flagship of new market design and in 1998 was the first to develop a pricing scheme that accounts for transmission congestion. This pricing scheme has become widely known as locational marginal pricing (LMP). New England was the last of these three pools to shift to the LMP paradigm, having done so in March 2003.

Generators have had to adapt to operating in these new markets. The following provides a background of the LMP environment in which generators are conducting business.

LMP in the US Northeast

Like real estate, deregulated energy markets are all about location. In the competitive New England (Nepool), New York Control Area (NYCA), and Middle Atlantic/Midwest (PJM RTO) markets, generator location is a direct determinant of energy revenues via the LMP paradigm. The Federal Energy Regulatory Commission (FERC) has endorsed LMP heavily in the hope that it will be implemented in all US markets – based in large measure on its success in the Northeast. Other power pools may adopt LMP pricing such as the Midwest Independent System Operator (MISO), which commenced its new LMP operations on 1 April 2005 and the California ISO is hoping to implement LMP in 2007. The bottom line: even if your generation assets are not now operating in a competitive energy market, chances are good that they will be in the future.

In the heyday of deregulated power markets, before Enron's collapse, a plethora of national and regional energy marketers had active trading desks and, often, a fleet of assets behind them. Plant managers were given dispatch instructions according to marketing strategies and contract commitments. In most cases, it was not necessary for them to have a deep understanding of the critical linkages among their unit, the grid and the market. And if one generator within a company fleet suffered an unplanned outage, leaving the company liable for firm power delivery, another unit could simply pick up the slack.

However, in the post-Enron days of bankrupt national and regional energy companies, limited trading activity, and little market appetite for "risk", many plant managers are now finding themselves facing the open market alone. In many cases, operating companies have taken over for creditors who were suddenly handed the keys to generating facilities by defaulting energy companies.

It is critical for plant operators to understand the economic impact of FERC's standard market design (SMD) and LMP on their plants. Likewise, it is imperative that developers of new generating facilities respond to market signals provided under SMD to ensure that new capacity is sited where revenue is highest.

Plant managers currently operating in deregulated and locationally priced electricity markets find themselves competing to sell energy in a complex environment, where, in addition to efficiency and cost competitiveness, location has become the primary determinant of future revenue flows from energy sales. Pre-SMD wholesale market design provided a pool-wide clearing price – for example, one price for Nepool in New England – based solely on generator bids for energy, with the lowest-cost units (based on heat rate and fuel cost) typically dispatched first. This clearing price applied to all market participants dispatched to supply energy, regardless of their unit location on the grid. As such, a generator in rural northern Maine would receive the same payment as a generator in urban Boston.

Under SMD, however, there are three location-specific components that make up the price any one generator will receive for its energy. In addition to a unit's energy price (generator energy bid), the ultimate LMP a generator receives factors in the cost of electrical losses incurred when power is transported from source to sink (meaning there's a penalty for remote generators), as well as congestion costs associated with transmission constraints or limitations created by the generator dispatch. Specifically, LMP = energy price − (congestion + losses).

Under the LMP paradigm, there are three types of "locational" pricing within the pool: at the node, the zone or hub.

❑ A node is defined by a specific bus location on the grid, to which any number of generators is connected to the transmission network. There can be thousands of nodes within a pool, each of which has a uniquely calculated nodal price. Generators are paid the nodal price for their energy supply.
❑ A zone is defined by a physical area. A zonal price is defined by the load-weighted average of the nodes located within that zonal area – for example, Zone J (New York City) and Zone K (Long Island) in the NYCA. Typically, buyers pay the zonal price for their energy purchases.
❑ A hub is meant to provide a representative price for the pool or a particular area. Examples include the Mass Hub in New England and the PJM Western Hub. The hub price is determined by a specific set of nodes that may or may not be located within the same zone, and often is the defining price in executing energy trades.

Congestion and losses
Congestion occurs when the flow of power increases along a transmission corridor and becomes constrained, making it necessary to dispatch more expensive local generation at the receiving end. This results in congestion-pricing differences between locations. Under the rules of SMD, generators receive payment according to their location as defined by the LMP. The system operator (for example, independent

system operators such as ISO-NE and NYISO, or ERCOT) uses a computer program to dispatch system generation resources in a least-cost manner, while calculating congestion and loss discounts (or premiums) specific to location (economic dispatch).

Contingency planning has an indirect but significant impact on locational pricing. Using "security-constrained dispatch", the system operator for a pool runs dispatch software with "contingency plans" built in to account for possible losses of major system generators and/or transmission lines. The calculated dispatch configuration allows the system to operate effectively, given multiple contingencies or system losses.

Fluctuations in generator availability or transmission capabilities impact nodal LMP values, and the security-constrained dispatch programs calculate various dispatch solutions for each outage scenario. Unplanned outages (especially large-base-load units) will increase energy costs pool-wide if higher-cost generation is dispatched to make up for the loss. Increased congestion or losses will change nodal LMPs. For example, a reduction in transmission capacity going into an urban load zone may reduce the LMP for a generator located in another zone by increasing its congestion costs and increasing the penalty on the source side of the congestion. Increased flows along transmission corridors in the direction of higher demand also increases the level of system losses. Higher levels of losses tend to reduce the LMP for remote generators, while increasing LMPs for urban generators.

A constrained urban area is known as a load pocket. Often, the availability of efficient generation capacity in densely populated urban areas is limited by space and environmental concerns. The import of power into urban areas from more remote locations bridges the gap in the shortfall between competitive urban generation and urban load. Cheaper power will flow to urban load pockets, but flows will be limited by transmission constraints. When transfer capacity has reached its limits, it is necessary to dispatch less efficient "in-city" capacity to ensure reliability, creating load-pocket LMPs that are often higher than surrounding areas or zones.

Day-ahead and real-time markets
Aside from any bilateral supply-contract obligations, there are generally two types of market in which plant managers can submit bids for their unit's energy output. The first is the day-ahead market (DAM). On the day before the energy is to be supplied, bidders into the DAM will have a deadline – noon, for example – before which energy bids are submitted to the system operator, made in increments of no less than one hour. The system operator collects all bids for generation supply along with load bids for energy purchases from the load-serving entities (LSEs). All bids are then entered into a dispatch program that calculates the energy clearing price and the LMPs for each nodal location in the

> pool. If a generator bid is equal to or lower than the LMP at its node, the unit will be notified by the system operator that it has been chosen for dispatch. If a generator bid is higher than the LMP at its node, however, it will not be dispatched.
>
> In the event that the generator is not dispatched into the DAM, it has the opportunity to bid into the real-time (RT) market, which offers the plant a second chance to receive revenue for providing electricity to the system. Bids into the RT market must be received no later than one hour prior to the operating hour and must be in minimum increments of one hour.
>
> DAM and RT prices can end up looking very similar if the transmission system conditions and generator availability are relatively constant over a 48-hour period. However, given the dynamic nature of electricity grids, there is a chance that the loss of a generator or transmission line prior to the DAM bid deadline could result in higher DAM prices, while the return to service of that same generator or transmission line by the next morning would result in comparatively lower RT prices. Likewise, DAM prices can clear and then a generator or transmission outage can cause RT prices to rise above the corresponding DAM prices.
>
> In general, a significantly greater portion of generator megawatts is bid into the DAM than into the RT market, because securing dispatch (and revenue) the day before is safer than tempting fate on the RT market. It is also generally easier for gas-fired generators to line up gas supplies the day before actual dispatch. A generator's type will help determine strategy as well. A large, base-load steam unit would want to secure the majority of its output into the DAM at a competitive cost to secure dispatch. At the same time, a smaller, more expensive fossil-fired peaking unit may wish to take its chances and bid aggressively into the RT market on a day that is expected to be extremely hot or cold during peak hours.

portion of the approximately 300 bids submitted into the DAM each hour is negative. This is due to bids from base-load facilities that wish to remain dispatched and generators with bilateral contracts that must submit bids for scheduling, but are not selling into the market. Conversely, in any given hour, there may be 2,500 to 3,000 bids that are greater than or equal to US$999/MW. The combination of negative and positive bids at US$999/MW and higher leaves only about 175 to 200 bids per hour that are between US$2.00/MW and US$999.00/MW. Yet, these bids represent only 58% to 66% of the bidders in the market, which may have implications for market-wide price levels.

Table 1 Bid profiles

	Bin 1, US$/MW	Bin 2, US$/MW	Bin 3, US$/MW	Bin 4, US$/MW	Bin 5, US$/MW	Bin 6, US$/MW
12PM	0	10	15	20	25	50
1PM	0	10	15	20	25	50
2PM	0	10	15	20	25	50
3PM	10	15	20	25	50	75
4PM	15	20	25	50	75	100
5PM	0	10	15	20	25	50
Plotted Average:	4.17	12.50	17.50	25.83	37.50	62.50

Table 2 Peak temperatures: August 7–10, 2001

Date	Daily Peak Temperature
August 7, 2001	98°
August 8, 2001	98°
August 9, 2001	101°
August 10, 2001	97°

The focus of the study was on pricing and flexibility, and responsiveness of bidders to market-level information and forecasts in addition to their own firm-specific factors. If suppliers to the New York City power market were unresponsive to market demand, we would anticipate the bids to look very similar from day to day regardless of temperature and load forecasts. If suppliers of power were highly responsive to incoming data, we would expect to see bids that look different from day to day as fuel prices, load forecasts and outage levels changed. Figures 10 and 11 show a series of three-dimensional plots representing the topology of the DAM for several days in August 2001. DAM bids for all the generators were averaged within each bin across the hours of 12 noon to 5 pm for 7–10 August. For example, if a supplier made the following bid from 12 noon to 5 pm for a given day, a bin-specific average was plotted in ascending order according to the first bin average for a supplier.

The primary focus of this exercise is to ask the question, "If there was a way to 'look at' bids, do bids look the same across days of varying temperature?" These plots provide a purely visual

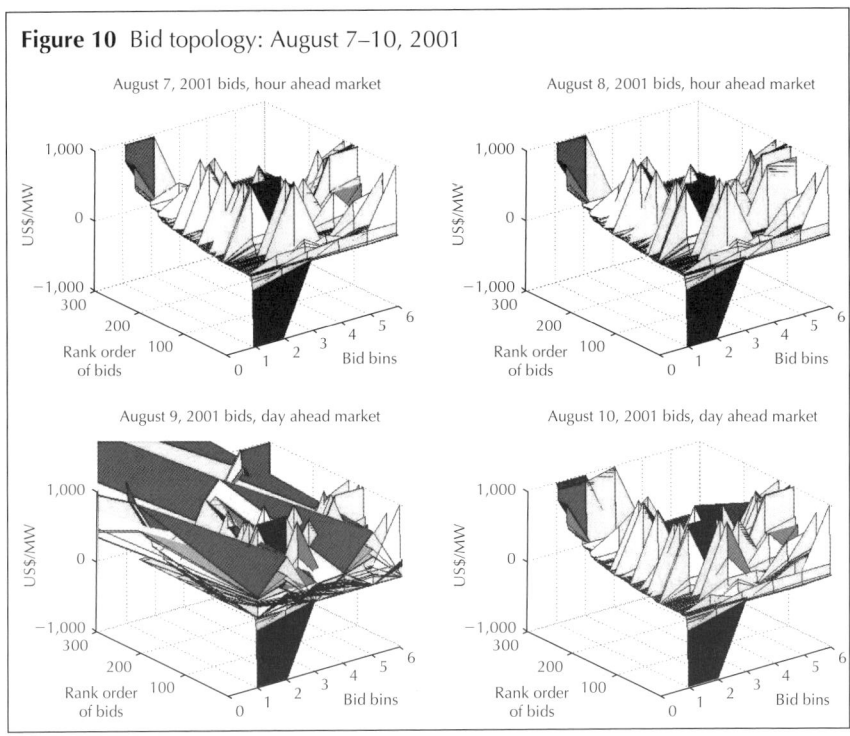

Figure 10 Bid topology: August 7–10, 2001

inspection of that question. Further, if a difference were to be discernible it should be most starkly visible when loads and temperatures are very high. August 9 was the hottest day of summer 2001, with a peak temperature of 101°F. However, the entire week was very hot with temperatures holding in the 90s.

The bids demonstrate that the suppliers do in fact respond to anticipated changes in demand, albeit only at extremes. Days with peak temperatures of 93° to 98° fail to motivate significant changes in bid behaviour. The topology of 7, 8 and 10 August appear nearly identical. Only with extreme temperatures, such as occurred on 9 August, did a visually discernible change in bid behaviour occur.

The differences in the DAM bids prompt the question, "If suppliers change their bids in the DAM under extreme circumstances, do the bids between the DAM and HAM look different or do they look the same?" The three-dimensional plot performs the same averaging of bids within bins from the hours of 12 noon to 5 pm for the DAM and for the HAM, plotting them in rank order by the first

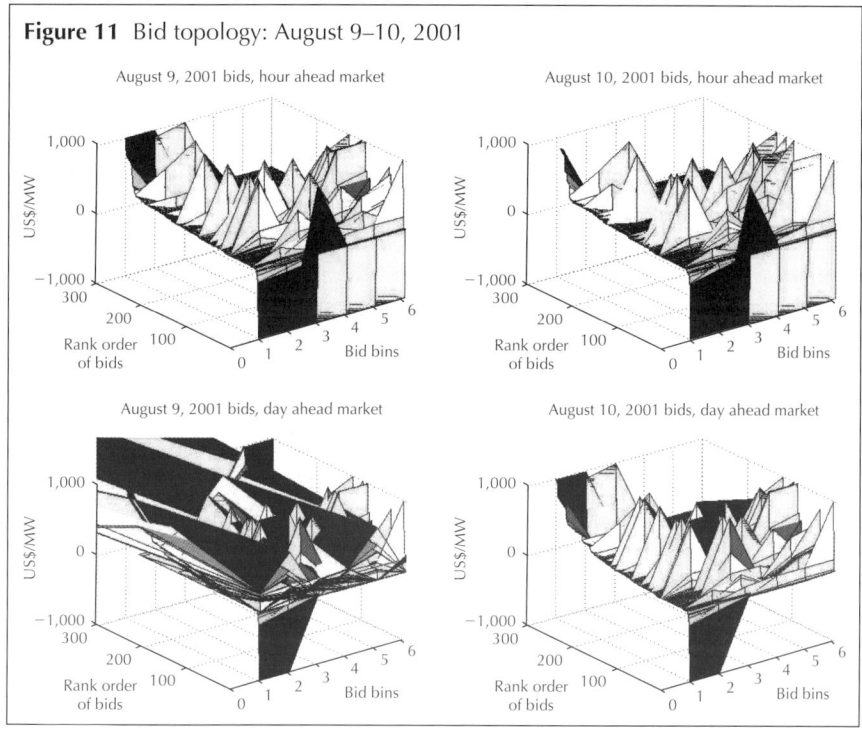

Figure 11 Bid topology: August 9–10, 2001

averaged bin for each of the generator IDs. Surprisingly, the 9 and 10 August HAM do not appear to demonstrate different bid strategies. Perhaps the even more surprising characteristic of these plots is how similar the HAM and DAM bids appear for 9 and 10 August. The only significant difference between the HAM and DAM resides in bids where generators may have been able to lock in bilateral contracts, which appear as large negative bids at the right-front face of each of the HAM plots. These visual inspections of the bids provide useful clues for thinking about bid behaviour, but one must measure and quantify what constitutes "different".

An obvious first question regarding the market behaviour of firms is simply, "How often did bidders change their bids?" Three weeks in summer 2001 were selected, and all generator bids for all peak hours (8 am–7 pm) across five week days. The results indicate that bidders change their bids more significantly across days than across hours. This is not surprising given that any adjustments to the change in output are more easily made over larger intervals of

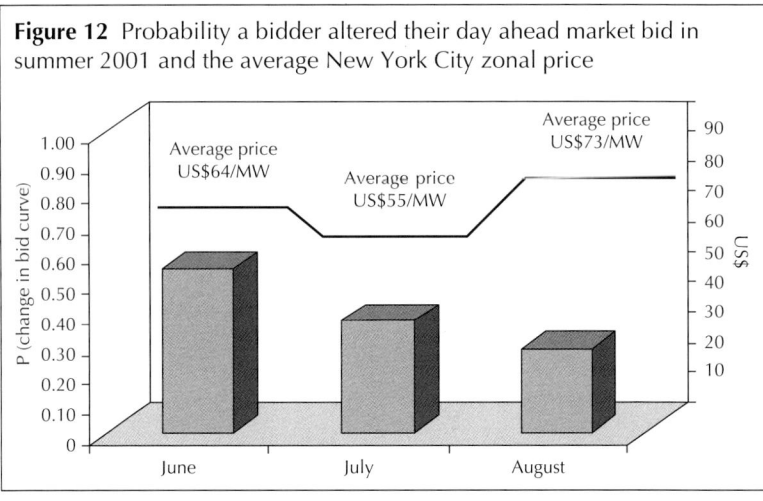

Figure 12 Probability a bidder altered their day ahead market bid in summer 2001 and the average New York City zonal price

time, particularly for base-load generators. Figure 12 shows the probability that a generator changed at least one of its bid bins into the DAM for given week. By a change we mean, specifically, any change in the price or megawatt level for any bin in their curve. Bids for the weeks of 11 June to 16 July, and 13 August, were matched by generator ID, and the probability that a generator changed the dollar value of at least one of the bins was calculated. Half or fewer of the bidders in the market changed their bids, and in August fewer than a third of the bidders made some modification to their bids.

The statistics and visual inspection clearly show the individual suppliers in the market are unresponsive, or, to use a term favoured by economists, the bidders are "sticky" in their exogenous sources of new information.

---

1 Split-curve, straddled and pure price taking bidders are highly correlated with the Day Ahead Zonal LBMP for 17% of the on-peak hours in 2001. Since each of these clusters acts as a pure price taker for all or some portion of their bid curve, almost by definition they could not set the price for that portion of their bid. So, these three clusters were grouped into a single cluster and the correlation was tested as a single pool of bids.
2 Bin – A grouping of like data elements, commonly seen in histogram charts.

Index

A
AGC (automated generation control) 279
Ampere's Law 171, 173
area control error (ACE) 185
auto regressive moving average (ARMA)-type process 161
available transfer capability (ATC) 181

B
Ball and Torus (1985) 88
Barone
 (2004a) 366
 (2004b) 387
Barraquand and Martineau (1995) 123
Barz (1999) 44
Basel Committee on Banking Supervision
 (1996a) 367
 (1996b) 369, 372, 388
 (1996c) 388
 (2004) 367–368, 385
Basic Generation Service (BGS) 337
Bjork (2004) 316
Black (1976) 128
Black and Scholes (1973) 124
Black–Scholes hedge ratio 93
Black–Scholes–Merton formula 65
Borenstein and Bushnell (1999) 233
Boyle (1977) 124
Brennan and Schwartz (1985) 63
Brownian motion 4, 33–34, 36, 63–64, 68, 71, 79, 103, 115, 124, 223, 297, 312, 376

Buy *et al* (1998) 350

C
California–Oregon border (COB) 11
Carmona and Durrleman (2003) 313
central limit theorem 90
Chaplin (1993) 149
Chibisov and Wolyniec (2004) 319, 326, 329
Cholesky's decomposition 139, 145
Clewlow and Strickland (1999) 78, 87
Climate Prediction Center (CPC) 255, 271
combined cycle gas turbine capacity (CCGT) 394, 396
combined-heat-and-power (CHP) 188
Committee of Chief Risk Officers (2003) 375
contract-for-difference (CfD) price 235
cooling degree days (CDD) 154
Cortazar and Schwartz (1994) 78, 352
Cournot game 233
Crampes and Fabra (2005) 244
Crank–Nicholson methods 106

D
Davidson (2003) 340
day-ahead market (DAM) 397, 411
DeGroot (1986) 374

Deng, Johnson and Sogomonian (1998) 43
Duffie and Lando (2001) 345, 348

E
East Center Area Reliability Coordination Agreement (ECAR) region 60
Edison Electric Institute (EEI) 334
El Niño cycles 259–261
Electric Reliability Council of Texas (ERCOT) 175, 186
Environmental Protection Agency (EPA) 107
Eustache (1998) 363
Evans and Green (2003) 226
Eydeland (2002) 292
Eydeland and Geman (1995) 64
Eydeland and Wolyniec
  (2002) 293, 313–314, 322, 328–329
  (2006) 300, 326, 329

F
Faraday's Law 171, 173
"fat tail" property 164
Federal Energy Regulatory Commission (FERC) 76, 175, 198, 361, 409
Federico and Rahman (2003) 226
financial transmission rights (FTR) 162, 198, 205
first-contingency incremental transfer capability (FCITC) 181
Fons (1994) 345

G
Game theory 229–230, 236
Gao and Wolyniec (2004) 293–294, 300
Gas Industry Standards Board (GISB) 334
Geman and Yor
  (1993) 64
  (1997) 71
Geman (1994) 63
Giesecke (2004) 345, 347

Giesecke and Goldberg (2004) 345
Global Forecast System (GFS) 269
Green and Newbery (1992) 233
Greenland block 262
Guldmann (1995) 372

H
Haug
  (1997) 312
  (1998) 131
Heath, Jarrow and Morton (1992) 352
Heath–Jarrow–Morton (HJM) extension 78
heating degree days (HDD) 154
Herfindahl–Hirschmann index 226
HFLI (high-frequency, low-impact) 383
Holton (2003) 369–370
hour-ahead market (HAM) 397
Hsieh (2003) 386
Hull (1997) 150
Hurricane Andrew (1992) 251
Hurricane Ivan (2004) 251

I
independent power producers (IPPs) 189
internal ratings-based (IRB) approach 367
International Swap Derivatives Association Inc (ISDA) 334

J
Jaillet, Ronn and Tompaidis (1998) 143
Jorion (2001) 388
"jump diffusion" model 134

K
Kamal and Derman (1999) 91
Kirchoff's Current Law (KCL) 192
Kirchoff's Voltage Law (KVL) 192
Klemperer (2003) 226
Koza (1992) 148

L

La Niña cycles 258–261
LFHI (low-frequency, high-impact) 383
Load-serving entities (LSEs) 203, 411
location marginal pricing (LMP) 153, 198, 282, 393, 409
location-based marginal prices (LBMP) 194
lognormal distribution 134, 137, 164, 318
Lopez (2004) 388
Loss-given default (LGD) 381
Louisiana region 61

M

Margrabe (1978) 137
Margrabe models 312
mark-to-market (MTM) 331
Market-implied heat rate (IHR) 316
Markov process 223
Merton
  (1973) 64
  (1976) 71, 134
  (1990) 79
Monte Carlo approaches 100
Monte Carlo method 6, 122–123, 138–139, 149

N

Nash equilibrium 230, 236, 246
National Centers for Environmental Prediction (NCEP) 269
National Electricity Reliability Council (Nerc) regions 102
National Oceanic and Atmospheric Administration (NOAA) 271
National Weather Service (NWS) 272
Nested Grid Model (NGM) 269
net present value (NPV) analysis 218
NITS (network integration transmission service) 279
non-utility generators (NUGs) 188
North American Electric Reliability Council (NERC) regions 77
North American Energy Standards Board (NAESB) 334
North American Mesoscale model (NAM) 269

O

Ocaña and Romero (1997) 233
Ohm's Law 171–172, 193
open-access same-time information system (OASIS) 188
Orstein–Uhlenbeck mean reversion 25

P

"parallel shift" method 53
partial differential equation (PDE) 100
Pearson models 312
Pennsylvania, New Jersey and Maryland (PJM) 19, 63, 100
PennWell (2005) 364
"peso problem" 112
Philipson (1999) 363
Pilipovic (1998) 135, 388
Poisson jump-diffusion process 88
Poisson processes 84–85
"power stack" function 68
Press (1967) 89
Press et al (1992) 89, 125, 139, 149
"price to beat" (PTB) 281
Pricing energy options 122
pricing power contingent claims (PCCs) 99
Probability of default (PD) 381
Public Utility Regulatory Policies Act (PURPA) 363

Q

"quanto" option 95

R

Rebonato (2004) 328
regional transmission organisation (RTO) 168, 198
Reiner (1992) 96

S

Samuelson hypothesis 85
Schwartz (1997) 87
simultaneous feasibility test (SFT) 200
Southern Oscillation Index (SOI) 260
Spark-spread options 123, 136, 138–139
standard market design (SMD) 393, 409
Stephen Figlewski (2003) 388
successive over-relaxation (SOR) methods 106

T

Telser (1958) 119
Tiedmann (1996) 95
Tikhonov and Arsenin (1977) 119
Tilley (1993) 123
total transfer capability (TTC) 181
transmission congestion rights (TCRs) 205
transmission loading relief (TLR) 190

V

value at risk (VAR) 34, 121, 145, 341, 368, 369
Vasicek (1977) 134
Vorst (1996) 131

W

Western Electricity Coordinating Council (WECC 175, 187
Western Systems Power Pool (WSPP) 334
Wilmott *et al* (1993) 116
Wolinsky (2003) 373

Z

zero volatility 144